About the Authors

Thanks to science backgrounds and their numerous science publications, both Patricia Barnes-Svarney and Thomas E. Svarney have had much more than a passing acquaintance with mathematics.

Barnes-Svarney has been a nonfiction science and science-fiction writer for 20 years. She has a bachelor's degree in geology and a master's degree in geography/geomorphology, and at one time she was planning to be a math major. Barnes-Svarney has had some 350 articles published in magazines and journals and is the author or coauthor of more than 30 books, including the award-winning *New York Public Library Science Desk Reference* and *Asteroid: Earth Destroyer or New Frontier?*, as well as several international best-selling children's books. In her spare time, she gets as much produce and herbs as she can out of her extensive gardens before the wildlife takes over.

Thomas E. Svarney brings extensive scientific training and experience, a love of nature, and creative artistry to his various projects. With Barnes-Svarney, he has written extensively about the natural world, including paleontology (*The Handy Dinosaur Answer Book*), oceanography (*The Handy Ocean Answer Book*), weather (*Skies of Fury: Weather Weirdness around the World*), natural hazards (*A Paranoid's Ultimate Survival Guide*), and reference (*The Oryx Guide to Natural History*). His passions include martial arts, Zen, *Felis catus,* and nature.

When they aren't traveling, the authors reside in the Finger Lakes region of upstate New York with their cats, Fluffernutter, Worf, and Pabu.

Other Visible Ink Press titles distributed by Omnigraphics

Armageddon Now (0–7808–0923–8)

The Astrology Book (0–7808–0719–7)

Black Firsts, 2nd Ed. (1–57859–153–8)

Black Heroes (0–7808–0728–6)

Conspiracies and Secret Societies (0–7808–0921–1)

The Encyclopedia of Religious Phenomena (0–7808–0924–6)

The Fortune-Telling Book (0–7808–0720–0)

The Handy Biology Answer Book (0–7808–0778–2)

The Handy Dinosaur Answer Book (0–7808–0724–3)

The Handy History Answer Book, 2nd Edition (0–7808–0925–4)

The Handy Math Answer Book (0–7808–0962–9)

The Handy Ocean Answer Book (0–7808–0725–1)

The Handy Physics Answer Book (0–7808–0723–5)

The Handy Politics Answer Book (0–7808–0726–X)

The Handy Presidents Answer Book (0–7808–0773–1)

The Handy Science Answer Book (0–7808–0727–8)

Real Ghosts, Restless Spirits, and Haunted Places (0–7808–0721–9)

The Religion Book (0–7808–0722–7)

The Spirit Book (0–7808–0922–X)

The Superhero Book (0–7808–0772–3)

Unexplained! 2nd Ed. (0–7808–0715–4)

The Vampire Book (0–7808–0716–2)

The Werewolf Book (0–7808–0717–0)

The Witch Book (0–7808–0718–9)

www.omnigraphics.com

THE
HANDY
MATH
ANSWER
BOOK

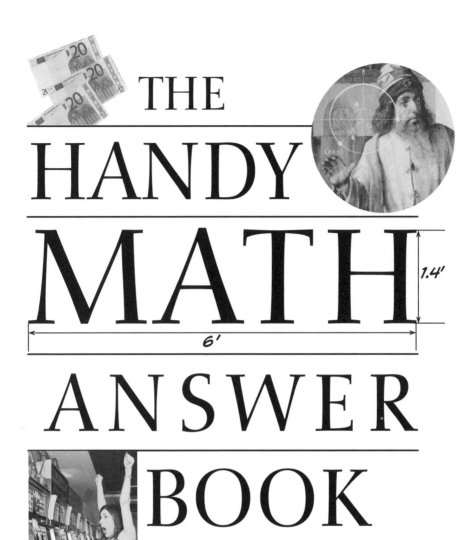

THE
HANDY
MATH

1.4'

6'

ANSWER
BOOK

Patricia Barnes-Svarney and Thomas E. Svarney

VISIBLE
INK
PRESS

Detroit

THE HANDY MATH ANSWER BOOK

Visible Ink Press®
43311 Joy Rd. #414
Canton, MI 48187-2075

Visible Ink Press is a registered trademark of Visible Ink Press LLC.

Art Director: Mary Claire Krzewinski
Line Art: Kevin Hile
Typesetting: The Graphix Group

ISBN 0-7808-0962-9

Cover images of euros used by permission of Photographer's Choice/Getty Images; Aristotle used by permission of The Bridgeman Art Library/Getty Images; woman at casino used by permission of Taxi/Getty Images; nautilus used by permission of The Image Bank/Getty Images.

Back cover images of mathematician used by permission of Stone/Getty Images; card game used by permission of The Image Bank/Getty Images.

Casebound editions published exclusively for Omnigraphics.

info@omnigraphics.com
1-800-875-1340
Editorial Office: 615 Griswold St.
Detroit, MI 48226
www.omnigraphics.com

Contents

HISTORY

THE BASICS

MATH IN SCIENCE AND ENGINEERING

Math All Around Us

Introduction

"As far as the laws of mathematics refer to reality, they are not certain; and as far as they are certain, they do not refer to reality."

Albert Einstein

We've all seen it. We've all experienced it, many times without knowing it. It's in the design of a beautiful stained-glass window in the middle of an Austrian cathedral. It's in the large and small workings of a car, computer, or space shuttle. It's in the innocent statement of a child asking, "How old are you?"

By now you've probably guessed what "it" is: mathematics.

Mathematics is everywhere. Sometimes it's as subtle as the symmetry of a butterfly's wings. Sometimes it's as blatant as the U.S. debt figures displayed on a sign outside the Internal Revenue Service building in New York City.

Numbers sneak into our lives. They are used to determine a prescription for eyeglasses; they reveal blood pressure, heart rate, and cholesterol levels, too. Numbers are used so you can follow a bus, train, or plane schedule; or they can help you figure out when your favorite store, restaurant, or library is open. In the home, numbers are used for recipes, figuring out the voltage on a circuit in an electric switchbox, and measuring a room for a carpet. Probably the most familiar connection we have to numbers is in our daily use of money. Numbers, for instance, let you know whether you're getting a fair deal on that morning cup of cappuccino.

The Handy Math Answer Book is your introduction to the world of numbers, from their long history (and hints of the future) to how we use math in our everyday lives. With more than 1,000 questions and answers in *The Handy Math Answer Book* (1,002, to be mathematically precise) and over 100 photographs, 70 illustrations, and dozens of equations to help explain or provide examples of fundamental mathematical principles, you'll cover a lot of ground in just one book!

Handy Math is split into four sections: "The History" includes famous (and sometimes infamous) people, places, and objects of mathematical importance; "The Basics"

explains the various branches of mathematics, from fundamental arithmetic to complex calculus; "Math in Science and Engineering" describes how relevant math is to such fields as architecture, the natural sciences, and even art; and "Math All Around Us" shows how much math is part of our daily lives, including everything from balancing a checkbook to playing the slots in Las Vegas.

The subject of math—and its many connections—is immense. After all, over two thousand years ago the Greek mathematician Euclid wrote thirteen books about geometry and other fields of mathematics (the famous *Elements*). It took him six of those volumes just to describe elementary plane geometry. Today, even more is known about mathematics, as you'll see in the list of resources described in the last chapter of this book. Here we've provided you with everything from recommended print sources to some of our favorite Web sites, such as "Dr. Math" and "SOS Math." In this way, *Handy Math* not only introduces you to the basics of math, but it also gives you the resources to continue on your own mathematical journey.

Be warned: This journey is an extensive one. But you'll soon learn that it's satisfying and rewarding in every way. Not only will you understand what math is all about, but you'll appreciate the mathematical beauty that surrounds you every day. Just as it has astounded us, we're sure you'll be amazed by how numbers, equations, and sundry other mathematical constructions continue to not only define, but also influence, the world around us.

Acknowledgments

Such a work as *The Handy Math Answer Book* could not have been completed without the help of many generous people. The authors would like to thank Roger Jänecke for originating the concept for this book; Kevin Hile for his patience, great editorial work, photo research, and line art design; Christa Gainor for always being there to answer our questions (and her amazing knowledge of topics); Roger Matuz for his friendly advice in helping us decide on content; Amy Keyzer and John Krol for proofreading; Lawrence Baker for the index; Mary Claire Krzewinski for design; Marco Di Vita of the Graphix Group for typesetting; Marty Connors for giving us the go-ahead for this project; and our agent and friend, Agnes Birnbaum, as always, for all her hard work.

Finally, the authors would like to thank the multitude of devoted mathematicians and those in other fields who use mathematics—past, present, and future. These people have, in a direct or indirect way, helped us all better understand our world.

It would also be nice to thank those people who first made up the numbering system so long ago, but that might be stretching our thanks a bit too much. After all, we wouldn't be using computers or cashing checks without numbers!

HISTORY

HISTORY OF MATHEMATICS

WHAT IS MATHEMATICS?

What is the **origin** of the **word "mathematics"**?

According to most sources, the word "mathematics" is derived from the Latin *mathematicus* and from the Greek *mathēmatikos,* meaning "mathematical." (Other forms include *mathēma,* meaning "learning," and *manthanein,* meaning "to learn.")

In simple terms, **what is mathematics**?

Mathematics is often referred to as the science of quantity. The two traditional branches of mathematics have been arithmetic and geometry, using the quantities of numbers and shapes. And although arithmetic and geometry are still of major importance, modern mathematics expands the field into more complex branches by using a greater variety of quantities.

Who were the **first humans** to **use simple forms of mathematics**?

No one really knows who first used simple forms of mathematics. It is thought that the earliest peoples used something resembling mathematics because they would have known the concepts of one, two, or many. Perhaps they even counted using items in nature, such as 1, represented by the Sun or Moon; 2, their eyes or wings of a bird; clover for 3; or legs of a fox for 4.

Archeologists have also found evidence of a crude form of mathematics in the tallying systems of certain ancient populations. These include notches in wooden sticks or bones and piles or lines of shells, sticks, or pebbles. This is an indication that certain prehistoric peoples had at least simple, visual ways of adding and subtracting things, but they did not yet have a numbering system such as we have today.

3

EARLY COUNTING AND NUMBERS

Early humans used all sorts of images to represent numbers, including the fox, the image of which was used to indicate the number 4. *Stone/Getty Images.*

What are some **examples** of **how early peoples counted**?

There were several different ways that early civilizations recorded the numbers of things. Some of the earliest archeological evidence of counting dates from about 35000 to 20000 BCE, in which several bones bear regularly spaced notches. Most of these marked bones have been found in western Europe, including in the Czech Republic and France. The purpose of the notches is unclear, but most scientists believe they do represent some method of counting. The marks may represent an early hunter's number of kills; a way of keeping track of inventory (such as sheep or weapons); or a way to track the movement of the Sun, Moon, or stars across the sky as a kind of crude calendar.

Not as far back in time, shepherds in certain parts of West Africa counted the animals in their flocks by using shells and various colored straps. As each sheep passed, the shepherd threaded a corresponding shell onto a white strap, until nine shells were reached. As the tenth sheep went by, he would remove the white shells and put one on a blue strap, representing ten. When 10 shells, representing 100 sheep, were on the blue strap, a shell would then be placed on a red strap, a color that represented what we would call the next decimal up. This would continue until the entire flock was counted. This is also a good example of the use of base 10. (For more information about bases, see "Math Basics.")

Certain cultures also used gestures, such as pointing out parts of the body, to represent numbers. For example, in the former British New Guinea, the Bugilai culture used the following gestures to represent numbers: 1, left hand little finger; 2, next finger; 3, middle finger; 4, index finger; 5, thumb; 6, wrist; 7, elbow; 8, shoulder; 9, left breast; 10, right breast.

Another method of counting was accomplished with string or rope. For example, in the early 16th century, the Incas used a complex form of string knots for accounting and sundry other reasons, such as calendars or messages. These recording strings were called *quipus,* with units represented by knots on the strings. Special officers of the king called *quipucamayocs,* or "keepers of the knots," were responsible for making and reading the quipus.

Why did the need for mathematics arise?

The reasons humans developed mathematics are the same reasons we use math in our own modern lives: People needed to count items, keep track of the seasons, and understand when to plant. Math may even have developed for religious reasons, such as in recording or predicting natural or celestial phenomena. For example, in ancient Egypt, flooding of the Nile River would wash away all landmarks and markers. In order to keep track of people's lands after the floods, a way to measure the Earth had to be invented. The Greeks took many of the Egyptian measurement ideas even further, creating mathematical methods such as algebra and trigonometry.

How did certain **ancient cultures** count **large numbers**?

It is not surprising that one of the earliest ways to count was the most obvious: using the hands. And because these "counting machines" were based on five digits on each hand, most cultures invented numbering systems using base 10. Today, we call these *base numbers*—or base of a number system—the numbers that determine place values. (For more information on base numbers, see "Math Basics.")

However, not every group chose 10. Some cultures chose the number 12 (or base 12); the Mayans, Aztecs, Basques, and Celts chose base 20, adding the ten digits of the feet. Still others, such as the Sumerians and Babylonians, used base 60 for reasons not yet well understood.

The numbering systems based on 10 (or 12, 20, or 60) started when people needed to represent large numbers using the smallest set of symbols. In order to do this, one particular set would be given a special role. A regular sequence of numbers would then be related to the chosen set. One can think of this as steps to various floors of a building in which the steps are the various numbers—the steps to the first floor are part of the "first order units"; the steps to the second floor are the "second order units"; and so on. In today's most common units (base 10), the first order units are the numbers 1 through 9, the second order units are 10 through 19, and so on.

What is the connection between **counting and mathematics**?

Although early counting is usually not considered to be mathematics, mathematics began with counting. Ancient peoples apparently used counting to keep track of sundry items, such as animals or lunar and solar movements. But it was only when agriculture, business, and industry began that the true development of mathematics became a necessity.

What is a **numeral**?

A numeral is a standard symbol for a number. For example, X is the Roman numeral that corresponds to 10 in the standard Hindu-Arabic system.

What were the **two fundamental ideas** in the development of **numerical symbols**?

There were two basic principles in the development of numerical symbols: First, a certain standard sign for the unit is repeated over and over, with each sign representing the number of units. For example, III is considered 3 in Roman numerals (see the Greek and Roman Mathematics section below for an explanation of Roman numerals). In the other principle, each number has its own distinct symbol. For example, "7" is the symbol that represents seven units in the standard Hindu-Arabic numerals. (See below for an explanation of Hindu-Arabic numbers; for more information, see "Math Basics.")

MESOPOTAMIAN NUMBERS AND MATHEMATICS

What was the **Sumerian oral counting system**?

The Sumerians—whose origins are debated, but who eventually settled in Mesopotamia—used base 60 in their oral counting method. Because it required the

Who were the Mesopotamians?

The explanation of who the Mesopotamians were is not easy because there are many historians who disagree on how to distinguish Mesopotamians from other cultures and ethnic groups. In most texts, the label "Mesopotamian" refers to most of the unrelated peoples who used cuneiform (a way of writing numbers; see below), including the Sumerians, Persians, and so on. They are also often referred to as Babylonians, after the city of Babylon, which was the center of many of the surrounding empires that occupied the fertile plain between the Tigris and Euphrates Rivers. But this area was also called Mesopotamia. Therefore, the more correct label for these people is probably "Mesopotamians."

In this text, Mesopotamians will be referred to by their various subdivisions because each brought new ideas to the numbering systems and, eventually, mathematics. These divisions include the Sumerians, Akkadians, and Babylonians.

memorization of so many signs, the Sumerians also used base 10 like steps of a ladder between the various orders of magnitude. For example, the numbers followed the sequence 1, 60, 60^2, 60^3, and so on. Each one of the iterations had a specific name, making the numbering system extremely complex.

No one truly knows why the Sumerians chose such a high base number. Theories range from connections to the number of days in a year, weights and measurements, and even that it was easier to use for their purposes. Today, this numbering system is still visible in the way we tell time (hours, minutes, seconds) and in our definitions of circular measurements (degrees, minutes, seconds).

How did the **Sumerian written counting system change** over time?

Around 3200 BCE, the Sumerians developed a written number system, attaching a special graphical symbol to each of the larger numbers at various intervals (1, 10, 60, 3,600, etc.). Because of the rarity of stone, and the difficulty in preserving leather, parchment, or wood, the Sumerians used a material that would not only last but would be easy to imprint: clay. Each symbol was written on wet clay tablets, then baked in the hot sunlight. This is why many of the tablets are still in existence today.

The Sumerian number system changed over the centuries. By about 3000 BCE, the Sumerians decided to turn their numbering symbols counterclockwise by 90 degrees. And by the 27th century BCE, the Sumerians began to physically write the numbers in a different way, mainly because they changed writing utensils from the old stylus that was cylindrical at one end and pointed at the other to a stylus that was flat. This change in writing utensils, but not the clay, created the need for new symbols. The

7

Who were the Akkadians?

The region of Mesopotamia was once the center of the Sumerian civilization, a culture that flourished before 3500 BCE. Not only did the Sumerians have a counting and writing system, but they were also a progressive culture, supporting irrigation systems, a legal system, and even a crude postal service. By about 2300 BCE, the Akkadians invaded the area, emerging as the dominant culture. As most conquerors do, they imposed their own language on the area and even used the Sumerians' cuneiform system to spread their language and traditions to the conquered culture.

Although the Akkadians brought a more backward culture into the mix, they were responsible for inventing the abacus, an ancient counting tool. By 2150 BCE, the Sumerians had had enough: They revolted against the Akkadian rule, eventually taking over again.

However, the Sumerians did not maintain their independence for long. By 2000 BCE their empire had collapsed, undermined by attacks from the west by Amorites and from the east by Elamites. As the Sumerians disappeared, they were replaced by the Assyro-Babylonians, who eventually established their capital at Babylon.

new way of writing numbers was called *cuneiform script,* which is from the Latin *cuneus,* meaning "a wedge" and *formis,* meaning "like."

Did any **cultures** use more than one **base number** in **their numbering system**?

Certain cultures may have used a particular base as their dominant numbering system, such as the Sumerians' base 60, but that doesn't mean they didn't use other base numbers. For example, the Sumerians, Assyrians, and Babylonians used base 12, mostly for use in their measurements. In addition, the Mesopotamian day was broken into 12 equal parts; they also divided the circle, ecliptic, and zodiac into 12 sections of 30 degrees each.

What was the **Babylonian numbering system**?

The Babylonians were one of the first to use a positional system within their numbering system—the value of a sign depends on the position it occupies in a string of signs. Neither the Sumerians nor the Akkadians used this system. The Babylonians also divided the day into 24 hours, an hour into 60 minutes, and a minute into 60 seconds, a way of telling time that has existed for the past 4,000 years. For example, the

What is the rule of position?

We are most familiar with the *rule of position,* or place value, as it is applied to the Hindu-Arabic numerals 1, 2, 3, 4, 5, 6, 7, 8, 9, and 0. This is because their values depend on the place or position they occupy in a written numerical expression. For example, the number 5 represents 5 units, 50 is 5 tens, 500 is 5 hundreds, and so on. The values of the 5s depends upon their position in the numerical expression. It is thought that the Chinese, Indian, Mayan, and Mesopotamian (Babylonian) cultures were the first to develop this concept of place value.

way we now write hours, minutes, and seconds is as follows: 6h, 20', 15"; the way the Babylonians would have written this same expression (as sexagesimal fractions) was 6 20/60 15/3600.

Were there any **problems** with the **Babylonian numbering system**?

Yes. One in particular was the use of numbers that looked essentially the same. The Babylonians conquered this problem by making sure the character spacing was different for these numbers. This ended the confusion, but only as long as the scribes writing the characters bothered to leave the spaces.

Another problem with the early Babylonian numbering system was not having a number to represent zero. The concept of zero in a numbering system did not exist at that time. And with their sophistication, it is strange that the early Babylonians never invented a symbol like zero to put into the empty positions in their numbering system. The lack of this important placeholder no doubt hampered early Babylonian astronomers and mathematicians from working out certain calculations.

Did the **Babylonians** finally use a **symbol** to indicate an **empty space** in their numbers?

Yes, but it took centuries. In the meantime, scribes would not use a symbol representing an empty space in a text, but would use phrases such as "the grain is finished" at the end of a computation that indicated a zero. Apparently, the Babylonians did comprehend the concepts of void and nothing, but they did not consider them to be synonymous.

Around 400 BCE, the Babylonians began to record an empty space in their numbers, which were still represented in cuneiform. Interestingly, they did not seem to view this space as a number—what we would call zero today—but merely as a placeholder.

9

What happened to the Babylonians?

After the Amorites (a Semitic people) founded Babylon, there were several dynasties that ruled the area, including those associated with the famous king and lawmaker, Hammurabi (1792–1750 BCE). It was periodically taken over, including in 1594 BCE by the Kassites and in the 12th century BCE by the Assyrians. Through all these conquests, most of the Babylonian culture retained its own distinctiveness. With the fall of the Assyrian Empire in 612 BCE, the Babylonian culture bloomed, at least until its conquest by Cyris of Persia in 539 BCE. It eventually died out a short time after being conquered by Alexander the Great (356–323 BCE) in 331 BCE (ironically, Alexander died in Babylon, unable to recover from a fever he contracted).

Who **invented** the symbol for **zero**?

Although the Babylonians determined there to be an empty space in their numbers, they did not have a symbol for zero. Archeologists believe that a crude symbol for zero was invented either in Indochina or India around the 7th century and by the Mayans independently about a hundred years earlier. What was the main problem with the invention of zero by the Mayans? Unlike more mobile cultures, they were not able to spread the word around the world. Thus, their claim as the first people to use the symbol for zero took centuries to uncover. (For more information about zero, see "Mathematics throughout History.")

What do we know about **Babylonian mathematical tables**?

Archeologists know that the Babylonians invented tables to represent various mathematical calculations. Evidence comes from two tables found in 1854 at Senkerah on the Euphrates River (dating from 2000 BCE). One listed the squares of numbers up to 59, and the other the cubes of numbers up to 32.

The Babylonians also used a method of division based on tables and the equation $a/b = a \times (1/b)$. With this equation, all that was necessary was a table of reciprocals; thus, the discovery of tables with reciprocals of numbers up to several billion.

They also constructed tables for the equation $n^3 + n^2$ in order to solve certain cubic equations. For example, in the equation $ax^3 + bx^2 = c$ (note: this is in our modern algebraic notation; the Babylonians had their own symbols for such an equation), they would multiply the equation by a^2, then divide it by b^3 to get $(ax/b)^3 + (ax/b)^2 = ca^2/b^3$.

If $y = ax/b$, then $y^3 + y^2 = ca^2/b^3$, which could now be solved by looking up the $n^3 + n^2$ table for the value of n that satisfies $n^3 + n^2 = ca^2/b^3$. When a solution was found

for y, then x was found by $x = by/a$. And the Babylonians did all this without the knowledge of algebra or the notations we are familiar with today.

What other **significant mathematical contributions** did the **Babylonians** make?

Throughout the centuries, the Babylonians made many mathematical contributions. They were the earliest people to know about the Pythagorean theorem, although it was not known by that name. In fact, Pythagoras, in his travels to the east, may have learned about the theorem that would eventually carry his name from the Babylonians. In addition, the Babylonians possessed all the theorems of plane geometry that the Greeks ascribed to Thales, including the theorem eventually named after him. They also may have been the most skilled algebraists of their time, even though the symbols and methods they used were much different than our modern algebraic notations and procedures.

Alexander the Great, depicted here in an 1899 painting of the Battle of Gaugamela, Iraq (331 BCE), by artist Benjamin Ide Wheeler, conquered much of the known world and brought an end to the Babylonian civilization. The rise and fall of civilizations throughout history did much to influence the development of mathematics over the centuries. *Library of Congress.*

EGYPTIAN NUMBERS AND MATHEMATICS

Who were the **Egyptians**?

The Egyptians rose to prominence around 3000 BCE in the area we now call Egypt, but their society was already advanced, urbanized, and expanding rapidly long before that time. Although their civilization arose about the same time that words and numbers were first written down in Mesopotamia, archeologists do not believe there was any sharing between the two cultures. The Egyptians already had writing and written numerals; plus, the Egyptian signs and symbols were taken exclusively from the flora and fauna of the Nile River basin. In addition, the Egyptians developed the utensils for writing signs about a thousand years earlier.

11

What type of **numerals** did the **Egyptians** use?

By about 3000 BCE, the Egyptians had a writing system based on hieroglyphs, or pictures that represented words. Their numerals were also based on hieroglyphs. They used a base-10 system of numerals: one unit, one ten, one hundred, and so on to one million. The main drawback to this system was the number of symbols needed to define the numbers.

Did the **Egyptians** eventually **develop different numerals**?

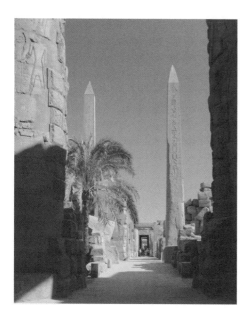

Hieroglyphs can often be found on such Egyptian structures as the Obelisks of Hatshepsut, Karnak Temple, near the ancient city of Thebes. *Robert Harding World Imagery/Getty Images.*

Yes, the Egyptians used another number system called hieratic numerals after the invention of writing on papyrus. This allowed larger numbers to be written in a more compact form. For example, there were separate symbols for 1 through 9; 10, 20, 30, and so on; 100, 200, 300, and so on; and 1,000, 2000, 3,000, and so on.

The only drawback was that the system required memorization of more symbols—many more than for hieroglyphic notation. It took four distinct hieratic symbols to represent the number 3,577; it took no less than 22 symbols to represent the same number in hieroglyphs, but most of those symbols were redundant (see illustration on p. 15).

Both hieroglyphic and hieratic numerals existed together for close to two thousand years—from the third to the first millennium BCE. In general, hieroglyph numerals were used when carved on such objects as stone obelisks, palace and temple walls, and tombs. The hieratic symbols were much faster and easier to scribe, and they were written on papyrus for records, inventories, wills, or for mathematical, astronomical, economic, legal—or even magical—works.

Even though it is thought that the hieratic symbols were developed from the corresponding hieroglyphs, the shapes of the signs changed considerably. One reason in particular came from the reed brushes used to write hieratic symbols; writing on papyrus differed greatly from writing using stone carvings, thus the need to change the symbols to fit the writing devices. And as kingdoms and dynasties changed, the hieratic numerals changed, too, with users having to memorize the many distinct signs.

What are some examples of Egyptian multiplication?

Egyptian multiplication methods did not require a great deal of memorization, just a knowledge of the two times tables. For a simple example, to multiply 12 times 16, they would start with 1 and 12. Then they would double each number in each row (1 × 2 and 12 × 2; 2 × 2 and 24 × 2; and so on) until the number 16, resulting in the answer 192:

1	12
2	24
4	48
8	96
16	192

Another example computes a number that is not a multiple in the row, such as 37 times 19:

1	19
2	38
4	76
8	152
16	304
32	608

First, do the usual procedure by starting with 1 and 19, then doubling the numbers until you get to 32 (if you double 32 [= 64], you've overshot the number 37). Because 37 is higher than 32, go back over the list on the left-hand side, figure out which numbers, with 32, add up to 37 (1, 4, and 32); then add the numbers that correspond to those numbers, to the right (19, 76, and 608), which equals the answer: 703. And you didn't even need a calculator!

Did the **Egyptians** use **fractions**?

Yes, the Egyptian numbering system dealt with fractions, albeit with symbols that do not resemble modern notation. Fractions were written by placing the hieroglyph for "mouth" over the hieroglyph for the numerical expression. For example, 1/5 and 1/10 would be seen as the first two illustrations represented in the box on p. 15. Other fractions, such as the two symbols for 1/2 (see illustration on p. 15), also have special signs.

What were the **problems** with the **Egyptian number system**?

The Egyptian number system had several problems, the most obvious being that it was not written with certain arithmetic calculations in mind. Similar to Roman

The Egyptian civilization did much to contribute to mathematics, including developing a numbering system and using geometry in architecture to create the famous pyramids and other buildings. *Photographer's Choice/Getty Images.*

numerals, Egyptian numbers could be used for addition and subtraction, but not for simple multiplication and division.

All was not lost, however, as the Egyptians devised a way to do multiplication and division that involved addition. Multiplying and dividing by 10 was easy with hieroglyphics—just replace each symbol in the given number by the sign for the next higher order. To multiply and divide by any other factor, Egyptians devised the tabulations based on the two times tables, or a sequence of duplications.

Why did the **Egyptians** need to **develop mathematics**?

Probably the most pressing reason for the development of Egyptian mathematics came from a periodic occurrence in nature: the flooding of the Nile River. With the advent of agriculture in the Nile River valleys, flooding was important, not only to provide fertile soil and water for the irrigated fields, but also to know when the fields would become dry. In addition, along with the growth of the Egyptian society came a need for a more complex way of keeping track of taxes, dividing property, buying and selling goods, and even amassing an army. Thus, the need for counting and mathematics arose, along with the development of a written system of numbers to complete and record the myriad of transactions.

Where does most of our knowledge of **Egyptian mathematics originate**?

Most of our knowledge of Egyptian mathematics comes from writings on papyrus, a type of writing paper made in ancient Egypt from the pith and long stems of the papyrus plant. Most papyri no longer exist, as the material is fragile and disintegrates over time. But two major papyri associated with Egyptian mathematics have survived.

Named after Scottish Egyptologist A. Henry Rhind, the *Rhind papyrus* is about 19 feet (6 meters) long and 1 foot (1/3 meter) wide. It was written around 1650 BCE by Ahmes, an Egyptian scribe who claimed he was copying a 200-year-old document (thus the original information is from about 1850 BCE). This papyrus contains 87 mathematical problems; most of these are practical, but some teach manipulation of the number system (though with no application in mind). For example, the first six problems of the Rhind papyrus ask the following: problem 1. how to divide n loaves between 10 men, in which $n = 1$; in problem 2, $n = 2$; in problem 3, $n = 6$; in prob-

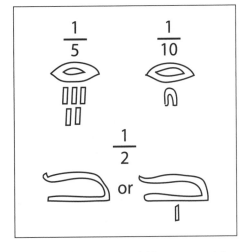

The number 3,577 is represented above using hiero-glyphs (top) and hieratic symbols (bottom). Notice these numbers are read from right to left.

The symbols for 1/5, 1/10, and 1/2 are represented above using hieroglyphs.

lem 4, $n = 7$; in problem 5, $n = 8$; and in problem 6, $n = 9$. In addition, 81 out of the 87 problems involve operating with fractions, while other problems involve quantities and even geometry. Rhind purchased the papyrus in 1858 in Luxor; it resides in the British Museum in London.

Written around the 12th Egyptian dynasty, and named after the Russian city, the mathematical information on the *Moscow papyrus* is not ascribed to any one Egyptian, as no name is recorded on the document. The papyrus contains 25 problems similar to those in the Rhind papyrus, and many that show the Egyptians had a good grasp of geometry, including a formula for a truncated pyramid. It resides in the Museum of Fine Arts in Moscow.

GREEK AND ROMAN MATHEMATICS

Why was **mathematics** so **important** to the **Greeks**?

With a numbering system in place and knowledge from the Babylonians, the Greeks became masters of mathematics, with the most progress taking place between the years of 300 BCE and 200 CE, although the Greek culture had been in existence long before that time. The Greeks changed the nature and approach to math, and they considered it one of the—if not the most—important subjects in science. The main reason for their proclivity towards mathematics is easy to understand: The Greeks preferred reasoning over any other activity. Mathematics is based on reasoning, unlike many scientific endeavors that require experimentation and observation.

15

The distance between the Moon and Earth was calculated by Hipparchus of Rhodes using basic trigonometry. *Stone/Getty Images.*

Who were some of the most **influential Ionian, Greek,** and **Hellenic mathematicians**?

The Ionians, Greeks, and Hellenics had some of the most progressive mathematicians of their time, including such mathematicians as Heron of Alexandria, Zeno of Elea, Eudoxus of Cnidus, Hippocrates of Chios, and Pappus. The following are only a few of the more influential mathematicians.

Thales of Miletus (c. 625–c. 550 BCE, Ionian), besides being purportedly the founder of a philosophy school and the first recorded western philosopher known, made great contributions to Greek mathematics, especially by presenting Babylonian mathematics to the Greek culture. His travels as a merchant undoubtedly exposed him to the geometry involved in measurement. Such concepts eventually helped him to introduce geometry to Greece, solving such problems as the height of the pyramids (using shadows), the distance of ships from a shoreline, and reportedly predicting a solar eclipse.

Hipparchus of Rhodes (c. 170–c. 125 BCE, Greek; also seen as Hipparchus of Nicaea) was an astronomer and mathematician who is credited with creating some of the basics of trigonometry. This helped immensely in his astronomical studies, including the determination of the Moon's distance from the Earth. Claudius Ptolemaeus (or Ptolemy) (c. 100–c. 170, Hellenic) was one of the most influential Greeks, not only in the field of astronomy, but also in geometry and cartography. Basing his works on Hipparchus, Ptolemy developed the idea of epicycles in which each planet revolves in a circular orbit, and each goes around an Earth-centered universe. The Ptolomaic way of explaining the solar system—which we now know is incorrect—dominated astronomy for more than a thousand years.

Diophantus (c. 210–c. 290) was considered by some scholars to be the "father of algebra." In his treatise Arithmetica, he solved equations in several variables for integral solutions, or what we call diophantine equations today. (For more about these equations, see "Algebra.") He also calculated negative numbers as solutions to some equations, but he considered such answers absurd.

What were **Archimedes's greatest contributions** to mathematics?

Historians consider Archimedes (c. 287–212 BCE, Hellenic) to be one of the greatest Greek mathematicians of the classic era. Known for his discovery of the hydrostatic

principle, he also excelled in the mechanics of simple machines; computed close limits on the value of "pi" by comparing polygons inscribed in and circumscribed about a circle; worked out the formula to calculate the volume of a sphere and cylinder; and expanded on Eudoxus's method of exhaustion that would eventually lead to integral calculus. He also created a way of expressing any natural number, no matter how large; this was something that was not possible with Greek numerals. (For more information about Archimedes, see "Mathematical Analysis" and "Geometry and Trigonometry.")

Ptolemy (center), depicted in this 1632 engraving discussing ideas with Aristotle (left) and Copernicus (right), discovered valuable concepts concerning cartography, geometry, and astronomy. *Library of Congress.*

What **Greek mathematician** made major **contributions to geometry**?

The Greek mathematician Euclid (c. 325–c. 270 BCE) contributed to the development of arithmetic and the geometric theory of quadratic equations. Although little is known about his life—except that he taught in Alexandria, Egypt—his contributions to geometry are well understood. The elementary geometry many of us learn in high school is still largely based on Euclid. His 13 books of geometry and other mathematics, titled *Elements* (or *Stoicheion* in Greek), were classics of his day. The first six volumes offer explanations of elementary plane geometry; the other books present the theory of numbers, certain problems in arithmetic (on a geometric basis), and solid geometry. He also defines basic terms such as point and line, certain related axioms and postulates, and a number of statements logically deduced from definitions, axioms, and postulates. (For more information on axioms and postulates, see "Foundations of Mathematics"; for more information about Euclid, see "Geometry and Trigonometry.")

What was **Pythagoras**'s importance to mathematics?

Although the Chinese and Mesopotamians had discovered it a thousand years before, most people credit Greek mathematician and philosopher Pythagoras of Samos (c. 582–c. 507 BCE) with being the first to prove the Pythagorean Theorem. This is a famous geometry theorem relating the length of a right-angled triangle's hypotenuse (h) to the lengths of the other two sides (a and b).

In other words, for any right triangle, the square of the length of the hypotenuse is equal to the sum of the squares of the lengths of the other two sides.

17

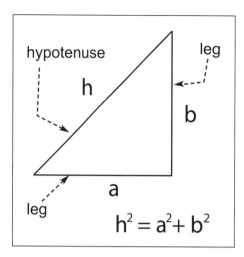

The Pythagorean Theorem is an easy way to determine the length of one side of a right triangle, given one knows the length of the other two sides.

$$h^2 = a^2 + b^2$$

What were **Pythagoras's** other **contributions**?

It is interesting that the Pythagorean Theorem was not Pythagoras's only contribution. He is considered the first pure mathematician. He also founded a school that stressed a fourfold division of knowledge, including number theory (deemed the most important of the pursuits at the school and using only the natural numbers), music, geometry, and astronomy (these subjects were called the *quadrivium* in the Middle Ages). Along with logic, grammar, and rhetoric, these studies collectively formed what was deemed the essential areas of knowledge for any well-rounded person.

Pythagoras not only taught these subjects, but also reincarnation and mysticism, establishing an order similar to, or perhaps influenced by, the earlier Orphic cult. The true lives of Pythagoras and his followers (who worshipped Pythagoras as a demigod) are a bit of a mystery, as they followed a strict code of secrecy and regarded their mathematical studies as something of a black art. The fundamental belief of the Pythagoreans was that "all is number," or that the entire universe—even abstract ethical concepts such as justice—could be explained in terms of numbers. But they also had some interesting non-mathematical beliefs, including an aversion to beans.

Although the Pythagoreans were influential in the fields of mathematics and geometry, they also made important contributions to astronomy and medicine and were the first to teach that the Earth revolved around a fixed point (the Sun). This idea would be popularized centuries later by Polish astronomer Nicolaus Copernicus (1473–1543). By the end of the 5th century BCE, the Pythagoreans had become social outcasts; many of them were killed as people grew angry at the group's interference with traditional religious customs.

Who was the **first** recorded **female mathematician**?

The first known female mathematician was Hypatia of Alexandria (370–415), who was probably taught by her mathematician and philosopher father, Theon of Alexandria. Around 400, she became the head of the Platonist school at Alexandria, lecturing on mathematics and philosophy. Little is known of her writings, and more legend is known of her than any true facts. It is thought that she was eventually killed by a mob.

What is the story behind "Archimedes in the bathtub"?

One of the most famous stories of Archimedes involves royalty: Hiero II of Syracuse, King of Sicily, wanted to determine if a crown (actually, a wreath) he had ordered was truly pure gold or alloyed with silver—in other words, whether or not the Royal Goldsmith had substituted some of the gold with silver. The king called in Archimedes to solve the problem. The Greek mathematician knew that silver was less dense than gold (in other words, silver was not as heavy as gold), but without pounding the crown into an easily weighed cubic shape, he didn't know how to determine the relative density of the irregularly shaped crown.

Perplexed, the mathematician did what many people do to get good ideas: he took a bath. As he entered the tub, he noticed how the water rose, which made him realize that the volume of the water that fell out of the tub was equal to that of the volume of his body. Legend has it that Archimedes ran naked through the streets shouting "Eureka!" ("I have found it!") He knew that a given weight of gold represented a smaller volume than an equal weight of silver because gold is much denser than silver, so not as much is needed to displace the water. In other words, a specific amount of gold would displace less water than an equal weight of silver.

The next day, Archimedes submerged the crown and an amount of gold equal to what was supposed to be in the crown. He found that Hiero's crown displaced less water than an equal weight of gold, thus proving the crown was alloyed with a less dense material (the silver) and not pure gold. This eventually led to the hydrostatic principle, as it is now called, presented in Archimedes's appropriately named treatise, *On Floating Bodies*. As for the goldsmith, he was beheaded for stealing the king's gold.

What is the **origin** of **Roman numerals**?

Because the history of Roman numerals is not well documented, their origin is highly debated. It is thought that the numerals were developed around 500 BCE, partially from primitive Greek alphabet symbols that were not incorporated into Latin. The actual reasons for the seven standard symbols are also argued. Some researchers believe the symbol for 1 (I) was derived from one digit on the hand; the symbol for 5 (V) may have developed because the outstretched hand held vertically forms a "V" from the space between the thumb and first finger; the symbol for 10 (X) may have been two Vs joined at the points, or it may have had to do with the way people or merchants used their hands to count in a way that resembled an "X." All the reasons offered so far have merely been educated guesses.

How ever the symbols were developed, they were used with efficiency and with remarkable aptitude by the Romans. Unlike the ancient Greeks, the Romans weren't

Why were early Greek calendars such a mess?

Unlike the Mesopotamian cultures, the early Greeks paid less attention to astronomy and more to cosmology (they were interested in studying where the Earth and other cosmic bodies stand in relation to the universe). Because of this, their astronomical observations were not accurate, creating confusing calendars. This also led to a major conundrum: Almost every Greek city kept time differently. In fact, during the Greek and Hellenistic times, most dates were given in terms of the Olympiads. This only created another time-keeping problem: If something happened during the 10th Olympiad, it meant the event occurred within a four-year span. Such notation creates headaches for historians, who end up making educated guesses as to the actual dates of Greek events, important people's deaths and births, and other significant historical occurrences.

truly interested in "pure" math, such as abstract geometry. Instead, they concentrated on "applied math," using mathematics and their Roman numerals for more practical purposes, such as building roads, temples, bridges, and aqueducts; for keeping merchant accounts; and for managing supplies for their armies.

Centuries after the Roman Empire fell, various cultures continued to use Roman numerals. Even today, the symbols are still in existence; they are used on certain timepieces, in formal documents, and for listing dates in the form of years. For example, just watch the end credits of your favorite movie or television program and you will often see the movie's copyright date represented with Roman numerals.

What are the **basic Roman numerals** and how are they used?

There are only seven basic Roman numerals, as seen in the following chart:

Number	Roman Numeral
1	I
5	V
10	X
50	L
100	C
500	D
1000	M

There are many rules, of course, to this method of writing numerals. For example, although the way to write a large number like 8,000 would be "MMMMMMMM," this is very cumbersome. In order to work with such large numbers, one rule was to write a

What was the "House of Wisdom"?

Around 786, the fifth Caliph of the Abbasid Dynasty began with Caliph Harun al-Rashid, a leader who encouraged learning, including the translation of many major Greek treatises into Arabic, such as Euclid's *Elements*. Al-Ma'mun (786–833), the next Caliph, was even more interested in scholarship, creating the House of Wisdom in Baghdad, one of several scientific centers in the Islamic Empire. Here, too, Greek works such as Galen's medical writings and Ptolemy's astronomical treatises were translated, not by language experts ignorant of mathematics, but by scientists and mathematicians such as Al-Kindi (801–873), Muhammad ibn Musa al-Khuwarizmi (see below), and the famous translator Hunayn ibn Ishaq (809–873).

bar over a numeral, meaning to multiply by 1,000. Thus, 8,000 would be VIII—equal to our Hindu-Arabic number 8—with a bar over the entire Roman numeral.

OTHER CULTURES AND EARLY MATHEMATICS

What did the **Chinese** add to the **study of mathematics**?

Despite the attention the Greeks have received concerning the development of mathematics, the Chinese were by no means uninterested in it. About the year 200 BCE, the Chinese developed place value notation, and 100 years later they began to use negative numbers. By the turn of the millennium and a few centuries beyond, they were using decimal fractions (even for the value of "pi" [π]) and the first magic squares (for more information about math puzzles, see "Recreational Math"). By the time European cultures had begun to decline—from about 530 to 1000 CE—the Chinese were contributing not only to the field of mathematics, but also to the study of magnetism, mechanical clocks, physical laws, and astronomy.

What is the most **famous Chinese mathematics book**?

The *Jiuzhang suanshu*, or *Nine Chapters on the Mathematical Art*, is the most famous mathematical book to come out of ancient China. This book dominated mathematical development for more than 1,500 years, with contributions by numerous Chinese scholars such as Xu Yue (c. 160–c. 227), though his contributions were lost. It contains 246 problems meant to provide methods to solve everyday questions concerning engineering, trade, taxation, and surveying.

21

Why is Omar Khayyám so famous?

Omar Khayyám is not as well known for his contributions to math as he is for being immortalized by Edward FitzGerald, the 19th-century English poet who translated Khayyám's own 600 short, four-line poems in the *Rubaiyat*. However, FitzGerald's translations were not exact, and most scholars agree that Khayyám did not write the line "a jug of wine, a loaf of bread, and Thou." Those words were actually conceived by FitzGerald. Interestingly enough, versions of the forms and verses used in the *Rubaiyat* existed in Persian literature long before Khayyám, and only about 120 verses can be attributed to him directly.

Who was **Aryabhata I**?

Aryabhata I (c. 476–550) was an Indian mathematician. Around 499 he wrote a treatise on quadratic equations and other scientific problems called *Aryabhatiya* in which he also determined the value of 3.1416 for pi (π). Although he developed some rules of arithmetic, trigonometry, and algebra, not all of them were correct.

What were some of the **contributions by the Arab world** to mathematics?

From about 700 to 1300, the Islamic culture was one of the most advanced civilizations in the West. The contributions of Arabic scholars to mathematics were helped not only by their contact with so many other cultures (mainly from India and China), but also because of the Islamic Empire's unifying, dominant Arabic language. Using knowledge from the Greeks, Arabian mathematics grew; the introduction of Indian numerals (often called Arabic numerals) also helped with mathematical calculations.

What are some **familiar Arabic terms** used in **mathematics**?

There are numerous Arabic terms we use today in our studies of mathematics. One of the most familiar is the term "algebra," which came from the title of the book *Al jabr w'al muqābalah* by Persian mathematician Muhammad ibn Musa al-Khuwarizmi (783–c. 850; also seen as al-Khowarizmi and al-Khwarizmi); he was the scholar who described the rules needed to do mathematical calculations in the Hindu-Arabic numeration system. The book, whose title is roughly translated as *Transposition and Reduction,* explains all about the basics of algebra. (For more information, see "Algebra.")

Another Arabic derivation is "algorithm," which stems from the Latinized version of Muhammad ibn Musa al-Khuwarizmi's own name. Over time, his name evolved from al-Khuwarizmi to Alchoarismi, then Algorismi, Algorismus, Algorisme, and finally Algorithm.

Who was **Omar Khayyám**?

Omar Khayyám (1048–1131), who was actually known as al-Khayyami, was a Persian mathematician, poet, and astronomer. He wrote the *Treatise on Demonstration of Problems of Algebra,* a book that contains a complete classification of cubic equations with geometric solutions, all of which are found by means of intersecting conic sections. He solved the general cubic equation hundreds of years before Niccoló Tartaglia in the 16th century, but his work only had positive roots, because it was completely geometrical (see elsewhere in this chapter for more about Tartaglia). He also calculated the length of the year to be 365.24219858156 days—a remarkably accurate result for his time—and proved that algebra was definitely related to geometry.

MATHEMATICS AFTER THE MIDDLE AGES

Who first introduced **Arabic notation** and the **concept of zero** to Europe?

Italian mathematician Leonardo of Pisa (c. 1170–c. 1250, who was also known as Fibonacci, or "son of Bonacci," although some historians say there is no evidence that he or his contemporaries ever used that name) brought the idea of Arabic notation and the concept of zero to Europe. His book *Liber abaci (The Book of the Abacus)* not only introduced zero but also the arithmetic and algebra he had learned in Arab countries. Another book, *Liber quadratorum (The Book of the Square)* was the first major European advance in number theory in a thousand years. He is also responsible for presenting the Fibonacci sequence. (For more information about Fibonacci and the Fibonacci sequence, see "Math Basics.")

What were the **major reasons** for **16th-century advances** in European mathematics?

There are several reasons for advances in mathematics at the end of the Middle Ages. The major reason, of course, was the beginning of the Renaissance, a time when there was a renewed interest in learning. Another important event that pushed mathematics was the invention of printing, which made many mathematics books, along with useful mathematical tables, available to a wide audience. Still another advancement was the replacement of the clumsy Roman numeral system by Hindu-Arabic numerals. (For more information about the Hindu-Arabic numerals, see "Math Basics.")

Who was **Scipione del Ferro**?

There were several mathematicians in the 16th century who worked on algebraic solutions to cubic and quartic equations. (For more information on cubic and quartic equa-

23

tions, see "Algebra.") One of the first was Scipione del Ferro (1465–1526), who in 1515 discovered a formula to solve cubic equations. He kept his work a complete secret until just before his death, when he revealed the method to his student Antonio Maria Fiore.

Who was **Adam Ries**?

Adam Ries (1492–1559) was the first person to write several books teaching the arithmetic method by the old abacus and new Indian methods; his books also presented the basics of addition, subtraction, multiplication, and division. Unlike most books of his time that were written in Latin and only understood by mathematicians, scientists, and engineers, Ries's works were written in his native German and were therefore understood by the general public. The books were also printed, making them more readily available to a wider audience.

Who was **François Viète**?

French mathematician François Viète (or Franciscus Vieta, 1540–1603) is often called the "founder of modern algebra." He introduced the use of letters as algebraic symbols (although Descartes [see below] introduced the convention of letters at the end of the alphabet [x, y, …] for unknowns and letters at the beginning of the alphabet [a, b, …] for knowns), and connected algebra with geometry and trigonometry. He also included trigonometric tables in his *Canon Mathematicus* (1571), along with the theory behind their construction. This book was originally meant to be a mathematical introduction to his unpublished astronomical treatise, *Ad harmonicon coeleste*. (For more about Viète, see "Algebra" and "Geometry and Trigonometry.")

What **century** produced the greatest **revolution** in **mathematics**?

Many mathematicians and historians believe that the 17th century saw not only the unprecedented growth of science but also the greatest revolution in mathematics. This century included the discovery of logarithms, the study of probability, the interactions between mathematics, physics, and astronomy, and the development of one of the most profound mathematical studies of all: calculus.

Who explained the **nature of logarithms**?

Scottish mathematician John Napier (1550–1617) first conceived the idea of logarithms in 1594. It took him 20 years, until 1614, to publish a canon of logarithms called *Mirifici logarithmorum canonis descripto* (*Description of the Wonderful Canon of Logarithms*). The canon explains the nature of logarithms, gives their rules of use, and offers logarithmic tables. (For more about logarithms, see "Algebra.")

What was the scandal between mathematicians working on cubic and quartic equations?

The early work on cubic equations was a tale of telling secrets, all taking place in Italy. No sooner had Antonio Maria Fiore (1526?–?)—considered a mediocre mathematician by scholars—received the secret of solving the cubic equation from Scipione del Ferro than he was spreading the rumor of its solution. A self-taught Italian mathematical genius known as Niccoló Tartaglia (1500–1557?; nicknamed "the stutterer") was already discovering how to solve many kinds of cubic equations. Not to be outdone, Tartaglia pushed himself to solve the equation $x^3 + mx^2 = n$, bragging about it when he had accomplished the task.

Fiore was outraged, which proved to be a fortuitous event for the study of cubic (and eventually quartic) equations. Demanding a public contest between himself and Tartaglia, the mathematicians were to give each other 30 problems with 40 to 50 days in which to solve them. Each problem solved earned a small prize, but the winner would be the one to solve the most problems. In the space of two hours, Tartaglia solved all Fiore's problems, all of which were based on $x^3 + mx^2 = n$. Eight days before the end of the contest, Tartaglia had found the general method for solving all types of cubic equations, while Fiore had solved none of Tartaglia's problems.

But the story does not end there. Around 1539, Italian physician and mathematician Girolamo Cardano (1501–1576; known in English as Jerome Cardan) stepped into the picture. Impressed with Tartaglia's abilities, Cardano asked him to visit. He also convinced Tartaglia to divulge his secret solution of the cubic equation, with Cardano promising not to tell until Tartaglia published his results.

Apparently, keeping secrets was not a common practice in Italy at this time, and Cardano beat Tartaglia to publication. Cardano eventually encouraged his student Luigi (Ludovico) Ferrari (1522–?) to work on solving the quartic equation, or the general polynomial equation of the fourth degree. Ferrari did just that, and in 1545 Cardano published his Latin treatise on algebra, *Ars Magna* (*The Great Art*), which included a combination of Tartaglia's and Ferrari's works in cubic and quartic equations.

Who originated **Cartesian coordinates**?

Cartesian coordinates are a way of finding the location of a point using distances from perpendicular axes. (For more information about coordinates, see "Geometry and Trigonometry.") The first steps toward such a coordinate system were suggested by French philosopher, mathematician, and scientist René Descartes (1596–1650; in

Latin, Renatus Cartesius); he was the first to publish a work explaining how to use coordinates for finding points in space. Around the same time, Pierre de Fermat developed the same idea independently (see below). Both Descartes's and Fermat's ideas would lead to what is now known as Cartesian coordinates.

Descartes is also considered by some to be the founder of analytical geometry. He contributed to the ideas involved in negative roots and exponent notation, explained the phenomenon of rainbows and the formation of clouds, and even dabbled in psychology.

Who was **Pierre de Fermat**?

French mathematician Pierre de Fermat (1601–1665) made many contributions to early methods leading to differential calculus; he was also considered by some to be the founder of modern number theory (see "Math Basics") and did much to establish coordinate geometry, eventually leading to Cartesian coordinates. He supposedly proved a theorem eventually called "Fermat's Last Theorem." It states that the equation $x^n + y^n = z^n$ has no non-zero integer solutions for x, y, and z when n is greater than 2. But there is no proof of Fermat's "proof," making most mathematicians skeptical about his supposed discovery.

Was **Fermat's last theorem** finally **solved**?

Just before the end of the 19th century, German industrialist and amateur mathematician Paul Wolfskehl, on the brink of suicide, began to explore a book on Fermat's Last Theorem. Enchanted with the numbers, he forgot about dying and instead believed that mathematics had saved him. To repay such a debt, he left 100,000 marks (about $2 million in today's money) to the Göttingen Academy of Science as a prize to anyone who could publish the complete proof of Fermat's Last Theorem. Announced in 1906 after Wolfskehl's death, thousands of incorrect proofs were turned in, but no true proof was offered.

But people kept trying—and failing. Fermat's Last Theorem was finally solved in 1994 by English mathematician Andrew John Wiles (1953–). Wiles was offered the Wolfskehl prize in 1997. By that time, the original $2 million had been affected by not only hyperinflation but also the devaluation of the mark, reducing its value to $50,000. But for Wiles, it didn't matter; his proving the Last Theorem had been a childhood dream.

It is interesting to note that some mathematicians do not believe Wiles uncovered the true proof of Fermat's Last Theorem. Instead, because many of the mathematical techniques used by Wiles were developed within recent decades (some even by Wiles himself), Wiles's proof—although a masterpiece of mathematics—could not possibly be the same as Fermat's. Still other mathematicians wonder about Fermat's words in claiming that he had found a proof. Was it really a proven or flawed proof he was talk-

ing about? Or was he such a genius that he took the proof he was able to see, in his time, to his grave? Like so many mysteries of history, we may never know.

Who began the mathematical study of **probability**?

French scientist and religious philosopher Blaise Pascal (1623–1662) is known not only for the study of probability but for many other mathematically oriented advances, such as a calculation machine (invented at age 19 to help his father with tax calculations, but it performed only additions), hydrostatics, and conic sections. He is also credited (along with Fermat) as the founder of modern theory of probability. (For more information about probability, see "Applied Mathematics.")

Seventeenth-century scientist Blaise Pascal was the founder of mathematical probability, as well as other achievements, such as devising one of the first calculating machines.

Who was **Sir Isaac Newton**?

Sir Isaac Newton (1642–1727) was an English mathematician and physicist considered by some to be one of the greatest scientists who ever lived. He was credited with inventing differential calculus in 1665 and integral calculus the following year. (For more information about calculus, see "Mathematical Analysis.") The list of his achievements—mathematical and scientific—does not end there: He is also credited as the discoverer of the general binomial theorem, he worked on infinite series, and he even made advancements in optics and chemistry.

Some of Newton's greatest contributions include the development of the law of universal gravitation, rules of planetary orbits, and sundry other astronomical concepts. By 1687, Newton had written one of his most famous books, *The Principia* or *Philosophiae naturalis principia mathematica* (*The Mathematical Principles of Natural Philosophy*), which is often called the greatest scientific book ever written. In it Newton presents his theories of motion, gravity, and mechanics. Although he had developed calculus earlier, he still used the customary classical geometry to work out physical problems within the book.

Who was Baron **Gottfried Wilhelm Leibniz**?

A contemporary of Isaac Newton, German philosopher and mathematician Baron Gottfried Wilhelm Leibniz (1646–1716) is considered by some to be a largely forgotten

mathematician, although his contributions to the field were just as important as Newton's in many ways. He is often called the founder of symbolic logic; he introduced the terms *coordinate, abscissas,* and *ordinate* for the field of coordinate geometry; he invented a machine that could do multiplication and division; he discovered the well-known series for pi divided by 4 ($\pi/4$) that bears his name; and he independently developed infinitesimal calculus and was the first to describe it in print. Because his work on calculus was published three years before Isaac Newton's, Leibniz's system of notation was universally adopted.

Who was considered the **first statistician**?

English statistician and tradesman John Graunt (1620–1674) was the first true statistician and wrote the first book on statistics, although statistics in a simpler form was known long before that. Graunt, a draper by profession, was the first to use a compilation of data, which in this case involved the records of bills of mortality, or the records of how and when people died in London from 1604 to 1661. In his *Natural and Political Observations Made upon the Bills of Mortality,* he determined certain inclinations, such as more boys were born than girls, women tend to live longer than men, etc. He also developed the first mortality table, which showed how long a person might expect to live after a certain age, a concept very familiar to us today, especially in fields such as insurance and health.

Why was the **Bernoulli family** important to mathematics?

The Bernoulli (also seen as Bernouilli) family of the 17th and 18th centuries is synonymous with mathematics and science. One of the developers of ordinary calculus, calculus of variations, and the first to use the word "integral" was Jacob Bernoulli (1654–1705; also known as Jakob, Jacques, or James). He also wrote about the theory of probability, is often credited for developing the field of statistics, and discovered a series of numbers that bear his name: the coefficients of the exponential series expansion of $x/(1 - e^{-x})$.

Not to be outdone, his brother Johann (1667–1748; also known as Jean or John) contributed to the field of integral and exponential calculus, was the founder of calculus of variations, and worked on geodesics, complex numbers, and trigonometry. His son was not far behind: Daniel Bernoulli (1700–1782) was considered the first mathematical physicist, publishing *Hydrodynamica* in 1738, which included his now famous principle named in his honor (Bernoulli's principle); and he brought out two ideas that were ahead of his time by many years: the law of conservation of energy and the kinetic-molecular theory of gases.

The Bernoulli legacy did not end there, with family members continuing to make great mathematical and scientific contributions. There were two Nicolaus Bernoullis: one, the brother of Jacob and Johann (1662–1716), was professor of mathematics at St.

What was in Joseph-Louis Lagrange's letter to Jean le Rond d'Alembert?

Italian-French astronomer and mathematician Comte Joseph-Louis Lagrange (1736–1813) made significant discoveries in mathematical astronomy, including many functions, theories, etc. that bear his name (for example, Lagrange point, Lagrange's equations, Lagrange's theorem, Lagrangian function). His mentor was none other than French scientist Jean le Rond d'Alembert (1717–1783), a physicist who expanded on Newton's laws of motion, contributed to the field of fluid motion, described the regular changes in the Earth's axis, and was the first to use partial differential equations in mathematical physics. He even had time to edit, along with French philosopher Denis Diderot (1713–1784), the *Encyclopedié,* a 17-volume encyclopedia of scientific knowledge published from 1751 to 1772.

Apparently, living in the years of such mathematical enlightenment had its drawbacks. In 1781 Lagrange wrote a letter to d'Alembert about his greatest fear: that the field of mathematics had reached its limit. At that point in time, Lagrange believed everything mathematical had been discovered, uncovered, and calculated. Little did he realize that mathematics was only in its infancy.

Petersburg, Russia's Academy of Sciences; the other, the son of Johann and brother of Daniel (1695–1726), was also a mathematician. Another Johann Bernoulli (1710–1790) was another son of Johann (and brother of Daniel), who succeeded his father in the chair of mathematics at Basel, Switzerland, and also contributed to physics. The younger Johann also had a son named Johann (1746–1807), who was astronomer royal in Berlin and also studied mathematics and geography. Finally, Jacob Bernoulli (1759–1789), yet another son of the younger Johann, succeeded his uncle Daniel in teaching mathematics and physics at St. Petersburg, but he met an untimely death by drowning.

Who was one of the most **prolific mathematicians** who ever lived?

Swiss mathematician Leonhard Euler (1707–1783) is considered to be one of the most prolific mathematicians who ever lived. In fact, his accomplishments are beyond the scope of this text. Suffice it to say that his collected works number more than 70 volumes, with contributions in pure and applied mathematics, including the calculus of variations, analysis, number theory, algebra, geometry, trigonometry, analytical mechanics, hydrodynamics, and the lunar theory (calculation of the motion of the Moon). Euler was one of the first to develop the methods of the calculus on a wide scale. His most famous book, *Elements of Algebra,* rapidly became a classic; and he wrote a geometry textbook (Yale University was the first American college to use the text).

Although half-blind for much of his life—and totally blind for his last 17 years—he had a near-legendary skill at calculation. Among his discoveries are the differential equation named for him (a formula relating the number of faces, edges, and vertices of a polyhedron, although Euler's formula was discovered earlier by René Descartes); and a famous equation connecting five fundamental numbers in mathematics. Like many in the Bernoulli family, Euler eventually worked at the Academy of Sciences in St. Petersburg, Russia, a center of learning founded by Peter the Great.

Who was **Karl Friedrich Gauss**?

German mathematician, physicist, and astronomer Karl Friedrich Gauss (1777–1855; also seen as Johann Carl [or Karl] Friedrich Gauss) was considered one of the greatest mathematicians of his time; some have even compared him to Archimedes and Newton. His greatest mathematical contributions were in the fields of higher arithmetic and number theory. He discovered the law of quadratic reciprocity, determined the method of least squares (independently of French mathematician Adrien-Marie Legendre [1752–1833]), popularized the symbol "i" as the square root of negative 1 (although Euler first used the symbol), did extensive investigations in the theory of space curves and surfaces, made contributions to differential geometry, and much more. In 1801, after the discovery (and subsequent loss) of the first asteroid, Ceres, by Giuseppe Piazzi, he calculated the object's orbit with little data; the asteroid was found again thanks to his calculations. He further calculated the orbits of asteroids found over the next few years.

When was **non-Euclidean geometry** first announced?

Non-Euclidean geometry—or a system of geometry different from that developed by Euclid (see p. 17)—was first announced by Russian mathematician Nikolai Ivanovich Lobachevski (1792–1856; also seen as Lobatchevsky) in 1826. This idea had already been independently developed by the Hungarian János (or Johann) Bolyai (1802–1860) in 1823 and by Karl Friedrich Gauss (1777–1855) in 1816, but Lobachevski was the first to publish on the subject.

In 1854 German mathematician Georg Friedrich Bernhard Riemann (1826–1866) presented several new general geometric principles. His suggestion of another form of non-Euclidean geometry further established this new way of looking at geometry. Riemann was also responsible for presenting the Riemann hypothesis (or zeta function), a complex function that remains an unsolved issue in mathematics today. (For more information about geometry and Riemann, see "Geometry and Trigonometry.")

Who developed the first ideas on **symbolic logic**?

English mathematician George Boole (1815–1864) was the first to develop ideas on symbolic logic, that is, the use of symbols to represent logical principles. He proposed

> ### Why was non-Euclidean geometry important to Albert Einstein?
>
> **N**on-Euclidian geometry, especially the form suggested by Bernhard Riemann, enabled Albert Einstein (1879–1955) to work on his general relativity theory (1916), showing that the true geometry of space may be non-Euclidean. (For more information about mathematics and Einstein, see "Math in the Physical Sciences.")

this in his treatise, *An Investigation of the Laws of Thought, on Which Are Founded the Mathematical Theories of Logic and Probabilities* (1854). Today, this is called Boolean algebra. (For more information about Boole, see "Algebra"; for more information about symbolic logic, see "Foundations of Mathematics.")

MODERN MATHEMATICS

Who first developed **set theory**?

German mathematician George (Georg) Ferdinand Ludwig Philipp Cantor (1845–1918) was not only known for his work on transfinite numbers, but also for his development of set theory, which is the basis of modern mathematical analysis (for more information on set theory, see "Foundations of Mathematics"). His *Mathematische Annalen* was a basic introduction to set theory. Unlike most long evolutionary histories of mathematical subjects, Cantor's set theory was his creation alone. In the late 19th century, Cantor also developed the Continuum Hypothesis. He realized that there were many different sized infinities, further conjecturing that two particular infinities constructed by different processes were the same size.

What was the *Principia Mathematica*?

In 1910 the first volume of the *Principia Mathematica* was published by Welsh mathematician and logician Bertrand Arthur William Russell (1872–1970) and English mathematician and philosopher Alfred North Whitehead (1861–1947). This book was an attempt to put mathematics on a logical foundation, developing logic theory as a basis for mathematics. It gave detailed derivations of many major theorems in set theory, examined finite and transfinite arithmetic, and presented elementary measure theory. The two mathematicians published three volumes, but the fourth, on geometry, was never completed.

On their own, both men did a great deal to advance mathematics, too. Russell discovered the Russell paradox (see below), introduced the theory of types, and popularized first-order predicate calculus. Russell's logic consisted of two main ideas: that all

mathematical truths can be translated into logical truths (or that the vocabulary of mathematics constitutes a proper subset of the vocabulary of logic) and that all mathematical proofs can be recast as logical proofs (or that the theorems of mathematics constitute a proper subset logical theorems).

Whitehead excelled not only in mathematics and logic but also in the philosophy of science and study of metaphysics. In mathematics, he extended the known range of algebraic procedures, and he was a prolific writer. In philosophy, he criticized the traditional theories for their lack of integrating the direct relationship between matter, space, and time; thus, he created a vocabulary of his own design, which he called the "philosophy of organism."

Who was **Kurt Gödel**?

For about a hundred years, mathematicians such as Bertrand Russell were trying to present axioms that would define the entire field of mathematics on an axiomatic basis. Austrian-American mathematician and logician Kurt Gödel (1906–1978) was the first to suggest that any formal system strong enough to include the laws of mathematics is either incomplete or inconsistent; this was called "Gödel's Incompleteness Theorem." Thus, axioms could not define all of mathematics.

Gödel also stated that the various branches of mathematics are based in part on propositions that are not provable within the system itself, although they may be proved by means of logical (metamathematical) systems external to mathematics. In other words, nothing is as simple as it seems; and, interestingly enough, Gödel's idea also implies that a computer can never be programmed to answer all mathematical questions.

What did **David Hilbert propose** in 1900?

In 1900 German mathematician David Hilbert (1862–1943) proposed 23 unsolved mathematical problems for the new century, most of which only proved to bring up

What was the "Golden Age of Logic"?

Kurt Gödel's work led to what is often described as the Golden Age of Logic. Spanning the years from about 1930 to the late 1970s, it was a time when there was a great deal of work done in mathematical logic. From the beginning, mathematicians broke into many camps that worked on various phases of logic (for more information about logic, see "Foundations of Mathematics"), including:

Proof theory—In which the mathematical proofs started by Aristotle and continued by Boole (see p. 30) were extensively studied, resulting in branches of this mathematics being applied to computing (including artificial intelligence).

Model theory—In which mathematicians investigated the connection between the truth in a mathematical structure and propositions about that structure.

Set theory—In which a breakthrough in 1963 showed that certain mathematical statements were undeterminable, a direct challenge to the major set theories of the time. This showed that Cantor's Continuum Hypothesis (see p. 31) is independent of the axioms of set theory, or that there are two mathematical possibilities: one that says the continuum hypothesis is true, one that says it is false.

Computability theory—In which mathematicians worked out the abstract theorems that would eventually help lead to computer technology. For example, English mathematician Alan Turing proved an abstract theorem that established the theoretical possibility of a single computing machine programmed to complete any computation. (For more information about Turing and computers, see p. 34 and "Math in Computing.")

other problems. By the 1920s Hilbert gathered many mathematicians—called the formalists—to prove that mathematics was consistent. But all did not go well as mathematical complications set in. By 1931 Kurt Gödel's Incompleteness Theorem dashed any more efforts by the formalists by proving that mathematics is either inconsistent or incomplete. (For more about Hilbert, see "Foundations of Mathematics.")

When was **quantum mechanics** developed?

There was not one major year in which quantum mechanics was developed, or even one major scientist who proposed the idea. This modern theory of physics evolved over about 30 years, with many scientists contributing to it. Beginning about 1900 Max Planck proposed that energies of any harmonic oscillator (such as the atoms of a black body radiator) are restricted to certain values. Mathematics came into play here,

too, with each value an integral multiple of a basic, minimum value. Planck developed the equation $E = h\nu$ (or h times "nu"), in which E (the energy of the basic quantum) is directly proportional to the ν (the frequency of the oscillator) multiplied by h, or Planck's constant (6.63×10^{-34} joule-second).

From there, mainly with the use of rigorous mathematics, others expanded or added to Planck's idea, including German scientist Albert Einstein (1879–1955), who explained the photoelectric effect; New Zealand-born British physicist Ernest Rutherford (1871–1937) and Danish physicist Neils Bohr (1885–1962), who explained both atomic structure and spectra; Austrian physicist Erwin Schrödinger (1887–1961), who developed wave mechanics; and German physicist Werner Karl Heisenberg (1901–1976), who discovered the uncertainty principle. Out of these studies came quantum mechanics (in the 1920s), quantum statistics, and quantum field theory. Today, quantum mechanics and Einstein's theory of relativity form the foundation of modern physics. These theories continually change or are modified as we get closer to understanding more about the physics—and mathematics—of our universe.

Who was **Alan Turing**?

British mathematician Alan Mathison Turing (1912–1954) was the first person to propose the idea of a simple computer. Called the Turing machine, its operation was limited to reading and writing symbols on tape, moving the tape to the left or right to read the symbols one at a time. This invention is often considered the start of the computer age. In fact, the definition of the word "computable" is a problem that can be solved by a Turing machine. Turing was also instrumental in interpreting and deciphering encrypted German messages using the Enigma cipher machine. (For more information on computers, see "Math in Computing.")

What is **chaos theory**?

Chaos theory is one of the "newest" ideas in mathematics. Developed in the last half of the 20th century, it affects not only math, but also physics, geology, biology, meteorology, and many other fields. Modern ideas about chaos began when theorists in various scientific disciplines started to question the linear analysis used in classical applied mathematics, most of which presumes an orderly periodicity that rarely occurs in nature. In the search to discover regularities, the idea of disorder had been ignored. To overcome this problem, chaos theorists developed deterministic, nonlinear dynamic models that explain irregular, unpredictable behavior. By 1961, American meteorologist Edward Norton Lorenz (1917–) noticed that small variations in the initial values of variables in his primitive computer weather model resulted in major divergent weather patterns. His discovery of a simple mathematical system with chaotic behavior led to the new mathematics of chaos theory.

Chaos theory recognizes the unpredictability of life, including the world's highly complex weather system, which can be influenced by a myriad of factors ranging from changes in temperature and humidity to alterations in geology and agricultural development. *Taxi/Getty Images.*

The use of chaos theory has enabled scientists and mathematicians to reveal the structure in aperiodic, unpredictable dynamic systems. For example, it has been used to examine crystal growth, the expansion of pollution plumes in water and in the air, and even to determine the formation of storm clouds. One of the reasons chaos theory has come to the forefront of science and mathematics is because of advancements in computers; high-end computers allow for a plethora of variables to enter into the complex chaos equations.

Who invented **catastrophe theory**?

Catastrophe theory—or the study of gradually changing forces that lead to so-called catastrophes (or abrupt changes)—was popularized by French mathematician René Thom (1923–2002) in 1972. Unable to use differential calculus in certain situations, Thom used other mathematical treatments of continuous action to produce a discontinuous result. Although it is not as popular as it once was, it is often used in biological and optical applications.

Who is **Benoit Mandelbrot**?

Benoit B. Mandelbrot (1924–) is the Polish-born, French mathematician who invented a branch of mathematics called fractal geometry, which is designed to find order in

35

Why is there no Nobel Prize for mathematics?

The Nobel Prizes were established at the bequest of Swedish chemical engineer Alfred Bernhard Nobel (1833–1896), the discoverer of dynamite. First awarded in 1901, the Nobel Prizes honor innovators in the fields of chemistry, physics, physiology or medicine, literature, and peace; a prize in economics was added in 1969, but there is no award for mathematics.

The lack of a mathematics prize has many stories attached, including one that states that Nobel's wife jilted him for Norwegian mathematician Magnus Gösta Mittag-Leffler, a notion made less plausible by the fact that Nobel never married. Most historians agree, however, that the reason has to do with Nobel's attitude toward mathematics: He simply did not consider mathematics sufficiently practical. To fill the gap, the Fields Medal of the International Mathematical Congress was established in 1932; it has the equivalent prestige of the Nobel with the limitation that it is only awarded for work done by mathematicians younger than 40 years old, and the monetary value is a mere $15,000 in Canadian dollars (or about $12,000 in U.S. dollars at press time).

But mathematics has not been left out of award-winning ceremonies. In 2003 Norway created the Abel Prize for mathematic achievement. Named after Norwegian mathematician Niels Henrik Abel (1802–1829), who proved that solving fifth-degree algebraic equations (quintics) is impossible, the award gives the winner a prize of six million Norwegian kroners (about $935,000 in American currency).

apparently erratic shapes and processes. A largely self-taught mathematician who did not like pure logical analysis, he was a pioneer of chaos theory, developing and finding applications for fractal geometry. Unlike traditional geometry with its regular shapes and whole-number dimensions, fractal geometry uses shapes found in nature with non-integer (or fractal—thus the name) dimensions. For example, twigs, tree branches, river systems, and shorelines can be examined using fractals. Today, fractals are often applied not only to the natural world but also to the chemical industry, computer graphics, and even the stock market.

MATHEMATICS THROUGHOUT HISTORY

THE CREATION OF ZERO AND PI

How did the **concept of zero evolve** over time?

The concept of zero developed because it was necessary to have a placeholder—or a number that holds a place—to make it easier to designate numbers in the tens, hundreds, thousands, etc. For example, the number 4,000 implies that the three places to the right of the 4 are "empty"—with only the thousandths column containing any value. Because zero technically means nothing, at first few people accepted the concept of "nothing" between numbers. Not that all cultures ignored the possibility of such an idea. For example, Hindu mathematicians, who wrote their math in verse, used words similar to "nothing," such as *sunya* ("void") and *akasa* ("space"). It is thought that the Babylonians were the first to have a placeholder in their numbering system, but not a zero; instead, it appears they used other symbols, such as a double hash-mark (also called wedges) as a placeholder.

Archeologists believe a crude symbol for zero probably started in Indochina or India about the 7th century—and by the Mayans independently about a hundred years earlier. While the isolated Mayans could not spread the idea of the zero, the Indians seemed to have no problem. Around 650 CE, zero became a mathematically important number in Indian mathematics—although the symbol was a bit different than today's zero. The familiar Hindu-Arabic symbol for zero—the open circle—would take several more centuries to become more readily accepted. (For more about zero and Hindu-Arabic symbols, see "History of Mathematics" and "Math Basics.")

What is **pi** and why is it important?

Pi (pronounced "pie"; the symbol is π) is the ratio of the circumference to the diameter of a circle. Another way of looking at pi is by the area of a circle: pi times the

square of the length of the radius, or as it is often phrased "pi r squared." There are other ways to consider the value of pi: 2 pi (2π) in radians is 360 degrees; thus, pi radians is 180 degrees and 1/2 pi ($1/2\pi$) radians is 90 degrees. (For more about pi and radians, see "Geometry and Trigonometry.")

What is the importance of pi? It was used in calculations to build the huge cathedrals of the Renaissance, to find basic Earth measurements, and it has been used to solve a plethora of other mathematical problems throughout the ages. Even today it is used in the calculations of items that surround everyone. To give just a few examples, it is used in geometric problems, such as machining parts for aircraft, spacecraft, and automobiles; in interpreting sine wave signals for radio, television, radar, telephones, and other such equipment; in all areas of engineering, including simulations and the modeling of a building's structural loads, and even to determine global paths of aircraft (airlines actually fly on an arc of a circle as they travel above the Earth).

What is the **value of pi**?

Pi is a number, a constant, and to 20 decimal places it is equal to 3.141592653589 79323846. But it doesn't end there: Pi is an infinite decimal. In other words, it has an infinite number of numbers to the right of the decimal point. Thus, no one will ever know the "end" number for pi. Not that mathematicians will stop trying any time soon. Today's supercomputers continue to work out the value of pi, and to date, researchers have taken the number to more than two hundred billion places. (For more about pi and computers, see "Math in Computing.")

Who first determined the value of pi?

People have been fascinated by pi throughout history. It was used by the Babylonians and Egyptians; the Chinese thought it stood for one thousand years. Some even give

Advances in architecture during the European Renaissance would not have been possible without similar advances in mathematics and a knowledge of the value of π (pi). This cathedral in York, England, is a prime example of what can be accomplished with mathematics. *Taxi/Getty Images.*

the Bible credit for mentioning the concept of pi (in which it apparently equaled 3): In one Biblical version of I Kings 7: 23–26, it states "And he made a molten sea, ten cubits from the one brim to the other: it was round all about, and his height was five cubits: and a line of thirty cubits did compass it about." The same verse is found in II Chronicles 4: 2–5 in reference to a vessel ("sea") made in the temple of Solomon, which was built around 950 BCE.

No one truly knows the origins of pi, although many historians believe it was probably figured out long ago. There are some clues as to its discovery, though. For example, some people claim the Egyptian Rhind papyrus (also called Ahmes papyrus), which was transcribed about 1650 BCE by Ahmes, an Egyptian scribe who claimed he was copying a 200-year-old document, contains a notation that pi equals 3.16, which is close to the real value of pi. (For more about the Rhind papyrus, see "History of Mathematics.")

But it was the Greeks who promoted the idea of pi the most: They were very interested in the properties of circles, especially the ratio of a circle to its diameter. In particular, Greek mathematician Archimedes (c. 287–212 BCE, Hellenic) computed close limits of pi by comparing polygons inscribed in and circumscribed about a circle. He applied the *method of exhaustion* to approximate the area of a circle, which,

in turn, led to a better approximation of pi (π). Through his iterations, he determined that $223/71 < \pi < 22/7$; the average of his two numbers equals 3.141851 (and so on). (For more about Archimedes, see "History of Mathematics" and "Geometry and Trigonometry.")

How was the value of **pi determined arithmetically**?

One of the earliest mathematical formulas for pi was determined by English mathematician John Wallis (1616–1703), who wrote the notation as:

$$2/\pi = (1.3.3.5.5.7. \ldots) / (2.2.4.4.6.6. \ldots)$$

Another more commonly recognized notation for pi is often attributed to German philosopher and mathematician Baron Gottfried Wilhelm Leibniz (1646–1716), but it is more likely the work of Scottish mathematician James Gregory (1638–1675):

$$\pi/4 = 1 - 1/3 + 1/5 - 1/7 + \ldots$$

Both are amazing examples of something that was not only figured out using geometric methods but also arithmetic methods.

What are the **measurements** of a **circle using pi**?

There are many measurements of a circle. The perimeter of a circle is called the *circumference;* to calculate it, multiply pi (π) times the diameter, or $c = (\pi d)$, or pi (π) times twice the radius, ($c = 2\pi r$). The area (a) of a circle is calculated by multiplying pi (π) times the radius squared, or $a = \pi r^2$.

When did people first start using measurements?

No one knows for certain the who, where, or when of the first use of measurements. No doubt people developed the first crude measurement systems out of necessity. For example, knowing the height of a human, versus the height of a lion, versus the height of the grass in which a human hid were probably some of the first (intuitive and necessary) measurements.

The first indications of measurements being used date back to around 6000 BCE in what today encompasses the area from Syria to Iran. As populations grew and the main source of food became farmland rather than wild game, new ways of calculating crops for growing and storage became necessary. In addition, in certain cultures during times of plenty, each person—depending on their status (from adult men who received the most, to women, children, and slaves who were given less)—received a specific measurement of food. During a famine, in order to stretch supplies, a certain minimal measurement of food was divided between each person. It is thought that the first true measuring was done by hand—in particular, measuring grains by the handful. In fact, the half-pint, or the contents of two hands cupped together, may be the only volume unit with a natural explanation.

DEVELOPMENT OF WEIGHTS AND MEASURES

What is **measurement**?

Measurement refers to the methods used to determine length, volume, distance, mass (weight), or some other quantity or dimension. Each measurement is defined by specific units, such as inches and centimeters for length, or pounds and kilograms for weight. Such measurements are an integral part of our world, from their importance in travel and trade, to weather forecasting and engineering a bridge.

Is **measurement tied** to **mathematics**?

Yes, measurement is definitely tied to mathematics. In particular, the first steps toward mathematics was using units (and eventually numbers) to describe physical quantities. There had to be a way to add and subtract the quantities, and most of those crude "calculations" were based on fundamental mathematics. For example, in order to trade horses for gold, merchants had to agree on how much a certain amount of gold (usually as weight) was worth, then translate that weight measurement into their barter system. In other words, "x" amount of gold would equal "y" amount of horses.

The advent of agriculture in human civilization necessitated the development of mathematical concepts so that farmers could better predict times to plant and harvest. *Gallo Images/Getty Images.*

What are some **standard measurement units** and their definitions?

For a helpful list of standard measurement units and systems for converting them to other types of units, see Appendix 1 in the back of this book.

Upon what system were **ancient measurements** based?

Initially, people used different measurement systems and methods depending on where they lived. Most towns had their own measurement system, which was based on the materials the residents had at hand. This made it difficult to trade from region to region.

Measurements eventually became based on familiar and common items. But that did not mean they were accurate. For example, length measurements were often based on parts of the human body, such as the length of a foot or width of the middle finger; longer lengths would be determined by strides or distances between outstretched arms. Because people were of different heights and body types, this meant the measurements changed depending on who did the measuring. Still, they were close enough for the needs at the time. Even longer lengths were based on familiar sights. For example, an acre was the amount of land that two oxen could plow in a day.

What is the **historical significance** of the **barleycorn** in measurement?

The barleycorn (just a grain of barley) definitely had a significant historical role in determining the length of an inch and the English foot (for more about the inch and foot, see below). In addition, in traditional English law, the various pound weights all referred to multiples of the "grain": A single barleycorn's weight equaled a grain, and multiples of a grain were important in weight measurement. Thus, some researchers believe the lowly barleycorn was actually at the origin of both weight and distance units in the English system.

Grains of barley, a common crop that was easy to obtain, were a convenient but not very accurate standard to use for measuring items in England before better standards were developed. *Taxi/ Getty Images.*

What were some **early units** used for calculating **length**?

The earliest length measurements reach back into ancient time, and it is a convoluted history. Some of the earliest measurements of length are the *cubit, digit, inch, yard, mile, furlong,* and *pace.* One of the earliest recorded length units is the *cubit.* It was invented by the Egyptians around 3000 BCE and was represented by the length of a man's arm from his elbow to his extended fingertips. Of course, not every person has the same body proportions, so a cubit could be off by a few inches. This was something the more precision-oriented Egyptians fixed by developing a *standard royal cubit.* This was maintained on a black granite rod accessible to all, enabling the citizenry to make their own measuring rods fit the royal standard.

The Egyptian cubit was not the only one. By 1700 BCE the Babylonians had changed the measurement of a cubit, making it slightly longer. In our measurement standards today, the Egyptian cubit would be equal to 524 millimeters (20.63 inches), and the Babylonian cubit (cubit II) would be equal to 530 millimeters (20.87 inches; the metric unit millimeters is used here, as it is an easier way to see the difference between these two cubits).

As the name implies, a *digit* was measured by the width of a person's middle finger and was considered the smallest basic unit of length. The Egyptians divided the digit into other units. For example, 28 digits equaled a cubit, four digits equaled a palm, and five digits equaled a hand. They further divided three palms (or 12 digits) into a small span, 14 digits (or a half cubit) into a large span, and 24 digits into a small cubit. To get smaller measurements than a digit, the Egyptians used fractions.

A cubit was once a common standard of measurement. One cubit equals the distance from a person's fingertips to the elbow. Of course, because different people have different body sizes, the length of a cubit would vary from person to person. *Stone/Getty Images.*

Over time, the measurement of an *inch* was all over the measurement map. For example, one inch was once defined as the distance from the tip to the first joint on a man's finger. The ancient civilization of the Harappan in the Punjab used the "Indus inch"; based on ruler markings found at excavation sites, it measured, in modern terms, about 1.32 inches (3.35 centimeters; see below for more about the Harappan). The inch was defined as one-thirty-sixth of King Henry I of England's arm in the 11th century, and by the 14th century, King Edward II of England ruled that one inch equaled three grains of barleycorn placed end to end lengthwise. (See box on p. 46 for more about both kings.)

Longer measurements were often measured by such units as yards, furlongs, and miles in Europe. At first, the *yard* was the length of a man's belt (also called a girdle). The yard became more "standard" for a while, when it was determined to be the distance from King Henry I's nose to the thumb of his outstretched arm. The term *mile* is derived from the Roman *mille passus,* or "1,000 double steps" (also called paces). The mile was determined by measuring 1,000 double steps, with each double step by a Roman soldier measuring five feet. Thus, 1,000 double steps equaled a mile, or 5,000 feet (1,524 meters). The current measurement of feet in a mile came in 1595, when, during the reign of England's Queen Elizabeth I, it was agreed that 5,280 feet (1,609 meters) would equal one mile. This was mainly chosen because of the popularity of the *furlong*—eight furlongs equaled 5,280 feet.

Finally, the *pace* was once attached to the Roman mile (see above). Today, a pace is a general measurement, defined as the length of one average step by an adult human, or about 2.5 to 3 feet (0.76 to 0.19 meters).

What were the **ancient definitions** of a **foot**?

Not all feet (or the foot) are created equal. The term *foot* in measurement has had a long history, with many stories claiming the origin-of-the-first-foot status. In fact, it seems as if the foot has ranged in size over the years—from 9.84 to 13.39 inches (25 to 34 centimeters)—depending on the time period and/or civilization.

For example, the ancient Harappan civilization of the Punjab (from around 2500 to

1700 BCE) used a measurement interpreted by many to represent a foot—a very large

What was the first civilization to use a decimal system of weights and measures?

Between 2500 and 1700 BCE, the Harappa (or Harappan) civilization of the Punjab—now a province in Pakistan—developed the earliest known decimal system of weights and measures (for more about decimals, see "Math Basics"). The proof was first found in the modern Punjab region, where cubical (some say hexahedral) weights in graduated sizes were uncovered at Harappa excavations.

Archaeologists believe that these weights were used as a standard Harappan weight system, represented by the ratio 1:2:4:8:16:32:64. The small weights have been found in many of the regional settlements and were probably used for trade and/or collecting taxes. The smallest weight is 0.8375 grams (0.00185 pounds), or as measured by the Harappa, 0.8525; the most common weight is approximately 13.4 grams (0.02954 pounds), or in Harappa, 13.64, the 16th ratio. Some larger weights represent a decimal increase, or 100 times the most common weight (the 16th ratio). Other weights correspond to ratios of 0.05, 0.1, 0.2, 0.5, 1, 2, 5, 10, 20, 50, 100, 200, and 500.

There is also evidence that the Harappan civilization had some of the most advanced length measurements of the time. For example, a bronze rod found at an excavation was marked in units of precisely 0.367 inch (0.93 centimeter). Such a measuring stick was perfect to plan roads, to construct drains for the cities, and even to build homes. An ivory scale found at Lothal, once occupied by the Harappan civilization, is the smallest division ever recorded on any measuring stick yet found from the Bronze Age, with each division approximately 0.06709 inch (0.1704 centimeter) apart.

foot, at about 13.2 inches (33.5 centimeters; see above for more about the Harappan). Around 1700 BCE, the Babylonians put their foot forward: A Babylonian foot was two-thirds of a Babylonian cubit. There are even records from Mesopotamia and Egypt showing yet another measurement system that included a foot of 11.0238 inches (300 millimeters). This was also known as the Egyptian foot, and it was standard in Egypt from predynastic times to the first millennium BCE. The Greek foot came close to today's foot, measuring about 12.1 inches (30.8 centimeters); a Roman foot measured in at 11.7 inches (29.6 centimeters). The list goes on, depending on the country and time period.

How was the **standard foot determined**?

Whatever the true story, the foot we are familiar with today is equal to 12 inches (30.48 centimeters). The true standardization of the foot came late in the 19th century, after the United States and Britain signed the "Treaty of the Meter." In this treaty,

Who developed the idea of a foot in measurement?

There is quite a mystery about who first developed the foot as a measurement unit. One story, which most scholars believe is a legend, is that a foot was the length of Charlemagne's (742–814) foot. Charlemagne (also known as Charles the Great) was King of the Franks and Emperor of the Holy Roman Empire. Standing at six feet four inches tall, he probably had a really big foot.

Still another story involves England's King Henry I (1068–1135), in which the length of an *arm* became important. Henry I ruled that the standard foot would be one-third of his 36-inch-long arm. This thus became the origin of our standardized unit of 12 inches to a foot, the inch being one thirty-sixth of a yard. According to the *Oxford English Dictionary,* the first confirmed usage of the word "foot" as a unit of measurement also occurred during the reign of Henry I. In honor of his arm, he ordered that an "Iron Ulna" (the ulna being the longer, inner bone in the forearm) be made. This iron stick represented the master standard yard for the entire kingdom.

But around 1324, in response to his subjects' cries for an even more standard measurement, England's King Edward II (1284–1327) changed things again. Recognizing the "Iron Ulna" was not universally available, he declared that "3 barleycorns, round and dry make an inch," and that 12 inches (or 36 barleycorns) would equal one foot.

It's interesting to note that even shoe sizes were tied to King Edward II and barleycorns. He declared that the difference between one shoe size to the next was the length of one barleycorn.

the foot was officially defined in terms of the new metric standards being adopted overseas. In the United States, the Metric Act of 1866 further defined the foot as equal to exactly 1200/3937 meter, or about 30.48006096 centimeters; this unit of measurement is still used for geodetic surveying purposes in the United States, and is called the survey foot. By 1959, the United States National Bureau of Standards redefined the foot to equal exactly 30.48 centimeters, or about 0.999998 survey feet. This definition was also adopted in Britain by the Weights and Measures Act of 1963; thus, a foot, or 30.48 centimeters, is also called an international foot.

What were some **early measurements of weight**?

Some of the early measurements of weight include the *grain, pound,* and *ton.* Ancient peoples used stones, seeds, and beans to measure weight, but grain (such as wheat or barleycorn) was a favorite. In fact, the *grain* (abbreviated "gr") is still one of the smallest units of weight used today (to compare, one pound equals 7,000 grains).

Why was the troy pound so historically important to weight measurement?

One of the oldest English weight systems was based on the 12-ounce troy pound. It was the basis by which coins were minted, and gold and silver weighed for trade and commerce. (The troy pound equaled 5,760 grains, and thus, in ounces, was 5,760/12 or 480 grains; 20 pennies weighed an ounce, and thus, a pennyweight equaled 480/20 or 24 grains.) The troy pound—and the entire system of connected weights—was used until the 19th century, mostly by jewelers and druggists. One holdover of the troy ounce (a portion of the troy pound) is found in today's pharmaceutical market to measure certain drugs. It is also seen in the financial markets, where it is used to interpret gold and silver prices.

The traditional *pound* as a unit of weight was used throughout the Roman Empire. But like many other measurements over time, the number of ounces in a pound seemed to shift and change. For example, the number of ounces in the Roman pound was 12; European merchants used 16 ounces to the pound. Eventually, 16 ounces in a pound became standard (for more about ounces, see below).

Back in the 19th century, the Americans—who did not like the British larger weights—decided that a hundredweight would equal 100 pounds (the British hundredweight was 112). This meant a ton was equal to 20 hundredweight for the American ton (or the American's *short ton* was 2,000 pounds), while the British *long ton* of 20 hundredweight was equal to 2,240 pounds. There were, of course, debates, but not everyone disagreed with the American short ton. It became the favorite of British merchants, who called it a *cental*. Eventually, the ton on the international market "went metric," and today a *metric ton* is close to the original British long ton. It is equal to 1,000 kilograms, or approximately 2,204 pounds, and is officially called a *tonne*. Although the International System (SI; see p. 50) standard uses tonne, the United States government recommends using the metric ton.

Where did the **pound** (and its abbreviation, "lb.") **originate**?

The origin of the word "pound" comes from *libra pondo,* or "pound of weight." The common abbreviation for pound (lb.) originated from letters in the Latin word *libra,* or balanced scales.

What is the difference between the **various pounds** and **ounces**?

The story behind the ounce is long and convoluted because historically people have been dissatisfied with the unit. For example, in medieval times English merchants

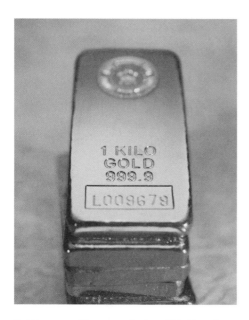

Gold bars are still measured using old-fashioned troy ounces. Thus, a pound of gold is equal to 12 troy ounces, not the usual 16 ounces per pound people typically use to measure other weights. *The Image Bank/Getty Images.*

were not happy with the troy pound, as it was less than the commercial pound in most of Europe. In response, the merchants developed an even larger pound, which was called the *libra mercatoria,* or *mercantile pound*. But by 1300 the complaints about the mercantile pound grew, because 15 troy ounces (or 7,200 grains) were easily divided by 15 and its divisors, but this was not as convenient as dividing by 12 troy ounces.

Soon, another type of pound was born in English commerce: the 16-ounce *avoirdupois* (roughly translated from the Old French as "goods of weight"). Modeled on a common Italian pound unit of the late 13th century, the avoirdupois pound weighed exactly 7,000 grains, which is easily divided for use in sales and trade. But because it was difficult to convert between the troy and avoirdupois units (the avoirdupois ounce is 7,000/16, or 437.5 grains, and 1 grain equals 1/7,000 avoirdupois pound, or 1/5,760 troy or apothecaries' pound; the troy ounce is 5,760/12,480 grains, or 31.1035 grams in metric), the standard soon shifted to using mostly the avoirdupois unit.

The avoirdupois is currently used in the United States and Britain. It is equal to 1/16th of a pound, or 28.3495 grams (in metric); the avoirdupois ounce is further divided into 16 drams (or drachm). The troy ounce hasn't been totally forgotten, though. Today, it is used mainly as units for precious metals and drugs, where it is often called the *apothecaries' ounce* (with its subdivisions of the *scruple,* or 20 grains, and the drachm, or 60 grains). In turn, the avoirdupois—our "ounce" for short—is used for almost everything else.

What were some **early measurements of volume** in terms of the **gallon**?

The names of the traditional volume units are the names of standard containers. Until the 18th century, the capacity of a container was difficult to accurately measure in cubic units. Thus, the standard containers were defined by the weight of a particular substance—such as wheat or beer—that the container could carry. For example, the basic English unit of volume, or the gallon, was originally defined as the volume of eight pounds of wheat. Other volumes were measured based on this gallon, depending on the different standard sizes of the containers.

But like most measurements over time, not all gallons were alike. During the American colonial period, the gallons from British commerce were based on dry and liquid commodities. For dry, the gallon was one-eighth of a *Winchester bushel* (defined by the English Parliament in 1696 as a cylindrical container 18.5 inches in diameter by 8 inches deep), which held 268.8 cubic inches of material. It was also called a "corn gallon" in England. For liquid, in England the gallon measurement was based on Queen Anne's wine gallon (also called the traditional British wine gallon), which measured exactly 231 cubic inches. This is why volume measurements in the United States include both the dry and liquid units, the dry units being about one-sixth larger than the corresponding liquid units.

By 1824, the British weren't as satisfied with the gallon divisions as the Americans. In response, the British Parliament abolished all the traditional gallons and established a system based on the *Imperial gallon*. It is still in use today, measuring 277.42 cubic inches, with the container holding exactly 10 pounds of water under specific (such as temperature and pressure) conditions.

What is a **rate**?

The rate is often used in measurement. It is defined as the comparison by division (similar to a ratio). For example, when measuring miles or kilometers per hour in your car, the rate equates the pairs of miles (kilometers) with hours. The translation for "a rate of 65 miles per hour" is that for each hour one will travel 65 miles as long as the speed remains the same for that hour.

What is **accuracy in measurement**?

Accuracy in measurement is based on *relative error* and number of *significant digits*. Relative error is the absolute error divided by the calculated (or estimated) value. For example, if a person expects to spend $10 per week at the local espresso bar, but he or she actually spends $12.50, the absolute error is $12.50 - 10.00 = 2.50$; the relative error then becomes $(2.50/10) = 0.25$ (to find out the percent, multiply by 100, or $0.25 \times 100 = 25$ percent of the original estimate).

Significant digits refers to a certain decimal place that determines the amount of rounding off to take place in the measurement. In most cases, this means that there are more numbers to the right of the decimal point. But beware. Accuracy in measurement does not mean the actual measurement taken was accurate. It only means that if there are a large number of significant digits, or if the relative error is low, the measurement is more accurate.

What are some **common modern measurement systems**?

There are several measurement systems in use today. The *English customary system* is also known as the standard system, U.S. customary system (or units), or English

units. It actually consists of two related systems: the U.S. customary units and the British Imperial System. The background of the units of measurement is historically rich and includes modern familiar terms, such as foot, inch, mile, and pound, as well as less well-known units, such as span, cubit, and rod. The official policy of the United States government is to designate the metric system as the preferred system for trade and commerce, but customary units are still widely used on consumer products and in industrial manufacturing.

In order to link all systems of weights and measures, both metric and non-metric, there is a network of international agreements supporting what is known as the *International System of Units*. It is abbreviated as SI (but *not* S.I.), in reference to the first two initials of its French name, *Système International d'Unités*. It was developed from an agreement signed in Paris on May 20, 1875, known as the Treaty of the Meter (Convention du Mètre). To date, 48 nations have signed the treaty. The SI is maintained by a small agency in Paris, the International Bureau of Weights and Measures (BIPM, or Bureau International des Poids et Mesures). Because there is a need to change or update the precision of measurements over time, the SI is updated every few years by the international General Conference on Weights and Measures (or CGPM, or Conférence Générale des Poids et Mesures), the two most recent meetings being in 2003 and 2007. SI is also referred to as the metric system, which is based on the meter. The word can also be used in mathematics (for example, metric space) or even computing (fontmetric file). It is often referred to incorrectly as "metrical." (See below for more about the metric system.)

What are the **base SI units**?

There are several base units at the heart of the International System (SI). The following lists the seven base units:

What countries have not officially adopted the metric system?

To date, there are only three countries that have not officially adopted the metric system: the United States, Liberia (western Africa), and Myanmar (formerly Burma, in Southeast Asia). All other countries—and the scientific world as a whole—have either used the metric system for many years or have adopted the measurement system in the past several decades. It's a bit of historical irony to note that the United States has hung on to such measurements as the foot, the standard measurement originated by the English who now use metric.

- meter (distance)
- kilogram (mass; related to weight)
- second (time)
- ampere (electric current)
- Kelvin (temperature)
- mole (amount of substance)
- candela (intensity of light)

Still other SI units—called *SI derived units*—are defined algebraically in terms of the above fundamental units. All the base units are consistent with the metric system called the *MKS,* or *mks, system,* which stands for meter, kilogram, and second. Another metric system is the *CGS,* or *cgs, system,* which stands for centimeter, gram, and second.

What are some of the **common metric/SI prefixes**?

The common metric and SI prefixes have been around for a while, but some were only recently added. In 1991, in order to apply standard units (SI units; see above) to a wide range of phenomena (especially in the scientific world), the Nineteenth General Conference on Weights and Measures lengthened the list to accommodate larger (and smaller) metric numbers—with the list now reaching from yotta- to yocto-. The following lists the American system (the name for large numbers) and the corresponding metric prefix and numerical equivalent (for comparison with prefixes and the power of ten, see "Math Basics"):

Common Metric/SI Prefixes

American system	metric prefix/symbol	number
1 septillion	yotta- / Y-	10^{24}
1 sextillion	zetta- / Z-	10^{21}
1 quintillion	exa- / E-	10^{18}
1 quadrillion	peta- / P-	10^{15}

51

American system	metric prefix/symbol	number
1 trillion	tera- / T-	10^{12}
1 billion	giga- / G-	10^9
1 million	mega- / M-	10^6
1 thousand	kilo- / k-	10^3
1 hundred	hecto- / h-	10^2
1 ten	deka- / da-	10
1 tenth	deci- / d-	10^{-1}
1 hundredth	centi- / c-	10^{-2}
1 thousandth	milli- / m-	10^{-3}
1 millionth	micro-/ μ	10^{-6}
1 billionth	nano- / n-	10^{-9}
1 trillionth	pico- / p-	10^{-12}
1 quadrillionth	femto- / f-	10^{-15}
1 quintillionth	atto- / a-	10^{-18}
1 sextillionth	zepto- / z-	10^{-21}
1 septillionth	yocto- / y-	10^{-24}

It is interesting to note that "deca-" is the recommended spelling by the International System (SI), but the United States National Institute of Standards and Technology spells the prefix "deka-." Thus, either one is considered by most references to be correct. There are also spelling variations between countries; for example, in Italy, hecto- is spelled etto- and kilo- is spelled chilo-. But the symbols remain standard through all languages. As for other numbers in the metric system—such as 10^5 or 10^{-5})—there are no set names or prefixes.

Why are some **prefix names different** in measurements?

The main reason why a prefix name would differ has to do with pronunciation and vowels: If the first letter of the unit name is a vowel and the pronunciation is difficult, the last letter of the prefix is omitted. For example, a metric measurement of 100 ares (2.471 acres) is a hectare (not hectoare) and 1 million ohms is a megohm (not

> ## Is it possible to convert international units seen on such items as vitamin bottles to milligrams or micrograms?
>
> No, there is no direct way to convert international units (IU) to mass units, such as milligrams. Most familiar to people who read vitamin and mineral bottles, an IU has nothing to do with weight; it is merely the measure of a drug or vitamin's potency or effect. Although it is possible to convert some items' IUs to a weight measurement, there is no consistent number. This is because not all materials weigh the same and the preparation of substances vary, making the total weights of one preparation differ from another.
>
> But there are some substances that can be converted, because for each substance there is an international agreement specifying the biological effect expected with a dose of 1 IU. For example, for vitamins, 1 IU of vitamin E equals 0.667 milligrams (mg); 1 IU of Vitamin C is equal to 0.05 mg. In terms of drugs, 1 IU of standard preparation insulin represents 45.5 micrograms; 1 IU of standard preparation penicillin equals 0.6 micrograms.

megaohm). There are exceptions, though, especially if the resulting prefix and unit sound fine, such as a milliampere. There are even times that another letter is added to make it easier to roll off the tongue. For example, the letter "l" is added to the term for 1 million ergs, making it a megalerg, not a megaerg or megerg.

How did the **metric system originate**?

In 1791 the French Revolution was in full swing when the metric system was proposed as a much needed plan to bring order to the many conflicting systems of weights and measures used throughout Europe. It would eventually replace all the traditional units (except those for time and angle measurements).

The system was adopted by the French revolutionary assembly in 1795; and the standard meter (the first metric standard) was adopted in 1799. But not everyone agreed with the metric system's use, and it took several decades before many European governments adopted the system. By 1820 Belgium, the Netherlands, and Luxembourg all required the use of the metric system. France, the originator of the system and its standards, took longer, finally making metric mandatory in 1837. Other countries such as Sweden were even slower. They accepted the system by 1878 and took another ten years to change from the old method to the metric.

How did the **first standard metric measurements evolve** over time?

The first standard metric units were developed by 1799: The meter was defined as one ten-millionth of the distance from the equator to the North Pole; the liter was defined

53

as the volume of one cubic decimeter, and the kilogram was the weight of a liter of pure water.

The standards metamorphosed over the years. For example, the first physical standard meter was in the form of a bar defined as a meter in length. By 1889 the International Bureau of Weights and Measures (BIPM, or Bureau International des Poids et Mesures) replaced the original meter bar. The new bar not only became a standard in France, but copies of the newest bar were distributed to the 17 countries that signed the Convention of the Meter (Convention du Mètre) in Paris in 1875. The accepted distance became two lines marked on a bar measuring 2 centimeters by 2 centimeters in cross-section and slightly longer than one meter; the bar itself was composed of 90 percent platinum and 10 percent iridium. But it was only a "standard meter" when it was at the temperature of melting ice.

By 1960 the BIPM decided to make a more accurate standard; mostly, this was done to satisfy the scientific community's need for precision. The new standard meter was based on the wavelength of light emitted by the krypton-86 atom (or 1,650,763.73 wavelengths of the atom's orange-red line in a vacuum). An even more precise measurement of the meter came about in 1983, when it became defined as the distance light travels in a vacuum in 1/299,792,458 second. This is the currently accepted standard.

What is **scientific notation**?

Scientific notation is a way of making larger and smaller numbers used in the scientific field easier to write, read, and take up less space in calculations. Scientists generally pick the power of ten that is multiplied by a number between 1 and 10 to express these numbers. For example, it is easier to write 0.00023334522 as 2.3334522×10^{-4}. (For more about scientific notation and power of ten, see "Math Basics.")

How is **temperature measured**?

Temperature is measured using a thermometer (*thermo* meaning "heat" and *meter* meaning "to measure"). The inventor of the thermometer was probably Galileo Galilei (1564–1642), who used a device called the thermoscope to measure hot and cold.

Temperatures are determined using various scales, the most popular being *Celsius, Fahrenheit,* and *Kelvin.* Invented by Swedish astronomer, mathematician, and physicist Anders Celsius (1701–1744) in 1742, *Celsius* used to be called the Centigrade scale (it can be capitalized or not; centigrade means "divided into 100 degrees"). He used 0 degrees Celsius to mark the freezing point of water; the point where water boils was marked as 100 degrees Celsius. Because of its ease of use (mainly because it is based on an even 100 degrees), it is the scale most used by scientists; it is also the scale most associated with the metric system.

Fahrenheit is the scale invented by Polish-born German physicist Daniel Gabriel Fahrenheit (1686–1736) in 1724. His thermometer contained mercury in a long, thin

What are the methods for converting temperatures between the various scales?

The following lists ways to convert from one temperature scale to another using, of course, simple mathematics:

Fahrenheit to Celsius: C° = (F° − 32) / 1.8; also seen as (5/9)(F° − 32)
Celsius to Fahrenheit: F° = (C° × 1.8) + 32; also seen as ((9/5)C°) + 32
Fahrenheit to Kelvin: K° = F° − 32 / 1.8 + 273.15
Kelvin to Fahrenheit: F° = (K° − 273.15) × 1.8 + 32
Celsius to Kelvin: K° = C° + 273.15
Kelvin to Celsius: C° = K° − 273.15

tube, which responded to changes in temperatures. He arbitrarily decided that the difference between water freezing and boiling—32 degrees Fahrenheit and 212 degrees Fahrenheit, respectively—would be 180 degrees.

The *Kelvin* scale was invented in 1848 by Lord Kelvin (1824–1907), who was also known as Sir William Thomson, Baron Kelvin of Largs. His scale starts at 0 degrees Kelvin, a point that is also known as *absolute zero,* the temperature at which all molecular activity ceases and the coldest temperature possible. His idea was that there was no limit to how hot things can get, but there was a limit to how cold. Kelvin's absolute zero is equal to −273.15 degrees Celsius, or −459.67 degrees Fahrenheit. So far, scientists believe nothing in the universe can get that cold.

TIME AND MATH IN HISTORY

How are **mathematics** and the study of **time connected**?

Mathematics is definitely tied to time. There has long been a need in human civilizations to record many sequences of events, especially those in nature that affected people. For example, the changing of the seasons was important to know, as it influenced the planting and growing of crops, when rivers would flood, and even when weather would change—from monsoon rains and harsh droughts to potential blizzards. The first such timekeepers counted the changing days and years by the movement of stars, the Sun, and the Moon across the sky, all of which are activities that included simple mathematical calculations.

What is some of the **earliest evidence** of **keeping time**?

No one agrees which culture(s) first invented timekeeping. Some historians and archeologists believe that marks on sticks and bones made by Ice Age hunters in

How did our present day become divided into hours, minutes, and seconds?

Divisions into hours, minutes, and seconds probably began with the Sumerians around 3000 BCE. They divided the day into 12 periods, and the periods into 30 sections. About one thousand years later, the Babylonian civilization, which was then in the same area as the Sumerians, broke the day into 24 hours, with each hour composed of 60 minutes, and each minute having 60 seconds.

It is unknown why the Babylonians chose to divide by 60 (also called a base number). Theories range from connections to the number of days in a year, weights and measurements, and even that the base-60 system was somehow easier for them to use. Whatever the explanation, their methods proved to be important to us centuries later. We still use 60 as the basis of our timekeeping system (hours, minutes, seconds) and in our definitions of circular measurements (degrees, minutes, seconds). (For more information about the Sumerian counting system, see "History of Mathematics.")

Europe around 20,000 years ago recorded days between successive new moons. Another hypothesis states that the measurement of time dates back some 10,000 years, which coincides with the development of agriculture, especially in terms of when to best plant crops. Still others point to timekeeping evidence dating back 5,000 to 6,000 years ago around the Middle East and North Africa. Whatever the true beginnings, most researchers agree that timekeeping is one of those subjects whose history will probably never be accurately known.

What **culture** took the **first steps toward timekeeping**?

Around 5,000 years ago, the Sumerians in the Tigris-Euphrates valley (today's Iraq) appear to have had a calendar, but it is unknown if they truly had a timekeeping device. The Sumerians divided the year into months of 30 days; the day was then divided into 12 periods (each corresponding to two of our modern hours) and the periods into 30 parts (each corresponding to four of our minutes).

Overall, many researchers agree that the Egyptians were the first serious timekeepers. Around 3500 BCE, they erected obelisks (tall, four-sided monuments), placing them in specific places in order to cast shadows as the Sun moved overhead. This thus created a large, crude form of a sundial. This sundial time was broken into two parts: before noon and after noon. Eventually, more divisions would be added, breaking down the time units even more into hours. Based on the length of the obelisks' shadows, the huge sundials could also be used to determine the longest and shortest days of the year.

What was one of the **first devices** used to **measure time**?

One of the first devices—smaller than the obelisks mentioned above—to measure time was a crude sundial. By about 1500 BCE, the true, small sundial (or shadow clock) was developed in Egypt. It was divided into ten parts, with two "twilight" hours marked. But it could only tell time for half a day; after noon, the sundial had to be turned 180 degrees to measure the afternoon hours.

More refinements of measuring time occurred later. In order to correct for the Sun's changing path over the sky throughout the year, the gnomon—or object that creates the shadow on the

A sundial, which uses shadows from the Sun to mark the passage of time, is one of the oldest time-keeping devices. *National Geographic/Getty Images.*

sundial—had to be set at the correct angle (what we call latitude). Eventually, the sundial was perfected. Multiple designs were used. For example, shortly before 27 BCE the Roman architect Marcus Vitruvius Pollio's (c. 90–20 BCE) *De architectura* described 13 different designs of sundials.

How does a **sundial** work?

The sundial tracks the apparent movement of the Sun across the sky. It does this by casting a shadow on the surface of a usually circular dial marked by hour and minute lines. The gnomon—or the shadow-casting, angular object on the dial—becomes the "axis" about which the Sun appears to rotate. To work correctly, it must point to the north celestial pole (near the star Polaris, also called the North Star); thus, the gnomon's angle is determined by the latitude of the user. For example, New York City is located at about 40.5 degrees north latitude, so a gnomon on a sundial in that city would be at a 40.5 degree angle on a sundial.

The sharper the shadow line, the greater the accuracy; in addition, larger sundials are more accurate, because the hour line can be divided into smaller units of time. But the sundial can't be too large. Eventually, diffraction of the sunlight around the gnomon causes the shadow to soften, making the time more difficult to read.

What is the definition of a **clock**?

A clock (from the Latin *cloca,* or "bell") is an instrument we use for measuring time. There are actually two main qualities that define a clock: First, it must have a regular,

constant, or repetitive action (or process) that will effectively mark off equal increments of time. For example, in the old days before our battery-driven, analog and digital clocks and watches, "clocks" included marking candles in even increments, or using a specific amount of sand in an hourglass to measure time.

Second, there has to be a way to keep track of the time increments and easily display the results. This eventually led to the development of watches, large clocks such as Big Ben in London, England, and even the clocks that count down the New Year. The most accurate clocks today are atomic clocks, which use an atomic frequency standard as the counter.

One of the most famous clocks on Earth is Big Ben in London, England. Although today's digital and atomic clocks are much more accurate, the charm of an old-fashioned analog clock still has its appeal. *Lonely Planet Images/Getty Images.*

What was the **driving force** behind the **development** of accurate **clocks**?

The true driving force behind accurate clocks began around the 16th century in relation to finding longitudinal measurement. As countries began to explore the world, an accurate way of telling a ship's position became a critical problem. With one time standard around the world (and clocks to tell those times), longitude, and thus position, could be determined. This would not only mean an increase in exploration but also wealth for the sponsoring country.

How was (and is) **one second defined**?

A second was once defined as 1/86,400 of a mean solar day. By 1956 this definition was changed by the International Bureau of Weights and Measures to 1/31,556,925.9747 of

> ## Where was the mechanical clock first invented?
>
> It is thought that the first mechanical clock was invented in medieval Europe and used most extensively by churches and monasteries (mainly to tell when to ring the bells for church attendance). The clocks had an arrangement of gears and wheels, which were all turned by attached weights. As gravity pulled on the weights, the wheels would turn in a slow, regular manner; as the wheels turned, they were also attached to a pointer, a way of marking the hours, but not yet minutes.
>
> The precursor to accurate time keeping came around 1500 with the advent of the "spring-powered clock," an invention by German locksmith Peter Henlein (1480–1542). It still had problems, though, especially with the slowing down of the clock as the spring unwound. But it became a favorite of the rich because of its small size, easily fitting on a mantle or shelf.

the length of the tropical year 1900. But like most measurements, the second definition changed again in 1964, when it was assigned to be the equivalent of 9,192,631,770 cycles of radiation associated with a particular change in state of a cesium-133 atom.

Interestingly enough, by 1983 the second became the "definer" of the meter: Scientists defined a meter as 1/299,792,458 the distance light travels in one second. This was done because the distance light travels in one second was more accurate than the former definition of the standard meter.

MATH AND CALENDARS IN HISTORY

What is the **connection** between **calendars and math**?

A calendar is essentially a numbering system that represents a systematic way of organizing days into weeks, months, years, and millennia, especially in terms of a human lifespan. It was the necessity to count, keep track of, and organize days, months, and so on that gave rise to calendars, all of which also entails the knowledge of mathematics to make such calculations.

When were the **first calendars invented**?

Although the first crude types of calendars may have appeared some 30,000 years ago—they were based on the movements of the Moon and indicated as marks on bones—the Egyptians are given credit for having the first true calendars. Scientists believe that around 4500 BCE the Egyptians needed such a tool to keep track of the Nile River's flooding. From about 4236 BCE, the beginning of the year was chosen as

the heliacal rising (when a star is finally seen after being blocked by the Sun's light) of the star Sirius, the brightest star in the sky located in the constellation of Canis Major. This occurred (and still occurs) in July, with the Nile flooding shortly after that, which made it a perfect starting point for the Egyptian calendar. The Egyptians divided the calendar into 365 days, but it was not the only calendar they used. There was also one used for planting and growing crops that was dependent on the lunar month.

What is a **lunar-based calendar**?

A lunar calendar is a calendar based on the orbit of our Moon. The new moon (when you can't see the Moon because it is aligned with the Sun) is usually the starting point of a lunar calendar. From there, the various phases seen from Earth include crescent, first quarter, and gibbous (these phases after a new moon are also labeled waxing, such as waxing crescent). When the entire face is seen, it is called a full moon; from there, the phases are seen "in reverse," and are labeled waning, such as waning crescent. Overall, the entire moon cycle takes about 29.530589 days. This cycle was used by many early cultures as a natural calendar.

What was the **problem** with the **lunar-based calendars**?

Nothing is perfect, especially a lunar month. The biggest drawback with using a lunar calendar is the fractional number of days, which makes a lunar calendar quickly go out of synch with the actual phases of the Moon. The first month would be off by about a half a day; the next month, a day; the next month, a day and a half; and so on. One way to help solve the problem was to alternate 30 and 29 day months, but this, too, eventually made the calendars go out of synch.

To compensate, certain cultures added (intercalations) or subtracted (extracalations) days from their calendar. For example, for more than a thousand years, the Muslims' lunar calendar has had an intercalation of 11 extra days over a period of 30 years, with each year being 12 lunar months. This calendar is only out of synch about one day every 2,500 years: To see this, mathematically speaking, the average length of a month over a 30-year period is figured out with the following equation: $(29.5 \times 360) + 11/360 = 29.530556$ days, in which 11 is the number of intercalated days, 360 is the number of months in a 30-year cycle (12 months \times 30 years), and 29.5 is the average number of days in the calendar month, or $(29 + 30) / 2$.

What is a **solar-based calendar**?

A solar-based calendar is one based on the apparent movement of the Sun across the sky as we orbit around our star. More than 2,500 years ago, various mathematicians and astronomers were basing a solar year on the equinoxes (when the Sun's direct rays are on the equator—or the beginning of fall and spring) and solstices (when the

Why does the Western calendar start with the birth of Christ?

The story behind the Western calendar—the one that developed into the calendar used most often today—started in the middle of the 6th century. Pope John I asked Dacian monk and scholar Dionysius Exiguus ("Dennis the Small," c. 470–c. 540; he was born in what is now Romania) to calculate the dates on which Easter would fall in future years. Dionysius, often called the inventor of the Christian calendar, decided to abandon the calendar numbering system that counted years from the beginning of Roman Emperor Diocletian's reign. Instead, being of Christian persuasion, he replaced it with a system that started with the birth of Christ. He labeled that year "1," mainly because there was no concept of zero in Roman numerals.

Sun's direct rays are on the latitudes marked tropic of Capricorn [winter in the Northern Hemisphere; summer in the Southern Hemisphere] or tropic of Cancer [winter in the Southern Hemisphere; summer in the Northern Hemisphere]).

As the measurement of the solar (and lunar) cycle became more accurate, calendars became increasingly sophisticated. But no calendar dominated until the last few centuries, with many cultures deriving their own calendars—some even combining lunar and solar cycles in a type of Moon-Sun or luni-solar calendar. This is why although there is one "standard calendar" used by most countries around the world, certain cultures still use their traditional calendars, including the Chinese, Jewish, and Muslim calendars.

How did **some ancient cultures refine** their **calendars**?

There were many different ways that various ancient cultures refined their calendars, and all of them entailed some type of mathematical calculation. One way to measure the length of a year was by using a *gnomon,* or a structure that casts a shadow (for more about gnomons and sundials, see p. 57). This was based on the apparent motion of the Sun across the sky, with the shadow not only used to tell daily time but also to determine the summer solstice, when the shadow created by the gnomon would be at its shortest at noon. By measuring two successive summer solstices and counting the days in between, various ancient cultures such as the Egyptians developed a more detailed calendar—and, as a bonus, determined the exact times of the solstice.

Around 135 BCE, Greek astronomer and mathematician Hipparchus of Rhodes (c. 170–c. 125 BCE) decided to compare his estimate of the vernal (spring in the Northern Hemisphere, occurring in March) equinox with that made by another astronomer about 150 years earlier. By averaging the number of days, he estimated that a year was equal to 365.24667 days, a number only off by about six minutes and 16 seconds.

61

What was the **Roman calendar**?

The first Roman calendar, according to legend, appeared when Rome was founded around 750 BCE. When it actually started is still up for discussion, but the calendar was based on the complexity of the solar-lunar cycles. At first, the calendar had ten months, starting in March; January and February were added as the calendar was modified. Politics entered into the determination of this calendar, too, with certain officials deciding to add days whenever they desired, and even what to name certain months.

What was the **Julian calendar**?

By the time of Julius Caesar (100–44 BCE), Roman calendar-keeping was a mess. Caesar decided to reform the Roman calendar, asking help from astronomer and mathematician Sosigenes of Alexandria (lived c. 1 BCE; not to be confused with Sosigenes the Peripatetic [c. 2nd century], an Egyptian philosopher). The year 46 BCE would consequently have 445 days—a time appropriately called "the year of confusion."

Sosigenes began the reformed year on January 1, 45 BCE, a year with 365 days, and proposed an additional day for every fourth year in February (leap day). The alternate months of the year (January, March, May, July, September, November) had 31 days; the other months would have 30 days. In the Julian Calendar, there was only one rule: Every year divisible by four was a leap year.

The vain heir to Caesar, Augustus Caesar (63 BCE–14 CE; a.k.a. Gaius Octavius, Octavian, Julius Caesar Octavianus, and Caesar Augustus), would change the Julian calendar in a several ways. Not only did he name the month of August after himself, but he would change the number of days in many months to their present usage, adding more confusion to the calendar.

The Julian calendar would govern Caesar's part of the world until 1582. Not that the Julian year was perfect: A year's 364.25 days was too long by 11 minutes 12 seconds. Although the difference between today's measurement of the year and the Julian year was not great, it adds up to 7.8 days over 1,000 years. But as with many decrees and mandates, Caesar, Sosigenes, and Octavian left it up to future generations to fix the problem.

What is the **Gregorian calendar**?

By 1582 the discrepancies in the Julian calendar were not interfering with timekeeping, but they were beginning to infringe on dates of the church's ecclesiastical holidays. The powerful Catholic church was not amused: Pope Gregory XIII, on the advice of several of his astronomers, decided to reposition days, striking out the excess ten days that had accumulated on the then-present-day calendar. Thus, October 4, 1582, was followed by October 15, 1582.

To fix the extra-days problem, the pope made sure that the last year of each century would be a leap year, but only when it is exactly divisible by 400. That means that three leap years are suppressed every four centuries; for example, 1900 was not a leap

What are some interesting facts about the Julian and Gregorian calendars?

An interesting fact about the Julian calendar is that it designates every fourth year as a leap year, a practice that was first introduced by King Ptolemy III of Egypt in 238 BCE. A quirk about the Gregorian calendar is that the longest time between two leap years is eight years. The last time such a stretch was seen was between 1896 and 1904; it will happen again between 2096 and 2104.

year, but 2000 was a leap year. (Today, the Gregorian calendar "rules" state that every year divisible by four is a leap year, except for years that are both divisible by 100 and not divisible by 400.)

Some countries eliminated the ten extra days, starting "fresh" with the Gregorian calendar. But not everyone agreed with the new calendar, especially those who distrusted and disliked the Catholic church. Eventually, by 1700 those who had not changed their calendars had collected too many extra days. In 1752, the English Parliament decreed that 11 days would be omitted from the month of September. England and its American colonies began to follow the Gregorian calendar, with most other countries following close behind. It is now the standard calendar used around the world.

What is a **problem** with our **modern calendar**?

The modern calendar could use some small changes, such as making sure we don't have to keep changing calendars each year (see below). But the real problem with the modern calendar isn't the human factor; it's nature. As our Earth orbits around the Sun, it wobbles like a spinning top in a process called precession. Because scientists can measure the planet's movements more accurately now than in the past, they know that the wobble is increasing. This is because the tides caused by the pull of the Sun and Moon are slowing the Earth's spin. And, like a top, as the spinning slows, the wobble increases and the length of the year decreases.

What does this mean for our calendar? It is already known that our calendar and the length of a year were only off by 24 seconds (0.00028 days) in 1582—a very small discrepancy that will eventually be noticed. But when you add in the slowing down of the Earth's rotation, it will make the year even shorter. In fact, since 1582, the year has decreased from 365.24222 days to 365.24219 days, or an actual decline of about 2.5 seconds.

Can we **change** the **calendars** now in use?

The present calendar is an annual one and changes every year—much to the happiness of calendar publishers. This is because 365 days in a year is not evenly divisible by

the number of days in the week: 365/7 = 52, with a remainder of 1 (or 52.142857…). This means that a given year usually begins and ends on the same weekday; and it also means that the next year bumps January 1 (and all following dates) to the next weekday, and a new calendar is born each year. But because the calendar we now have is so ingrained in everything we do, it is doubtful that there will be any changes soon.

Not that there haven't been suggestions. One is called the *World* (or *Worldsday*) *Calendar,* in which each date would always fall on the same day of the week, and all the holidays occur on the same day of the year. With this calendar, each year begins on Sunday, January 1, and each working year begins on Monday, January 2. The reason why the calendar is called "perpetual" or "perennial" is that the year ends with a 365th day following December 30, which is marked with a "W" for "Worldsday" (our current "December 31"). Leap year days would still have to be added, such as at the end of June (some suggest a June 31 be added). Both extra days could act as world holidays.

The drawbacks? Besides the obvious—no one wanting to change an already entrenched system—the superstitious would revolt. After all, on the World Calendar there are four Friday the 13ths.

THE BASICS

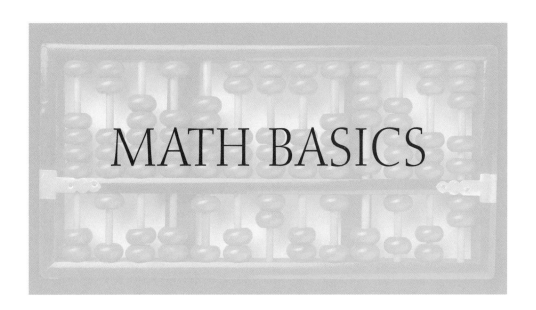

MATH BASICS

BASIC ARITHMETIC

What is **arithmetic**?

Arithmetic is a branch of mathematics that deals with numerical computation; specifically, it includes computation using integers, rational numbers, real numbers, or complex numbers. The word "arithmetic" has its roots in the Greek word for "to count" (*arithmeein;* also *arithmos,* or "number").

Arithmetic contains all the rules for combining two or more numbers. In most cases, when mathematicians talk about elementary arithmetic, they are speaking of those subjects most of us learned in grade school: addition, subtraction, multiplication, and division being the most common; and fractions, geometry and measurements, ratios and proportion, simple probabilities, and algebra examined in more advanced levels. For even more advanced students, such arithmetic lessons as congruence calculation, root extraction, power computations, and advanced factorizations are often presented.

Are there more **advanced concepts** in **arithmetic**?

Yes, arithmetic can even be more advanced than the ideas mentioned above. For example, higher arithmetic is the archaic term for *number theory,* which is the study of the properties of integers, or whole numbers ($0, \pm 1, \pm 2 \ldots$). It can include anything from simpler arithmetic concepts to the more complex, such as diophantine equations (for more information about these equations, see "Algebra"), prime numbers (see below), and functions such as the Riemann hypothesis (for more about Friedrich Bernhard Riemann, see "History of Mathematics" and "Geometry and Trigonometry").

There are other more advanced ideas in arithmetic, too. For example, *modular arithmetic* is known as the arithmetic of congruences (see below). The *model theory*

In 1556 the first arithmetic book was published in North America by Brother Juan Diez Freyle, a Franciscan friar. The name of the book was *Sumario compendioso de las quentas de plata y oro que in los reynos del Piru son necessarias a los mercaderes y todo genero de tratantes: Con algunas reglas tocantes al Arithmética*. The title translates as *Comprehensive Summary of the Counting of Silver and Gold, Which, in the Kingdoms of Peru, Are Necessary for Merchants and All Kinds of Traders*. The book explained the conversion of gold ore into value equivalents in different types of coinage in the Old World, problems that required the use of ratios and proportions. Diez also included a short chapter on algebra.

The first English-language mathematics book written in North America was published in 1729 by Isaac Greenwood and titled *Arithmetik, Vulgar and Decimal* ("vulgar" refers to the common people). Greenwood's life was also somewhat vulgar: He was appointed to the first Hollis Professorship of Mathematics and Natural Philosophy at Harvard University in Massachusetts when it was founded in 1727. By 1737 he was removed for "intemperance." Reportedly, he drank too much, and more than likely his views, philosophical and otherwise, differed greatly from those of his colleagues at the university.

discusses the existence of "non-standard" models of arithmetic. And *floating-point arithmetic* is performed on real numbers by computers or other automated devices.

What is **arithmetic progression**?

Arithmetic progression is one of the more simple types of series in mathematics. It is usually in the form of a, $a + d$, $a + 2d$, $a + 3d$, and so on, in which a is the first term and d is the constant difference between the two successive terms. A progression is also seen as these numbers are added, as in $a + (a + d) + (a + 2d) + (a + 3d) +, ..., (a + (n + 1)d)$. An example of an arithmetic progression would be $2 + 6 + 10 + 14 + ...$, in which d is equal to 4.

What do **computers and arithmetic** have in **common**?

Computers and arithmetic have a great deal in common. *Arithmetical operations* are actually digital computer operations in which the numerical quantities are computed, either through adding, subtracting, multiplying, dividing, or otherwise comparing them. *Arithmetical instructions* give a computer program direction to perform an arithmetic operation on specific types of data, such as addition, subtraction, multiplication, and division. The sections of the computer that carry out these computations and

other logic operations are called *arithmetical units* (or arithmetic sections). (For more information about computers and math, see "Math in Computing.")

ALL ABOUT NUMBERS

What is a **number**?

The term *number* can be defined in many ways, including a sizable collection of people or things, and even an indefinite quantity or collection. In mathematics a number, or numeral, is usually defined as a symbolic representation of a specific quantity or place in a sequence; to most people, the most familiar numbers are 1, 2, 3, 4, 5, and so on.

Computers have become such an everyday part of our lives that we hardly think about them. Yet the mathematical concepts that lie behind their operations are staggering. *Taxi/Getty Images.*

What is a **decimal system**?

The *decimal system* uses the base 10 notation system to represent real numbers. A *decimal expansion* is the expression of a number within the decimal system, such as 1, 15, 359, 18.7, and 3.14159. Each number within the system is called a *decimal digit*. (Such decimal notation—or a numbering notation based on decimals—was first used in India around the year 594.) The *decimal point* is represented by a period placed to the right of a unit's place in a decimal number. It is interesting to note that a comma is used in continental Europe to denote a decimal point, such as 3,25 (translated as 3.25), which in this case would logically be called the *decimal comma*.

What is currently the most **common numeration system**?

The most common numeration system in use today is the Hindu-Arabic. This set of numerals has ten digits in a place value decimal system, which is a fancy way of saying that a decimal system—one based on tens—is an integral part of the system and that each number has a certain value depending on its place in the list of numbers.

How did the **Hindu-Arabic** numerals spread to **Europe**?

Hindu-Arabic numerals (often less accurately called Arabic numerals or numbers) had their roots in India before 300 BCE. From there, the use of Indian numerals followed the

69

One of the earliest and most common devices developed for making everyday calculations was the abacus, which was still being used in Europe as late as the 15th century. *Photographer's Choice/Getty Images.*

western trade routes to Spain and Northern Africa that were taken by the Arabic/Islamic peoples; this consequently resulted in the expanded use of these symbols.

It took several more centuries to spread the idea to Europe. Although the Spanish used some Hindu-Arabic symbols as early as the late 900s, records of a more extensive use of these symbols occurred around 1202. Italian mathematician Leonardo of Pisa (also known as Fibonacci, c. 1170–c. 1250; for more about Fibonacci, see p. 77 and "History of Mathematics") introduced the Hindu-Arabic numbers in his book *Liber Abaci* (*The Book of the Abacus*). The acceptance of such a numbering system was difficult. For example, in some places in Italy it was forbidden to use anything but Roman numerals. By the late 15th century, most people in Europe were still using an abacus and Roman numerals.

The 16th century was the turning point, with European traders, surveyors, bookkeepers, and merchants spreading the use of the Hindu-Arabic numerals. After all, it took longer to record data using Roman numerals than with Hindu-Arabic numbers. The advent of the printing press also helped by standardizing the way the Hindu-Arabic numbers looked. By the 18th century, the "new" numeration system was entrenched, establishing a system that dominates the way we work with and perceive numbers in the 21st century. (For more information about Hindu-Arabic and Roman numerals, see "History of Mathematics.")

How did the **Hindu-Arabic numbers evolve**?

The evolution of the Hindu-Arabic numbers was not a straight line from India to Arabia and on to Europe. In between, the Arabic cultures had more than one number sys-

	1	2	3	4	5	6	7	8	9	10	0
Babylonian	▼	▼▼	▼▼▼	▼▼▼	▼▼▼	▼▼▼	▼▼▼	▼▼▼	▼▼▼	◀	
Egyptian	∣	∣∣	∣∣∣	∣∣∣∣	∣∣∣ ∣∣	∣∣∣ ∣∣∣	∣∣∣∣ ∣∣∣	∣∣∣∣ ∣∣∣∣	∣∣∣ ∣∣∣ ∣∣∣	∩	
Greek	A	B	Γ	Δ	E	F	Z	H	Θ	I	
Ancient Chinese	—	=	☰	▱	丘	大	七	人	九	十	
Maya	•	••	•••	••••	—	•̄	•̈̄	•••̄	••••̄	=	◯
Hindu	?	?	?	?	?	ε	?	?	?	?	○
Hindu-Arabic 15th century	/	2	3	ℓ	4	6	∧	8	9	/0	○
Modern Hindu-Arabic	1	2	3	4	5	6	7	8	9	10	0

Different cultures generally employed one of two strategies when creating symbols for numbers: multiple marks that indicated single numbers or multiples of fives or tens (e.g., Babylonian, Egyptian, Mayan), or a more abstract system using a single symbol for the numbers one through nine, with numbers then being shifted over one or more places to indicate multiples of ten, hundreds, etc. (e.g., Hindu, Hindu-Arabic).

tem to contend with, including at least three different types of arithmetic: finger-reckoning arithmetic (counting on fingers), a sexagesimal system with numbers written in letters of the Arabic alphabet, and Indian numeral arithmetic.

The evolution of the Hindu-Arabic numbers continued throughout time and includes some good reasons for why our numbers look as they do today. For example, historians believe that between 970 and 1082, the numbers 2 and 3 changed significantly, rotating 90 degrees from their original written position. This is thought to be due to how scribes worked: Sitting cross-legged, they wrote on a scroll they wound from right to left across their body. This caused them to write from top to bottom, not our usual left to right; the script was then rotated when the scroll was read.

How are **numbers classified**?

The set of natural numbers are also called integers—or counting or whole numbers—which are usually defined as the positive and negative whole numbers, along with zero (0). But many times mathematicians do not use the term "natural numbers," and instead define numbers based on the following terminology and/or symbols:

71

Group	Name	Symbol
...,−3,−2,−1, 0, 1, 2, 3...	integers	Z (after the German word *Zahl*, for "number")
1, 2, 3...	positive integers (or often referred to as natural numbers)	Z+ or N
−1, −2, −3...	negative integers	Z−
0, 1, 2, 3 ...	nonnegative integers (or often referred to as whole numbers)	Z* (Z-star)
0, −1, −2, −3 ...	nonpositive integers	(no symbol)

What do **integers include**?

Integers include the whole numbers (also called positive integers or natural numbers), negative whole numbers (also called negative integers or the negatives of the naturals), and zero. Numbers such as 3/4, 5.993, 6.2, −3.2, and pi (π; or 3.14 ...) are *not* considered integers. Only integers are used when speaking of odd and even numbers (zero is considered to be an even number; for more about zero, see p. 90).

What is a **place value**?

The *place value,* or "rule of position," are numbers whose value depends on the place or position they occupy in a written numerical expression. In the Hindu-Arabic counting system, the numerals 1, 2, 3, 4, 5, 6, 7, 8, 9, and 0 have certain specific place values. For example, the number 7 represents 7 (7 units), 70 (as 7 tens), 700 (as 7 hundreds), and so on—the 7's values based on their position in the numerical expression. (For more information about place values, see "History of Mathematics.")

What are some of the **highest numbers** we know?

Numbers can go on forever—in other words, there is an infinite number of numbers. The highest numbers we are most familiar with include million, billion, and trillion, which we often see when referring to such quantities as the number of miles to the outer planets or the federal budget deficit. These larger numbers are usually separated by commas at the thousands, millions, trillions, etc., place values. For example, for 3,490 the comma is after the thousand place value; for 1,384,993 it is put after the million and thousand place value. Larger numbers written for technical papers are sometimes expressed with spaces; for example, 11 384 443 is equivalent to 11,384,443.

Not every country labels the highest numbers in the same way. For example, in the American system denominations above 1,000 million (or the American billion) are 1,000 times the preceding one (for example, one trillion is 1,000 billion; one

What is a googol?

The "googol" is the invention of Milton Sirotta, the eight-year-old nephew of mathematician Edward Kasner, who once asked the young boy to name the number 1 followed by 100 zeros (10 to the 100th power or 10^{100}). A googol is an incredibly large number, so there is little it can represent. Although you might guess it would represent some collection of astronomical entities, such as the number of elementary particles in the universe, it does not: Scientists estimate only about 10 to the 80th power (10^{80}) such particles exist.

Googol was soon followed by *googolplex,* a name invented by another mathematician, and said to equal 10 to the power of googol (or 1 followed by 10 to the power of 100 zeros). No one has ever seen such a large number printed out. And although it is true that computer processing power doubles about every one to two years, it is still too early to print the number represented by a googolplex. Thus, many ask why begin at all, since attempts to do so will soon be overtaken by faster processors? In fact, it is estimated that it will take another 500 years before such an endeavor is achieved.

quadrillion is 1,000 trillion). But in the British system, the first denomination above 1,000 milliards (the British billion) is 1,000,000 times the preceding one (for example, one trillion is 1,000,000 billion; one quadrillion is 1,000,000 trillion). The American system is based on an early French system, which the French ironically no longer follow. Their larger number names now correspond to the British system, as do those of many other countries in Europe.

American and British Naming Systems for Large Numbers

American Name	British Name	Number (in powers of ten)
billion	milliard	10^9
trillion	billion	10^{12}
quadrillion	—	10^{15}
quintillion	trillion	10^{18}
sextillion	—	10^{21}
septillion	quadrillion	10^{24}
octillion	—	10^{27}
nonillion	quintillion	10^{30}
decillion	—	10^{33}
undecillion	sextillion	10^{36}
duodecillion	—	10^{39}
tredecillion	septillion	10^{42}

American Name	British Name	Number (in powers of ten)
quattuordecillion	—	10^{45}
quindecillion	octillion	10^{48}
sexdecillion	—	10^{51}
septdecillion	nonillion	10^{54}
octodecillion	—	10^{57}
novemdecillion	decillion	10^{60}
vigintillion	—	10^{63}
—	undecillion	10^{66}
—	duodecillion	10^{72}
—	tredecillion	10^{78}
—	quattuordecillion	10^{84}
—	quindecillion	10^{90}
—	sexdecillion	10^{96}
—	septendecillion	10^{102}
—	octodecillion	10^{108}
—	novemdecillion	10^{114}
—	vigintillion	10^{120}
centillion	—	10^{303}
—	centillion	10^{600}

What are **non-vanishing** and **vanishing numbers**?

A non-vanishing number means just what the term implies: A quantity that is non-zero everywhere. For example, in the expression $x^4 + 1$, the answer will never be zero (even when x is zero or a negative number). The answer for the expression x^2 is called "vanishing" because if $x = 0$, the expression's answer "vanishes" to zero.

What are **rational, irrational,** and **real numbers**?

Rational, or fractional, numbers are most often regarded as divisions (or ratios) of integers. By creating a fraction (dividing one integer by another), a rational number produces either a number that "ends" or repeats decimals. For example, 1/4 equals the decimal equivalent of 0.25; 1/3 is equivalent to 0.33333.... Both of these are rational numbers. (For more information about fractions, see p. 98.)

On the other hand, *irrational numbers* are all the numbers that can be written as non-repeating, non-finite (or non-terminating) decimals. Also called non-rational numbers, they include the decimal equivalent of "pi" (or 3.141592 ...). Finally, if you put the rational and irrational numbers together, they form the *real numbers.* Most numbers we use in our everyday lives are real numbers.

What are imaginary numbers?

The "opposite" of real numbers are (logically enough) called *imaginary numbers*. In particular, they are all non-zero multiples (real numbers) of the square root of -1, which is also represented as i, with the formula defined as follows:

$$i = \sqrt{-1}$$

$$\text{then } i^2 = (\sqrt{-1})^2 = -1$$

The $\sqrt{-1}$ does not have a position on a number line; and no number can be squared to get -1. (If you square a positive number, the result is positive; if you square a negative number, the result is also a positive number.) Thus, in order to square a number to get a negative one, mathematicians invented the imaginary number, i.

Can there be **more than one type of number**?

Yes, numbers can be classified as more than one type, and it's not always easy to keep them straight. The following lists some ways to better understand the plethora of number types:

- A rational number is not always an integer: 4/1 is an integer, but 2/3 is not; but an integer is always a rational number because it can be represented by a fraction by putting the integer over 1, or /1, such as 2/1 or 234/1

- A number can either be rational or irrational but not both

- The number for pi (3.141592 …) is irrational (the decimal does not repeat) and real

- 0.25 is considered rational (the numbers terminate) and real

- The fraction 5/3 is rational (it's a fraction) and real

- The number 10 can be explained using many terms, including a counting number, whole number, integer, rational, and real

How do **regular** and **non-regular numbers** differ?

Regular and non-regular numbers are actually other terms for rational numbers. Regular numbers are positive integers that have a finite decimal expansion. In other words, a number that seems to "end." For example, one quarter (1/4) is equal to the decimal equivalent of 0.25 in which the numbers end with "5." A non-regular number is one that includes repeating decimals—numbers that seem to go on forever. For example, one third (1/3) is equal to the decimal equivalent of 0.3333… in which the 3s go on indefinitely.

How do you perform **imaginary number computations**?

Imaginary numbers come in handy to do many computations, especially something called *simplification*. Here are some "simple" examples of how to use imaginary numbers:

To simplify the square root (or *sqrt*) of -25:
$$\sqrt{-25} = \sqrt{-25} \times -1 = \sqrt{-25} \times \sqrt{-1} = 5i$$

To simplify $2i + 4i$:
$$2i + 4i = (2 + 4)i = 6i$$

To simplify $21i - 5i$:
$$21i - 5i = (21i - 5i) = 16i$$

To multiply and simplify $(2i)(4i)$:
$$(2i)(4i) = (2 \times 4)(ii) = (8)(i^2) = (8)(-1) = -8$$

Who first came up with the **idea for imaginary numbers**?

The origin of *i* is difficult to trace. Some historians give credit to Italian physician and mathematician Girolamo Cardano (1501–1576; in English, known as Jerome Cardan). In 1545, he is said to have started modern mathematics, first mentioning not only

negative numbers but imaginary numbers in his Latin treatise *Ars Magna* (*The Great Art*). But Cardano did not consider the imaginary numbers as the real mathematical objects we do today. To him, they were merely convenient "fiction" to classify certain polynomial properties, describing how their roots would behave when he pretended they even *had* roots.

Most agree that around 1777, Swiss mathematician Leonhard Euler (1707–1783) used "i" and "−i" (negative i) for the two different square roots of −1, thus eliminating some of the problems associated with notation when putting polynomials into categories. (He is also credited with originating the notation *a* + *bi* for complex numbers.) Much to the consternation of many past and present mathematicians, *i* and −*i* were called "imaginary," mainly because the number's function at the time of Euler was not clearly understood. When German mathematician, physicist, and astronomer Johann Friedrich Carl Gauss (1777–1855)

Eighteenth-century Swiss mathematician Leonhard Euler, who published over 70 volumes on mathematics in his lifetime, was one of the greatest contributors to the discipline that ever lived. Euler developed important concepts in such areas as geometry, calculus, trigonometry, algebra, hydrodynamics, and much more. He also created the concept of the imaginary number that is the square root of −1. *Library of Congress.*

used them for the geometric interpretation of complex numbers as points in a plane, the usefulness of imaginary numbers became apparent. (For more information on Gauss, Cardano, and Euler, see "History of Mathematics.")

What are **complex numbers**?

Complex numbers have two parts: a "real" part (any real number) and an "imaginary" part (any number with an *i* in it). The standard complex number format is "*a* + *bi*," or a real number plus an imaginary number. It is also often seen as *x* + *iy* because while real numbers are viewed on a line, complex numbers are viewed graphically on an Argand (or polar) coordinate system: The imaginary numbers make up the vertical (or *y*) axis as *iy,* while the horizontal (or *x*) axis is occupied by real numbers. (For more information about coordinate systems, see "Geometry and Trigonometry.")

What is the **polar form** of a **complex number**?

The polar form of a complex number is equal to a real number expressed as an angle's cosine, and the imaginary number (*i*) times the same angle's sine, with the angle

expressed in radians (for more about angles, sines, cosines, and polar forms, see "Geometry and Trigonometry"). This is seen in equation form as: $r(\cos \theta + i \sin \theta)$, in which r is the radius vector, θ is the angle, and i is the imaginary number.

Is there such a thing as a **perfect number**?

Yes, there is such a thing as a perfect number, but it is not what we think of as true perfection. To mathematicians, perfect numbers are somewhat rare. They are defined as a natural number (or positive integer) in which the sum of its positive divisors (or the bottom number in a fraction that divides the number to equal another whole number, and includes 1 but not the number itself) is the number itself. For example, 6 is considered a perfect number because its divisors are 1, 2,

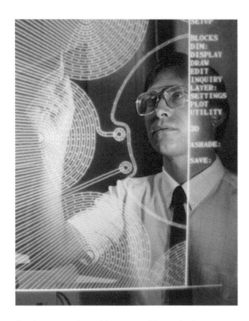

Engineers, such as this man making calculations on a CAD (computer aided design) of an electric motor core, employ complex mathematics every day to perform their jobs. *Taxi/Getty Images.*

and 3—or $1 + 2 + 3 = 6$. The next perfect numbers are 28 ($1 + 2 + 4 + 7 + 14$), 496; 8,128; 33,550,336; 8,589,869,056; 137,438,691,328; 2,305,843,008,139,952,128, and so on. Larger and larger perfect numbers are still being discovered, especially with the help of today's faster and more memory-packed computers.

What is meant by **one-to-one correspondence**?

A one-to-one correspondence means just what it implies: that the number of objects, numbers, or whatever is the same as the set of other objects, numbers, or whatever. (In set theory, the one-to-one correspondence means something different; for more about set theory, see "Foundations of Mathematics.")

Everyone has no doubt had contact with one-to-one correspondence without even thinking about it. For example, there is a one-to-one correspondence of the number 10 to the number of fingers on both hands (ten). Counting a deck of cards is a one-to-one correspondence—each number, from 1 to 52, representing a card in the deck. When you compare two decks of cards, putting the cards side-by-side to equal 52 in each deck can also be considered a one-to-one correspondence.

Not everything is counted in such a way. For example, when mathematicians want to know the size of an unknown quantity, they put the unknown quantity in a one-to-one correspondence with a known quantity.

Anyone who has played a card game has used their knowledge of one-to-one correspondence, though most of us would not know enough to call it that. *The Image Bank/Getty Images.*

What are **ordinal and cardinal numbers**?

In common, arithmetic terms, *cardinal numbers* are those that express amounts; they are also used in simple counting or to answer the question of quantity (how many). They can be nouns (try counting to *ten*); as pronouns (*ten* were discovered); or adjectives (*ten* cats). Specifically, the term is from the Latin *cardin,* meaning "stem" or "hinge," referring to the most important or principal numbers, with others depending (hinging) on those numbers. We are most familiar with the cardinal numbers as our counting numbers, or the Hindu-Arabic numeration system—1, 2, 3, and so on.

Ordinal numbers are much different. In common, arithmetic terms, ordinal numbers are adjectives that describe the numerical position of an object, such as first, second, third, and so on. They are used to show the order of succession for objects (*second* chair), names (*second* month), or periods of time (*2nd* century).

Note that cardinal and ordinal numbers are easily divided. For example, in the Hindu-Arabic numeration system, the cardinal numbers may be read as ordinals, such as May 10 being read as "May tenth." Such differences are even harder to distinguish when it comes to Roman numerals. Most of the time, these numerals are considered cardinal numbers (I, II, III, etc.), but they can also be ordinal numbers in certain con-

Do numbers continue into infinity?

When most of us think of infinity, we envision the universe continuing on forever; and in mathematics, we often think of numbers that are never-ending. Sometimes it is difficult to understand infinity, since our own lives—and most of our experiences—are finite (they eventually end). Infinity is a mind-boggling concept.

There are several rules to mathematical infinity. The three most important are: No matter how high you count numbers, you can always count higher; no matter what length you draw parallel lines, they will never meet; and when starting with a line, dividing it in half, then dividing that in half, and so on, you will never stop dividing the resulting line segment.

Even though scientists and mathematicians agree that infinity exists theoretically, it is often a difficult concept to understand and accept. Is it true that the number of particles in the universe are infinite? Does the universe continue on forever? Do parallel lines eventually meet at a place we have yet to discover? If a particle is infinitely divided, just how small can an atomic particle become? And to add to the unimaginable explanations, German mathematician George (Georg) Ferdinand Ludwig Philipp Cantor (1845–1918) mathematically reasoned out that not only do infinities come in different sizes, but there are an infinite number of infinities.

texts, such as Henry VIII (Henry the Eighth). Roman numerals can even contain ordinal suffixes, such as the IXth Dynasty.

Ordinal and Cardinal Numbers and Symbols

Hindu-Arabic Symbol	Roman Symbol	Cardinal Number Name	Ordinal Number Name	Ordinal Symbol
0	n/a	zero/naught/cipher		
1	I	one	first	1st
2	II	two	second	2d/2nd
3	III	three	third	3d/3rd
4	IV	four	fourth	4th
5	V	five	fifth	5th
6	VI	six	sixth	6th
7	VII	seven	seventh	7th
8	VIII	eight	eighth	8th
9	IX	nine	ninth	9th

Hindu-Arabic Symbol	Roman Symbol	Cardinal Number Name	Ordinal Number Name	Ordinal Symbol
10	X	ten	tenth	10th
11	XI	eleven	eleventh	11th
12	XII	twelve	twelfth	12th
13	XIII	thirteen	thirteenth	13th
14	XIV	fourteen	fourteenth	14th
15	XV	fifteen	fifteenth	15th
16	XVI	sixteen	sixteenth	16th
17	XVII	seventeen	seventeenth	17th
18	XVIII	eighteen	eighteenth	18th
19	XIX	nineteen	nineteenth	19th
20	XX	twenty	twentieth	20th
30	XXX	thirty	thirtieth	30th
40	XL	forty	fortieth	40th
50	L	fifty	fiftieth	50th
60	LX	sixty	sixtieth	60th
70	LXX	seventy	seventieth	70th
80	LXXX	eighty	eightieth	80th
90	XC	ninety	ninetieth	90th
100	C	one hundred	one hundredth	100th
400	CD	four hundred	four hundredth	400th
500	D	five hundred	five hundredth	500th
900	CM	nine hundred	nine hundredth	900th
1,000	M	one thousand	one thousandth	1,000th

Why are the terms **cardinal and ordinal** sometimes **confused** in mathematics?

Cardinal and ordinal numbers are sometimes confused because they have two distinct mathematical definitions. The cardinal numbers in the numbering system should not be confused with cardinal numbers in set theory, in which any method of counting sets using a cardinal number gives the same result. Likewise, the ordinal numbers in arithmetic should not be confused with ordinal numbers in set theory: Such numbers, often called ordinals for short, are the order type of a well-ordered set. They are divided into two types: finite and transfinite ordinals. (For more about sets, ordinals, and cardinal numbers, see "Foundations of Mathematics.")

MORE ABOUT NUMBERS

We can look to the familiar face of a clock to illustrate the concept of "clock arithmetic" or modulo 12. With clocks, we can count through 11 before getting to 12, where we start over again at zero. Thus, if it is 7 o'clock and we add 6 hours, we get 1 o'clock, not 13 o'clock. *Taxi/Getty Images.*

What is **congruence**?

In reference to numbers, congruence is the property of two integers having the same remainder upon division by another integer. The term also is often used in geometry to describe a property of geometric formations (for more information about congruence in geometry, see "Geometry and Trigonometry"). Still another way of using congruence is in number theory, in which modular arithmetic is the arithmetic of congruences, which is sometimes informally called "clock arithmetic."

How does **modular arithmetic** work?

In modular arithmetic, numbers "wrap around" when they reach a fixed quantity. This is also called the *modulus*—thus the name modular arithmetic—with the standard way of writing the form as "mod 12" or "mod 2."

In this case, if the two numbers b (also called the base) and c (also called the remainder) are subtracted ($b - c$), and their difference is a number integrally divisible by m, or ($b - c)/m$, then b and c are said to be congruent modulo m. Mathematically, "b is congruent to c (modulo m)" is written as follows, with the symbol for congruence (\equiv):

$b \equiv c \pmod{m}$

But if $b - c$ is not integrally divisible by m, then it is said, "b is not congruent to c (modulo m)," or

$b \not\equiv c \pmod{m}$

More formally, modular arithmetic includes any "non-trivial homomorphic image of the ring of integers." We can interpret this interesting definition using a clock. The modulus would be the number 12 on the clock (arithmetic modulo 12), with an associated ring labeled C_{12} and the allowable numbers being 0, 1, 2, 3, 4, 5, 6, 7, 8, 9, 10, and 11. Another example is arithmetic modulo 2, with an associated ring of C_2, or allowable numbers of 1 and 2.

What are **prime and composite** numbers?

Prime numbers are positive integers (natural numbers) that are greater than 1 and have only 1 and the prime number as divisors (factors). Another way to define a prime number is an integer greater than 1 in which its only positive divisors are 1 and itself. For example, prime numbers less than 20 are 2, 3, 5, 7, 11, 13, 17, and 19. All other integers greater than 1 that are not prime are called *composite numbers*.

There are other rules: The number 1 is unique, and is not considered a prime or composite number. And one of the basic theorems of arithmetic is that any positive integer is either a prime or the product of a unique set of prime numbers. For example, the number 12 is not a prime, but it has a unique "prime calculation" written as: $2 \times 2 \times 3$.

What is the **Sieve of Eratosthenes**?

The smallest prime numbers—those less than 1 million—can be determined using something invented circa 240 BCE: the Sieve of Eratosthenes. This method was named after astronomer and mathematician Eratosthenes of Cyrene (276–196 BCE), who was actually more famous for calculating the circumference of the Earth than for his work with prime numbers.

To determine primes using this method, make a list of all the integers less than or equal to n (numbers greater than 1) and get rid of all the multiples of all primes less than or equal to the square root of n. The numbers that are left are all primes. For example, to determine primes less than 100, start with 2 as the first prime; then write all odd numbers from 3 to 100 (there is no need to write the even numbers). Take 3 as the first prime and cross out all its multiples in the numbers you listed. Take the next number, 5, and then 7, and cross out all their multiples. By the time you reach 11, many numbers will be eliminated and you will have reached a number greater than the square root of 100 (11 is greater than 10, the square root of 100). Thus, all the numbers you have left will be primes.

What are the **first 100 prime numbers**?

The first 100 prime numbers are as follows:

2	3	5	7	11	13	17	19	23	29
31	37	41	43	47	53	59	61	67	71
73	79	83	89	97	101	103	107	109	113
127	131	137	139	149	151	157	163	167	173
179	181	191	193	197	199	211	223	227	229
233	239	241	251	257	263	269	271	277	281
283	293	307	311	313	317	331	337	347	349
353	359	367	373	379	383	389	397	401	409
419	421	431	433	439	443	449	457	461	463
467	479	487	491	499	503	509	521	523	541

Are there **different** types of **prime numbers**?

Yes, there are different types of prime numbers, including the following:

Mersenne primes—See the boxed text for an explanation

Twin primes—Primes of the form p and $p + 2$ (in other words, they differ by two); discovering such a prime involves finding two primes.

Factorial/primorial primes—Primorial primes are of the form $n\# \pm 1$; factorial primes are of the form $n! \pm 1$.

Sophie Germain primes—This is an odd prime p for which $2p + 1$ is also a prime. It was named after Sophie Germain (1776–1831), who proved that the first case of Fermat's Last Theorem for exponents was divisible by such primes.

Other names for prime numbers are mainly for descriptive purposes. For example, in 1984, mathematician Samuel Yates defined a *titanic prime* to be any prime with at least 1,000 digits. In the past few decades since his definition, there have been over a thousand times more such primes discovered. Yates also coined the term *gigantic prime* to indicate a prime with at least 10,000 digits. A great deal has happened in the last few decades, so it is only a matter of time before the first ten-million-digit prime is found, although it is still unknown what name that prime number will be given.

What is the **Fibonacci sequence**?

Italian mathematician Leonardo of Pisa (c. 1170–c. 1250; also known as Fibonacci, or "son of Bonacci," although some historians say there is no evidence that he or his contemporaries ever used the name) may be known for helping to introduce Hindu-Arabic numerals to Europe (see p. 70), but he also is famous for the sequence of numbers he discovered. This sequence—initially pursued as an exercise to determine how fast a

What is the story behind the Mersenne primes?

Mersenne primes (or Mersenne numbers) are connected to prime numbers. They come in the form of $2^p - 1$, in which p is a prime; or, to put it another way, when $2^p - 1$ is prime, it is said to be a Mersenne prime.

Centuries ago, many mathematicians believed that numbers from the form $2^p - 1$ (they actually used the form $2^n - 1$, which is the same as the $2^p - 1$ used today) were prime for all primes p. By the 16th century, it was proven that $2^{11} - 1 = 2,047$ was not prime. By 1603 Pietro Cataldi (1548–1626) correctly discovered that $p = 17$ and $p = 19$ were both prime, but he was wrong to add 23, 29, and 37 to his prime numbers list. Soon, others discovered his errors, including French mathematician Pierre de Fermat (1601–1665) in 1640 and Swiss mathematician Leonhard Euler (1707–1783) in 1738.

The hunt for primes continued. The name "Mersenne" actually came from the French priest Father Marin Mersenne (1588–1648), who in 1644 referred to such numbers in the preface to his book *Cogitata Physica-Mathematica*. He believed that these special primes were: $p = 2, 3, 5, 7, 13, 17, 19, 31, 67, 127$, and 257. But like earlier attempts at determining prime numbers, many of Mersenne's numbers were in error. It took three centuries more to check Mersenne's range of numbers, and by 1947 the correct list of Mersenne primes were: $p = 2, 3, 5, 7, 13, 17, 19, 31, 61, 89, 107$ and 127. Interestingly enough, even though Mersenne incorrectly stated that certain numbers belonged to this group—he probably didn't verify all the numbers on his list—his name is still attached to these numbers.

Mersenne Prime	Year Discovered	Mersenne Prime	Year Discovered	Mersenne Prime	Year Discovered
2	—	1,279	1952	110,503	1988
3	—	2,203	1952	132,049	1983
5	—	2,281	1952	216,091	1985
7	—	3,217	1957	756,839	1992
13	1456	4,253	1961	859,433	1994
17	1588	4,423	1961	1,257,787	1996
19	1588	9,689	1963	1,398,269	1996
31	1772	9,941	1963	2,976,221	1997
61	1883	11,213	1963	3,021,377	1998
89	1911	19,937	1971	6,972,593	1999
107	1914	21,701	1978	13,466,917*	2001
127	1876	23,209	1979	20,996,011*	2003
521	1952	44,497	1979	24,036,583*	2004
607	1952	86,243	1982		

*These numbers have not yet been confirmed as Mersenne primes by the Great Internet Mersenne Prime Search at press time (GIMPS; see boxed text).

pair of rabbits can reproduce per year—is formed by adding the two preceding numbers to find the next number, starting with a pair of ones. Thus, the Fibonacci numbers in this sequence are 1, 1, 2, 3, 5, 8, 13, 21, 34, 55, 89, 144, 233, 377, and so on; or

$$1 + 1 = 2$$
$$2 + 1 = 3$$
$$3 + 2 = 5$$
$$5 + 3 = 8$$
$$8 + 5 = 13$$
$$13 + 8 = 21$$
$$21 + 13 = 34$$
$$34 + 21 = 55$$
$$55 + 34 = 89$$
$$89 + 55 = 144$$
$$144 + 89 = 233$$
$$233 + 144 = 377, \text{ and so on.}$$

What are **exponents**?

Exponents are actually shorthand for multiplications and represent the number of times a number is being multiplied (called the *base*). For example, $9 \times 9 = 9^2$, or $9 \times 9 \times 9 = 9^3$.

Another term for this process is "raising to a power," in which the exponent (a number as a superscript) is the "power." For example, 9^3 is "nine raised to the third power." It is also easier to write larger numbers with exponents. For instance, instead of writing *xxxx*, we can write x^4. (For more information about exponents, see "Algebra.")

How can we **simplify expressions** using **exponents**?

The following are ways to solve simple mathematical expressions with the same base:

To multiply two numbers with different exponents, add the exponents:

$(x^2)(x^4)$

$= (xx)(xxxx)$

$= xxxxxx$

$= x^{2+4}$

$= x^6$

To raise a number with a power, multiply the exponents:

$(x^3)^3$

$= (x^3)(x^3)(x^3)$

$= (xxx)(xxx)(xxx)$

$= xxxxxxxxx$

$= x^{(3 \times 3)}$

$= x^9$

What **common mistake** is often made in **exponential equations**?

Unlike multiplication, exponents do not "distribute" over addition. For example, $(2 + 5)^2$ *does not* mean $2^2 + 5^2 = 4 + 25 = 29$. In this case, you add the numbers in the parentheses first, then square that number. The *true* answer to this equation is $(2 + 5)^2 = (7)^2 = 49$.

What is a **base** in mathematics?

The term "base" has many meanings in the English language, including several that apply to the field of mathematics. When talking about sets, bases are the open sets whose union forms an abstract entity called a topological space. In geometry, the base represents the side of a polygon or polyhedron that is perceived as its bottom; when referring to an isosceles triangle, the base is the side that differs in length from the other two (thus, the base angles include the side that is thought of as the base). Algebraists also use the term base to describe either the number used with an exponent to create a power, such as $3^4 = 81$; or to write the same number as a subscript to a logarithm, such as $\log_3 81 = 4$. (For more information about logarithms, see "Algebra.")

One of the more familiar uses of the term base in mathematics deals with our numbering system, in which a base is a natural number whose powers are added to produce a specific number. For example, using 10 as a base, the number 2583.789 is actually $(2 \times 10^3) + (5 \times 10^2) + (8 \times 10^1) + (3 \times 10^0) + (7 \times 10^{-1}) + (8 \times 10^{-2}) + (9 \times 10^{-3})$.

How do you **convert** from the **binary** to the **decimal system** and vice versa?

Besides the decimal system, one of the most familiar number systems is the binary numeration system; this is mainly because of its use in computers (for more about computers, see "Math in Computing"). In a binary numeration system, only 1 and 0 are used—or a base 2 system. Converting between binary and decimal systems is fairly simple; just remember that each digit in the binary number represents a power of two.

The first column in the base 2 math is the *units column,* then the twos, fours, eights, etc. columns, all of which can only be filled with 0s or 1s. Since there is no single digit that stands for "2" in base 2, when you get to what stands for 2, you put a 1 in the 2's column and a 0 in the units column, creating one 2 and no 1s. Thus, the base ten "two" (2_{10}, or just 2 in decimal form) is written in the binary as 10_2; a 3 (3_{10}, or just 3 in decimal form) in base 2 is actually "one 2 and one 1," or 11_2. The number 4 is actually 2×2, so you eliminate the 2 and unit columns and put a 1 in the 4s column. Thus, 4_{10} (or just 4 in decimal form) is written in binary form as 100_2. To see how computers "translate" decimal to binary numbers, here are the first ten conversions:

Decimal	Binary	Explanation
0	0	no 1s
1	1	one 1
2	10	one 2 and no 1s
3	11	one 2 and one 1
4	100	one 4, no 2s, and no 1s
5	101	one 4, no 2s, and one 1
6	110	one 4, one 2, and no 1s
7	111	one 4, one 2, and one 1
8	1000	one 8, no 4s, no 2s, and no 1s
9	1001	one 8, no 4s, no 2s, and one 1
10	1010	one 8, no 4s, one 2, and no 1s

How do you **express** certain numbers in **powers of ten** notation?

A number in power of ten notation is represented as a base number (or *mantissa*) times ten raised to some power (or *exponent*). This means the mantissa is multiplied by ten times the number of times the power indicates. For example, in the equation 32×10^4, the 32 is the mantissa multiplied by $10 \times 10 \times 10 \times 10$. This would equal 320,000.

Power of ten notation can also be used to express the same number in various ways. For example, the distance from the Earth to the Sun averages 93,000,000 miles. This can be represented by the following:

93,000,000

93×10^6

9.3×10^7

or 0.93×10^8, and so on.

The distance between the Earth and the Sun is most easily expressed using powers of ten as 9.3×10^7. *The Image Bank/Getty Images.*

Why do mathematicians and scientists often use **scientific notation**?

In scientific notation, scientists generally pick the power of ten that is multiplied by a number between 1 and 10, which makes larger (and smaller) numbers easier to write and read because they take up less space.

For example, in the case of the average distance between Earth and Sun, 9.3×10^7 miles is mathematically much easier to work with than any larger or smaller exponent numbers. The following are several examples of large and small numbers in powers of ten notation similar to numbers used in scientific notation (note: numbers that are extremely small in scientific notation will have negative exponents):

- 748,000 can be represented by 7.48×10^5
- 245 can be represented by 2.45×10^2
- $-45,000$ can be represented by -4.5×10^4
- 0.025 can be represented by 2.5×10^{-2}
- -0.0036 can be represented by -3.6×10^{-3}
- 0.0000409 can be represented by 4.09×10^{-5}
- 0.0000000014 can be represented by 1.4×10^{-9}

What **prefixes** define the **powers of ten**?

Many standard prefixes are used to represent powers of ten, most of which are employed quite frequently to express various units. We are all familiar with some of them, including kilo- (from the Greek *chilioi,* or "a thousand"), milli- (from the Latin *mille,* or "thousand") , and micro- (from the Greek *mikros,* or "small"). The following chart defines some prefixes representing the powers of ten:

Prefixes for Powers of Ten

Multiple	Name
10^{18}	exa
10^{15}	peta
10^{12}	tera
10^{9}	giga
10^{6}	mega
10^{3}	kilo
10^{2}	hecto
10	deka
10^{-1}	deci
10^{-2}	centi
10^{-3}	milli
10^{-6}	micro
10^{-9}	nano
10^{-12}	pico
10^{-15}	femto
10^{-18}	atto

THE CONCEPT OF ZERO

What is a **placeholder**?

A *placeholder* is a number that, as the name implies, holds a place. Initially, various cultures used a placeholder—often a dot or a space—to show an empty spot or place not used in a numeral (or other way of counting). This was in place of what we now call a zero (0), a symbol that is not only a placeholder but also an essential number in our numeration system. (For more about the evolution of the symbol zero [0], see "History of Mathematics.")

What are the **definitions of zero** and **non-zero**?

Those who use the Hindu-Arabic numeration system are all familiar with the concept of "zero" (0), and its importance. The symbol "zero" represents a valuable placeholder; it is also the additive identity element of an algebraic system (when a number and its

additive inverse add up to 0); and, finally, it is the starting point in measurements. The zero symbol is also called a cipher (no relation to sending secret messages) or the symbol for the absence of quantity (although be aware that cipher can also mean any Hindu-Arabic numeral). In other words, zero is naught (nothing, or, from the Old English, *nāwiht,* meaning "not" [na] and "thing" [wiht]).

Mathematicians also use the word *non-zero* to represent a quantity that does not equal zero. A real non-zero number must either be positive or negative; a complex non-zero number can be the real or imaginary part of the equation. (For more information about complex numbers, see above.)

What is an **indeterminate number**?

As with many other fields, mathematics has terms that are sometimes confusing or overlapping. For example, it is interesting to note that there *is* such a number as 0/0, which is called an *indeterminate* number. But be careful: This is not the same as an undefined number. If an indeterminate number comes up somewhere, you never know the value for your specific case—and you can conceivably give it any number of values. Confusing? Don't worry, you're not alone. It is one of those quandaries that often baffles the best mathematicians.

BASIC MATHEMATICAL OPERATIONS

What does an **equal sign** represent?

In standard arithmetic terms, the equal sign (=) is a symbol that represents two amounts with the same value. For example $7 = 7$; or $3 + 4 = 7$. When something is not equal, the sign is ≠, as in $2 \neq 3$ or $3 + 7 \neq 12$.

In a line of computer code, the equal sign can mean something much different. For example, for codes used under certain conditions, such as a JavaScript-reading computer, a single equal sign in a line of script means "is". There is even a double equal sign (= =) that means "is equal to," an important difference when writing code for the computer. In addition—and confusing as it may seem—the computer codes for "not equal" are different, and include != and /=.

What is **addition**?

Addition is an operation in which two numbers, called the *addends,* produce a third number called the *sum.* Natural numbers are added by starting with the first addend and counting as many more numbers as the second addend. For example, for $2 + 4$, you would think 2, 3 (as the first number after 2), 4 (as the second number), 5 (as the third number), and 6 (as the fourth number), thus the numbers add up to 6. Not all numbers are added in the same way as natural numbers.

91

When was the equal sign introduced into mathematics?

The equal sign (=) is a relatively new invention in mathematics. It was first used by British mathematician Robert Recorde (1510–1558; also seen erroneously as Record) in his book *The Whetstone of Witte* (1557), the first algebra book introduced in England. In it he justifies using two parallel line segments "bicause noe 2 thynges can be moare equalle (sic)" ("because no two things can be more equal than parallel lines.") It was not an immediately popular symbol, though, with mathematicians continuing to use a range of symbols for equal, including the two parallel lines (||) used by Wilhelm Xylander in 1575, and *ae* or *oe* (both from the word *aequalis,* the Latin for "equal"). But for the most part, the word "equal" was written in an equation until around 1600, when Recorde's symbol became more readily accepted, and it continues to be so today.

What does **carry** mean in **addition**?

In addition, *carry* defines a way in which larger numbers are added. In this arithmetic operation, there is a shifting of leading digits into the next column to the left when the sum of that column exceeds a single digit. For example, carrying is evident in the following operation in the base 10 numbering system (adding addends 234 and 168, giving the sum 402):

$$\begin{array}{r} \overset{1\,1}{234} \\ +\,168 \\ \hline 402 \end{array}$$

By adding the same columns, starting in the right-hand column, 8 + 4 is 12, a number larger than 9, the highest number than can exist in that spot; carry the amount of ten to the next column, leaving 2; add 3 + 6 + 1 (which, when carried, represents 10 for this column), which equals 10, again a number larger than 9. Again, carry the one to the next column leaving 0; adding the 2 + 1 + 1 (carried) equals 4. All total, the sum is 402.

How do you **subtract numbers**?

Subtraction is the "opposite" (or, in set theory, the inverse) of addition: In its simplest form, one whole number is essentially taken away from another whole number. When you subtract numbers, you are answering the question of how many are left. For example, if 23 people leave a building that has 123 people (123 − 23), there would be 100 people left in the building.

What does **borrow** mean in **subtraction**?

As with "carry" in addition, to "borrow" in subtraction means to take amounts from one number and assign them to the next. In this procedure, the 10 is borrowed from

the next highest digit column in order to obtain a positive difference in the nearby column. For example, borrowing is evident in the following operation in the base 10 numbering system (1,234 minus 567, giving the result of 667):

$$1\overset{1}{\underset{}{2}}\overset{12}{\underset{}{3}}\overset{1}{4}$$
$$-567$$
$$\overline{667}$$

When you subtract the same columns, starting in the right-hand column, the 4 is too small to subtract from 7; therefore, you need to borrow 10 from the column to the left, boosting the number 4 to 14. The column to the left then loses 10, and 2 is left. Again, the 2 is too small to subtract from 6, so 10 is borrowed from the next column to the left, boosting the number 2 to 12. The last number to the left is dropped down by 10, giving 11; the 5 is subtracted from the 11, resulting in 6 in that column.

Where did the **symbols plus (+)** and **minus (−)** originate?

One of the first books to use the plus and minus signs was written in 1489 in Johann Widmann's (c. 1460–?) *Mercantile Arithmetic*. Originally, he used the signs + and − to indicate excesses and deficits (what we would call credit and debit) in business dealings. But some historians believe the + sign initially evolved from the French *et,* or "and," because the written "e" and "t" resemble the + sign.

Although they were probably used before in general mathematics, the first person known to have used the + and − symbols in writing algebraic expressions lived in the early 1500s: Dutch mathematician Vander Hoecke. The symbols finally went into general use in England when Robert Recorde's (1510–1558; also seen erroneously as Record) book *The Whetstone of Witte* was published; this is the same book responsible for bringing the equal (=) sign to the forefront of mathematics (see above).

What is **multiplication**?

The word "multiply" comes from the Latin roots *multi* ("many") and *pli* ("folds"). English poet Geoffrey Chaucer (1340?–1400) may have used the word first as a verb in his *A Treatise on the Astrolab* (1391). In multiplication, two natural numbers are multiplied together (the numbers are called *factors*; the less-used terms for two such numbers are multiplicand and multiplier), producing what is called a *product*. Multiplication is actually a form of repeated addition. For example, 2 × 3 means 2 + 2 + 2 (or 6).

What are **multiplication tables**?

Multiplication tables are just what the name implies: a table of multiplication. Most of these tables have a specific purpose: mainly, to multiply numbers in the rows and columns by each other to find a product. One of the simplest multiplication tables

deals with the whole numbers. Here is one example:

0	1	2	3	4	5
1	1	2	3	4	5
2	2	4	6	8	10
3	3	6	9	12	15
4	4	8	12	16	20

Where did the **symbols** for **multiplication** originate?

The 17th century seems to be the century when the basic mathematical symbols were developed. The best reasons for the development of such symbols make sense: they were faster and easier to write, took up less written space, and helped the printing process. Although the use of these symbols would eventually be standardized—so everyone would understand the meaning of certain mathematical operations—it took a while for this to happen.

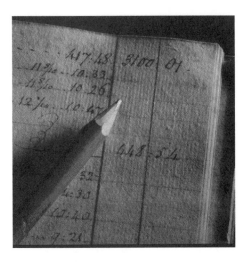

Ledgers are commonly used in business for accounting purposes. Debits and credits are added and subtracted to keep track of profits and expenses. *The Image Bank/Getty Images.*

For example, in 1686 German mathematician Gottfried Wilhelm Leibniz (1646–1716) was using the symbol ∩ for multiplication and ∪ for division. Eventually, English mathematician and scientist Thomas Harriot (1560–1621; also seen as Hariot) used the dot to indicate multiplication in his treatise *Artis anayticae praxis* (1631). (He also developed the greater than [>] and less than [<] symbols). That same year, English mathematician William Oughtred (1575–1660) used the symbol "×" for multiplication in his book *Clavis mathematicae,* in which he was also the first to mention the plus-minus symbol [±].)

Today, we use a number of symbols for multiplication operations. The most common symbols are ×, * , ·, and (), as in 2 × 3, 2 * 3, 2 · 3, and (2)(3).

How is the term **inverse** used in arithmetic operations?

Inverse operations are those that "undo" another operation. In particular, subtraction is the inverse of addition because a + b − b = a; division is also the inverse operation to multiplication.

The *inverse of a number* can be expressed as follows: The additive inverse of a real or complex number *a* is the number that, when added to *a,* equals 0. In multiplication, the multiplicative inverse of *a* is the number that, when multiplied by *a,* equals 1.

Why can't you divide by zero?

Dividing by zero is like the old saying "You can't get something from nothing." Mathematically speaking, it's the same way: You can't "divide by nothing." In fact, when something is divided by zero, the answer is always undefined.

Here are a few ways of looking at this: There is a rule in arithmetic that $a(b/a) = b$. So if we say that $1/0 = 5$, then $0(1/0) = 0 \times 5 = 0$. In other words, if you could divide by 0, this rule would not work. Another way to look at the "no to 0 as a divisor" problem is through multiplication: if $10/2 = 5$, we know that $5 \times 2 = 10$; the same for $5/1 = 5$, thus we know that $5 \times 1 = 5$. But if you take $5/0$, that would mean that the answer times 0 would equal 5, but anything times 0 is equal to zero. Because there is no answer to this dilemma, mathematicians say you can't divide by zero.

What is **division**?

The word "divide" comes from the Latin root *vidua* (referring to a separation; the word "divide" shares its major root with the word "widow") and *di*, a prefix that is a contraction of *dis*, meaning "apart" or "away." In division, the number being divided is called the *dividend*, while the number dividing it is called the *divisor*. The end result is called the *quotient*. For example, in $20/5 = 4$, 20 is the dividend, 5 the divisor, and 4 the quotient.

Division in mathematics is a relatively new concept for the masses; it was only taught at university levels after the 16th century. The first to offer division to the public was German mathematician Adam Ries (1492–1559; also seen as Risz, Riesz, Riese, or Ris) in his work *Rechenung nach der lenge, auff den Linihen vnd (sic) Feder,* often shortened to *Practica*. His work reached more people for an important reason: Instead of the usual practice of writing a mathematical book in Latin, he wrote his book in German, thus reaching a wider audience.

Where did the **symbols** for **division** originate?

The history behind the division symbols is long and complicated. The following lists how the major ones developed:

Closed parentheses—The arrangement "8)24," meaning 24 divided by 8 in this case, was used by Michael Stifel (1486 or 1487–1567) in *Arithmetica integra* (1544).

The obelus—By 1659 Swiss mathematician Johann Heinrich Rahn (1622–1676) introduced the division symbol (\div, called an *obelus*) in his book *Teutsche Algebra*. The symbol was a combination of ":" and "−". (This division symbol was used by many

writers before Rahn as a minus sign.) In 1668, when Rahn's book was translated into English with additions by English mathematician John Pell (1610–1685), the division symbol was retained. Some say Pell greatly influenced Rahn to develop the symbol, but most historians agree that there is little evidence of such a connection.

Slash—Another sign for division, the slash (/) was actually first used for fractions, such as 2/3 or 1/2. It can be extended into other, larger or smaller numbers, such as 123/112 and 0.112/0.334. Little is known about its origins, but it is known that this symbol was sometimes used for subtraction, until it became standard practice for representing division.

How did the **symbols** for **long division** develop?

In the 19th century, United States textbooks typically showed long division with the divisor, dividend, and quotient on the same line, separated by parentheses, as in 36)108(3. In the same century, in examples of short division, a vinculum (line) was placed under the dividend, with the vinculum almost attached to the bottom of the parenthesis. The quotient was written under the vinculum, as per the following:

$$5 \overline{)455}$$
$$\overline{91}$$

By late in the 19th century, the vinculum was almost attached to the top of the parenthesis and the quotient was written above the vinculum, as per the following:

$$\overline{91}$$
$$5 \overline{)455}$$

These symbols are similar to what we see in our elementary dealings with long division, but our vinculum is attached to the parenthesis. Interestingly enough, there is no name for the symbol used for long division ($\overline{)}$).

What are the **least common multiple** and **denominator**?

The smallest common multiple (whole number) of two or more whole numbers is called the *lowest (or least) common multiple (LCM)*. For example, for the numbers 3 and 8, the multiples of 8 are 8, 16, 24, 32, and so on; the multiples of 3 are 3, 6, 9, 12, 15, 18, 21, 24, 27, and so on. Therefore, the LCM of 3 and 8 is 24.

The *least common denominator* (LCD) is mainly used to carry out the addition or subtraction of fractions. In order to do these operations, the fractions need to have the same denominator. (For more information about fractions, see below.) The easiest way to work on such calculations is to determine the lowest number possible for the denominator—a number called the least common denominator (LCD)—which is actually the common factor by which two numbers are divisible. For example, to add 1/6 and 1/8, we have to find the least common multiple of the denominators. In this

case, the number is 24: Multiply $1/6 \times 4/4$ and $1/8 \times 3/3$, to change each addend to some number of "24ths"; or $1/6 \times 4/4 = 4/24$; and $1/8 \times 3/3 = 3/24$. Thus, $1/6 + 1/8 = 4/24 + 3/24 = (4+3)/24 = 7/24$.

What is the **root** of a number?

The root of a real or complex number is a number that, when raised to some exponent (or multiplied by itself the number of times indicated by the exponent), equals the original number. Most people are familiar with square and cube roots. There are actually many other such roots, including the real fourth roots, real fifth roots, and so on, and roots associated with complex numbers. For example, the real fourth roots of 16 are 2 and -2; the real fifth root of -32 is -2.

What are the **square** and **square roots**?

When you multiply a real or complex number by itself, you are actually squaring that number. Mathematicians express the square of a number using the superscript 2, or, for example, 2^2. The square of a real number is always positive, whether the number is 2^2 ($= 4$) or -2^2 ($= 4$; a negative times a negative equals a positive real number).

A square root is a number that when multiplied by itself equals a specific product. For example, if $t^2 = s$, then $t = \pm \sqrt{s}$, in which t is the square root and s is a positive number. For example, the two square roots of 16 ($\sqrt{16}$) are 4 and -4, as $4^2 = 16$ and $-4^2 = 16$.

What are the **cube** and the **cube root**?

Much like a square, a cube is when you multiply a real or complex number by twice itself (making a total of three numbers). Mathematicians express the cube of a number using the superscript 3; or, for example, 2^3, or $2 \times 2 \times 2$. Unlike a square of a number, the cube of a number will not always be positive, such as $-3 \times -3 \times -3$, which equals -9.

A cube root is a number that when multiplied by itself two more times has the product of s, in which $t^3 = s$. For example, the cube root of 125 (s) is 5 (t), or written as $\sqrt[3]{125} = 5$; the cube root of -125 is -5.

What is a **factor** and what does **factorization** mean?

A *factor* is a portion of a quantity that when multiplied by other factors gives the entire quantity (or product). In order to determine such factors (or divisors), you have to use *factorization* (also called *factoring* or *to factor*). When factoring an integer, it is referred to as prime factorization; when factoring a polynomial, it is called polynomial factorization.

What is **prime factorization**?

Many of us are most familiar with prime factorization, which is a way of taking a number and breaking it down into its constituent primes. An example of prime factorization is as follows: One finds the "simplest" representation of the given quantity in terms of smaller parts—in the case of 15, the factors would be 1, 3, 5, and 15 (essentially, all the numbers that will divide integrally into 15). Not that prime factorization is always that easy. Larger numbers make it more difficult to factor, and many sophisticated prime algorithms have been devised for larger—and different types—of numbers.

What does the **greatest common factor** mean?

The *greatest common factor* (or GCF; sometimes called highest common factor) of two whole numbers is the largest whole number that is a factor of both. Take, for example, the numbers 12 and 15: The factors of 12 are 1, 2, 3, 4, 6, and 12; the factors of 15 are 1, 3, 5, and 15. Therefore, the common factors—or numbers in both lists of factors—are 1 and 3; and the greatest (highest) common factor in this case is 3.

There is another method used to discover the GCF: listing the numbers' prime factors, then multiplying those numbers. For example, the prime factorizations of 12 and 15 are: $2 \times 2 \times 3 = 12$ and $3 \times 5 = 15$. Notice that the prime numbers have 3 in common; thus, the GCF is 3.

An example with larger numbers is to find the GCF of 36 and 54. Working it out by the first method, the factors of 36 are 1, 2, 3, 4, 6, 9, 12, 18, and 36; the factors of 54 are 1, 2, 3, 6, 9, 18, 27, and 54. The greatest (or highest) common factor of both numbers is 18. To work it out using prime factorization, the prime factorization of 36 is $2 \times 2 \times 3 \times 3$; the prime factorization of 54 is $2 \times 3 \times 3 \times 3$. Both these factorizations have one 2 and two 3s in common; thus, we multiply those common numbers, or $2 \times 3 \times 3 = 18$.

FRACTIONS

What are **proper** and **improper fractions**?

The word "fraction" to most of us means a part of something; in mathematics, it represents a type of numeral, in most cases the quotient of two integers, with the top number called the *numerator* (the number of parts) and the bottom number the *denominator* (how many parts the whole is divided into). Written out, the numerator and denominator are separated by a "/" or "—". A fraction is usually denoted by a/b, in which "a" and "b" are whole numbers and "b" is not equal to zero. (For the explanation of why you cannot divide by zero, see above).

A rational number between 0 and 1 can be represented by fractions (by the division of two numbers). If the quotient is less than one, such as 1/2 or 2/5, then it is called a *proper fraction;* if the quotient is greater than one—or, in other words, if the

Can a fraction's numerator be zero?

In the context of division (see above), we learned that it is not possible to divide by 0, which is labeled an undefined number. But you can have zero (0) as your numerator. And any allowed fraction (one that doesn't have a 0 in the denominator) that has 0 as its numerator will always be equal to zero. For example, 0/5 and 0/345 both equal 0.

numerator of a fraction is larger than the denominator, such as 23/7—it is called an *improper fraction*.

How are **fractions converted to decimals** and vice-versa?

In the most commonly used place value, the decimal system, numbers smaller than 1 can be expressed as fractions called *decimal fractions*. In this system, the decimal fractions are expressed in terms of tenths, hundredths, thousandths, and so on. For example, for the fraction 1/2, or 1 divided by 2, the decimal fraction is 0.5; and, vice versa, the decimal fraction 0.5, or 5/10ths, is equal to 1/2.

Not all fractions are so easily converted to decimals. It depends on the type of number, especially if it is an irrational or rational number. Some decimal fractions include an infinite number of decimal places to be expressed exactly; something that is not possible as far as we know (who can write an infinite number of numbers?) And some decimal fractions repeat forever, such as 1/3 = 0.3333....

What are the **rules** for **adding and subtracting fractions**?

When adding fractions, the denominators need to be the same, but you can't add the denominators to get the answer. Simply put, if the denominators are already the same, the fractions are simple to add, such as 1/3 + 1/3 = (1 + 1)/3 = 2/3. If the denominators are not the same, find the common denominator by multiplication. For example, 1/2 + 1/3 = 3/6 + 2/6 = (3 + 2)/6 = 5/6.

When subtracting fractions, the denominators again need to be the same; and again you can't add (or subtract) the denominators to get the answer. If the denominators are the same, subtract the fractions, such as 2/3 − 1/3 = (2 − 1)/3 = 1/3. If the denominators are not the same, find the common denominator by multiplication, such as 1/2 − 1/3 = 3/6 − 2/6 = (3 − 2)/6 = 1/6.

What are the **rules** for **multiplying and dividing fractions**?

As to be expected, there are rules for multiplying and dividing fractions. Multiplication of fractions is very straightforward—just multiply the numerators and denominators,

then simplify the resulting fraction, if needed (or if you can). For example, $2/5 \times 4/7 = (2 \times 4)/(5 \times 7) = 8/35$ (this number can't be simplified).

Division of fractions entails one main rule: You need to flip over, or invert, the "divisor" fraction (the fraction on the bottom) to get the result (this is also called the reciprocal of the fraction; see below). Here are the steps: First, change the division sign to a multiplication sign after inverting the fraction to the right of the sign. Multiply the numerators and denominators, and write the result. You can then simplify or reduce the fraction if needed. For example, $1/2 \div 1/4 = 1/2 \times 4/1 = 4/2$. This number can be simplified to 2.

How are **decimal fractions** calculated by **adding, subtracting, multiplying** and **dividing**?

Decimal fractions are added, subtracted, multiplied and divided much like whole numbers, but with decimal differences. The following gives some examples: Adding such numbers as $0.3 + 0.2$ is simple: $0.3 + 0.2 = 0.5$. Adding whole and decimal fractions is also easy: $2.4 + 5 = 7.4$. These numbers are also easy to subtract, such as $0.3 - 0.2 = 0.1$. Multiplication and division with fractions is also similar to doing so with regular numbers, although the placement of the decimal point is all important. For example, multiplying $24.45 \times 0.002 = 0.0489$; dividing the same numbers $24.45/.002 = 12,225$. (It's interesting to note here a "mathematical surprise": In the last example, dividing the small numbers equaled a much larger number—the opposite of what most of us would expect.)

How do you **reduce a fraction**?

To reduce a fraction, there are three general steps: factor the numerator, factor the denominator, and cancel out the *fractional mixes* that have the value of one. The leftover number is the reduced fraction. For example, to reduce 16/56, factor the numerator ($16 = 2 \times 2 \times 2 \times 2$) and factor the denominator ($56 = 2 \times 2 \times 2 \times 7$); then eliminate the 2s (2/2 equals 1):

$$\frac{16}{56} = \frac{2 \times 2 \times 2 \times 2}{2 \times 2 \times 2 \times 7}$$

The reduced fraction equals 2/7.

What do you calculate an **equivalent fraction**?

An equivalent fraction—also called "building fractions"—is the reverse of reducing the fraction: Instead of searching for the 1 in the fractional mix that you can reduce, you insert a 1 and build the fractions. For example, to find the equivalent fraction for 1/4, using the number 3, multiply the numerator and denominator by 3 (3/3 = 1); 1/4

What are unit fractions and how are they tied to ancient Egypt?

A *unit fraction* is one that has a numerator of 1, such as 1/2, 1/4, and 1/43545. One of the earliest discussions of unit fractions—a table of representations of 1/n—was found on the famous Rhind papyrus (also called the Rhind Mathematical Papyrus), dated to around 1650 BCE. This record—a table copied by the Egyptians from another papyrus dated 200 years earlier—represented a sum of distinct unit fractions for odd "n" numbers between 5 and 101. To write a certain fraction, they would add combinations of 1/n. For example, instead of writing 2/5, they wrote 1/3 + 1/15; for 2/29, they wrote 1/24 + 1/58 + 1/174 + 1/232.

Because of the Rhind papyrus discovery, the sums of unit fractions are now called *Egyptian fractions*. No one truly knows why the Egyptians chose this method for representing fractions, but some historians believe it was a "wrong turn" in Egyptian mathematical history. Whatever the reason, the Egyptians (apparently successfully) used this system for 2,000 years. (For more information about the Rhind papyrus, see "History of Mathematics.")

\times 3/3 = (1 \times 3)/(4 \times 3) = 3/12; therefore, the equivalent fraction in this case is 1/4 = 3/12 (the equal sign is used to represent equivalent fractions).

What is the **reciprocal** of a number?

The reciprocal of a number is obtained when a given number is divided into 1: the results are called the reciprocal of that number. For example, the reciprocal of 6 is 1 divided by 6, or 1/6. Reciprocals come in most handy when dividing fractions (see above to learn more about dividing fractions).

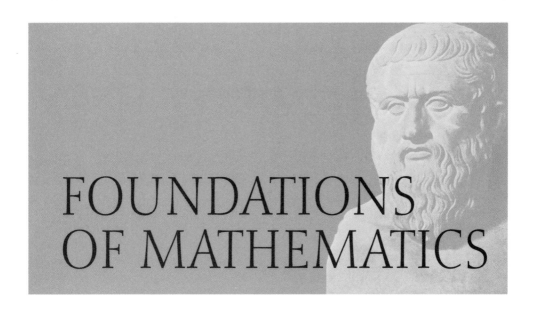

FOUNDATIONS OF MATHEMATICS

FOUNDATIONS AND LOGIC

What are the **foundations of mathematics**?

The foundations of mathematics include how to formulate and analyze the language (you have to "speak" the right mathematical language to make meaningful mathematical statements), axioms (a statement accepted as true without proof), and developing logical methods in all mathematical studies. The most basic mathematical concepts in the foundations include numbers, shapes, sets, functions, algorithms, axioms, definitions, and proofs.

Why do so many **philosophers study** the **foundations of mathematics**?

There are three underlying reasons why philosophers often study the foundations of mathematics. First, these foundations have always been a part of scientific thought, with the abstract nature of mathematical objects offering unusual and often unique philosophical quandaries. Second, the subject offers a high level of technical sophistication, allowing philosophers to develop connections between models and patterns, laying the groundwork for many other sciences. And finally, the foundations of mathematics provides ways for philosophers (and mathematicians) to try out general philosophical doctrines in a specific scientific context.

What is **logic**?

Although it closely resembles mathematics (and is sometimes used as a basis for it), logic is a branch of knowledge or inquiry that is separate from mathematics and the sciences, but it is still used by both fields in various ways. Simply put, logic is described as the systematic study of well-founded inference, in which there is a definite distinction

between logical validity (also known as the formal properties of the inference process) and truth. This also means that a true result may come from an invalid argument (see below for the definition of an argument). For example, "all cats are cute; Fluffer is a cat; therefore, Fluffer is cute," is a valid inference; whereas, "all cats are cute; Fluffer is cute; therefore, Fluffer is a cat," is an invalid inference, even if Fluffer really is a cat.

What is the **historical basis** for **mathematical logic**?

Most mathematicians believe that systematic logic began with Aristotle's collection of works titled *Organon* (*Tool*), in which he introduced his ideas on logic. In particular, Aristotle used general forms to describe logic, such as if all x are y; and all y are z; then all x are z. He presented three laws basic to all valid thought: the law of identity, or A is A (for example, an acorn will always yield an oak tree and nothing else); the law of contradiction, or A cannot be both A and not A (for example, an honest woman cannot be a thief); and the law of the excluded middle, or either or, in which A must either be A or not A (for example, a dog can be brown or not brown). Interestingly, author Ayn Rand divided her novel *Atlas Shrugged* into three parts after these three principles as a tribute to Aristotle.

Was **mathematics** always based on a **logical foundation**?

No, not all of mathematics was always based on a logical foundation, but some ancient cultures did develop certain aspects of logic in their thought. The Greeks were probably one of the first cultures to understand logic's role in mathematics and philosophy, and they studied the subject extensively. For example, geometry, as presented by Greek mathematician Euclid (c. 325–c. 270 BCE), had some foundations in logic. Greek scientist and philosopher Aristotle's (384–322 BCE) rules for syllogisms were also based on logic, and he wrote the first systematic treatise on logic. But his logic

works were based on ordinary language—making them a matter of interpretation and subject to ambiguities.

It was not until the development of calculus that most of mathematics was put on a logical foundation. By the 17th century, people such as German mathematician Gottfried Wilhelm Leibniz (1646–1716) began to demand a more regular and symbolic way to express logic. Logic truly became a part of mathematics around the mid-19th century, especially with the 1847 publication of English mathematician George Boole's (1815–1864) *The Mathematical Analysis of Logic* and English mathematician Augustus De Morgan's (1806–1871) *Formal Logic*. Thus, mathematics began to encompass symbolic logic with precise rules to manipulate those symbols (see below for more about symbolic logic).

Syllogisms, a simple exercise developed by Greek philosopher Aristotle, use basic concepts of logic also seen in mathematics. Aristotle also contributed to mathematics by originating the concept of proofs. *The Bridgeman Art Library/Getty Images.*

Of course, nothing is perfect, although mathematicians in the late 19th and early 20th centuries hoped it would be. They believed that all of mathematics could be described using symbolic logic and made purely formal. But in the 1930s, Austrian-American mathematician and logician Kurt Gödel (1906–1978) put a damper on such an idea by showing that not all truths could be derived by a formal logic system.

What were **Aristotle's syllogisms**?

Syllogisms, which are often attributed to Aristotle, are the verbal versions of the formal deductive rules for logic. Aristotle believed that any logical argument could be explained in these standard forms. He divided them into a major premise ("all squirrels eat nuts"), a minor premise ("Fred is a squirrel"), and a conclusion derived by a rule of logic ("Fred eats nuts"). The classical syllogism is, "All men are mortal. Socrates is a man. Therefore Socrates is mortal." This form of logic—called syllogistic logic—would dominate Western cultural thought for more than 2,000 years.

What are **subjects and predicates** in **Aristotelian logic**?

In Aristotelian logic there are grammatical distinctions between a subject and a predicate. The *subject* is usually an individual entity (an object, house, city, man, animal);

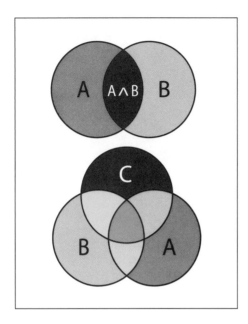

In these examples of Venn diagrams, the top illustration represents an order-two diagram, and the bottom is an order-three diagram.

or it may be a class of entities (objects, houses, cities, men, animals). The *predicate* is the property or mode of existence that does or does not exist with a given subject. For example, a singular plant (subject) may or may not be blooming (predicate); all houses (subject) may or may not have two stories (predicate).

Who invented a way of **analyzing syllogisms**?

In 1880 English logician John Venn (1834–1923) presented a method to analyze syllogisms, now known as Venn diagrams. Venn initially criticized such diagrams in works by his contemporaries, especially those of English mathematicians George Boole (1815–1864) and Augustus De Morgan (1806–1871). But in 1880 Venn introduced his own, now famous, version of the diagrams in his paper *On the Diagrammatic and Mechanical Representation of Prepositions and Reasonings*. By 1881, along with correcting Boole's work, Venn further elaborated on the diagrams in his book *Symbolic Logic*. Today we are most familiar with Venn diagrams in connection with understanding sets.

Although Venn is credited with the diagrams, he was not the first person to use such geometric methods to represent syllogistic logic. German mathematician Gottfried Wilhelm Leibniz (1646–1716) used such graphic representations in his work. And even Swiss mathematician Leonhard Euler (1707–1783) is known to have presented diagrams that had a definite "Venn-ish" look a century before John Venn.

What are some **examples** of **Venn diagrams**?

Venn diagrams are schematic illustrations used in logic theory to show collections of sets and the relationship between them. Overlapping circles represent the sets (or the subjects and predicates in syllogistic logic); the standard way of presenting such diagrams include the intersection of two (order-two diagram) to three (order-three diagram) circles. Based on what circles intersect and the areas shaded, a conclusion about the sets may then be read directly from the diagram. Such illustrations can include the union of two sets, the intersection of two sets, the complement of a set, and the complement of the union of two sets. (For more information about sets, see p. 122).

MATHEMATICAL AND FORMAL LOGIC

What is **mathematical logic**?

Mathematical logic is not the logic of mathematics, but is really the mathematics of logic composed of those parts of logic that can be modeled mathematically. Overall, it was invented to understand and present the work of Austrian-American mathematician and logician Kurt Gödel (1906–1978) and his interpretation of the foundations of mathematics in the early 20th century. Although mathematicians use mathematical logic to have rational and reasonable discussions of the many issues in the foundations of mathematics, not everything is agreed upon.

The human brain is more than just a remarkable calculating device; it also possesses the capacity for intuition. Some mathematicians have created the concept of "intuitionism" to reflect the idea that concepts of language and math are really just all in your head. *Visuals Unlimited/Getty Images.*

What is **intuitionism**?

There are some people within philosophy and mathematics who reject the formalism of mathematics and believe in *intuitionism,* which says that words and formulas have significance only as a reflection of the mind's activity. Intuitionists believe that a theorem is meaningful only if it represents a mental construction of a mathematical or logical entity. This is different from the classical approach that states that the existence of an entity can be proven by refuting its non-existence. For example, if you said "A or B" to an intuitionist, he or she believes that either A or B can be proved; but if you said, "A or not A," this is not allowed, since you cannot assume that it is always possible to either prove or disprove statement A.

What is a **proposition** in mathematical logic?

A *proposition* in mathematical logic is a statement that can be proven to be either true or false. For example, if you say, "The bear is black," that is a proposition; but the statement "the bear is x," can't be true or false until a particular value for x is chosen; therefore, it is not a proposition.

What is **symbolic logic**?

Symbolic logic (also called formal logic) is mainly concerned with the structure of reasoning. It determines the meaning and relationship of statements used to represent

107

German mathematician David Hilbert (1862–1943) contributed a great deal to mathematical logic, as well as mathematics in general. In 1890 his proof of the theorem of invariants replaced earlier work on the subject and paved the way for modern algebraic geometry; by 1897, his algebraic number theory led to many developments in that field. His contributions also included discoveries in number theory, mathematical logic, differential equations, multivariable calculus, Euclidean geometry, and even mathematical (theoretical) physics.

Hilbert is most well known for presenting "Hilbert's problems," which originally were a set of 23 unsolved mathematical problems that he hoped would eventually lead to many more disciplines within the field of mathematics. His idea worked: As mathematicians attempted to solve the problems their efforts led to mathematical discoveries in the 20th century, although a number of the problems have yet to be solved. (For more information about Hilbert, see "History of Mathematics.")

specific mathematical concepts and provides a means to compose proofs of statements. Symbolic logic draws most notably on set theory. It uses variables combined by operations such as *not* or *and,* and assigns symbols to them ("~" and "&", respectively).

What are **truth values** and **truth functions**?

As seen above, when discussing propositional calculus, a proposition is any declarative sentence that is either true (T) or false (F). Mathematicians refer to T or F as the *truth value* of the statement.

The combinations of such statements are known as *truth functions,* with their true values determined from the overall true values of their contents. Truth-functional analysis includes the following logical operators:

Negation—The negation of a statement is false if the original statement is true, and true if the original statement is false; it refers to "it is not the case that" or simply "not" in natural language.

Conjunction—The conjunction of two statements is true only if both are true and false in all other instances; it refers to "and" in natural language.

Alteration—Alteration (or disjunction) of two statements is false only if both are false and true in all other instances; it refers to "or" (and "either … or") in natural language.

Conditional—Conditional (or implication) is false only if the antecedent is true and the consequent is false, and is true in all other instances; it refers to "if … then" or "implies" in natural language.

Biconditional—Biconditional (double implication or bi-implication) is true only if the two statements have the same value, either true or false; it refers to "if and only if...." in natural language.

What is a **truth table**?

A truth table is a two-dimensional array of truth values derived by determining the validity of arguments through assigning all possible combinations of truth values to the statements. This simple form of logic depends on a combination of certain statements, using terms such as "not" or "and," along with the input values.

The first columns correspond to the possible input values and the last columns to the operations being performed; the rows list all possible combinations of true (T) or false (F) inputs, together with the corresponding outputs. The following is a truth table for the three most common binary operations of logic ("if ... then," "or," "and"), using s and t as the statements:

s	t	if s, then t	s or t	s and t
T	T	T	T	T
T	F	F	T	F
F	T	T	T	F
F	F	T	F	F

What are **logical operators** in truth tables?

Logical operators in truth tables include such words as "and" or "or," which are all represented by certain symbols (for more about logical operators in predicate calculus, see below). For example, "and" (also called the conjunction operator) is also referred to as a *binary operator*. It is one of the most useful logical operators, as in "p

AND q," represented by the symbols ∧ or &. The "or" (also called the disjunction operator) is also a binary operator, as in "p OR q", and represented by the symbols ∨ and |. The "not" (also called the negation or inversion) operator is known as a *unary operator,* and is represented by the symbols ~ or ¬ (in computer programming, NOT is often represented by the !). The "implies" (or implication operator) is also a binary operator; its symbols include ∴, ⊃, and Ã.

But note: Not all logical operators seem to represent words the way we are accustomed to using them, and many times they seem to contradict their proper definitions. But in a truth table, the logical operator means what it means—without the usual nuances of our English language.

What is a **formula**?

In mathematics, a formula is generally a rule, principle, or fact that is displayed in terms of mathematical symbols. (Although the Latin plural form of formula is "formulae," "formulas" is the accepted common use in mathematics.) These equations express a definite, fixed relationship between certain quantities (usually expressed by letters), with their relationship indicated by algebraic symbols. For example, scientist Albert Einstein's famous $E = mc^2$ is a formula representing energy (E) equal to mass (m) times the speed of light (c) squared. The word "formula" is also used in logic; it is written as a propositional or sentential formula, or "a formula in propositional calculus is one that uses 'and,' 'or,' and so on."

What is **predicate calculus**?

Predicate calculus (also called first-order logic, functional calculus, or quantification theory) is a theory in symbolic logic that uses statements such as "there exists an object such that ..." or "for all objects, it is the case that...." It is a much more solid theory than propositional calculus, taking the interrelationship between sentences

Who was responsible for expanding the ideas of predicate calculus?

The German philosopher and mathematician Friedrich Ludwig Gottlob Frege (1848–1925) presented a way to rearrange sentences to make their logic clearer and to show how the sentences relate in various ways in his 1879 treatise, *Begriffsschrift* (German for "Concept Script"). Before Frege began his work, formal logic (in the form of propositional or sentential calculus; see above) used such words as "and," "or," and so on, but the method could not break the sentences down into smaller parts. For example, formal logic could not show how the sentence "Cats are animals" actually entails "parts of cats are parts of animals."

Frege added words such as "all," "some," and "none," using variables and quantifiers to rearrange the sentences, therefore making them more precise in their meaning. He also developed two of the major qualifier symbols for predicate calculus, the upside-down A (∀) and the backward E (∃). Frege's work was the foundation for modern logical theory, even though his work was defective in several respects and was considered awkward to use. By the 1910s and 1920s, Frege's system was modified and streamlined into today's predicate calculus.

much farther, but it is weaker than certain branches of mathematics, such as arithmetic and set theory.

What is a **quantifier** in predicate calculus?

A sentence or many sentences containing a variable (such as x) can be made into true or false propositions simply by using a *quantifier*. The quantifier actually assigns a truth value to the sentence, depending on the set of values allowed for that variable. There are two major quantifiers: the existential and universal quantifiers, which are represented by the logical operator symbols of ∃ and ∀, respectively, although there are also more exotic types of logic that use different quantifiers.

How is **predicate calculus interpreted**?

Predicate calculus may be a general system of logic, but it accurately expresses a large variety of assertions and provides many types of reasoning. It is definitely more flexible than Aristotle's syllogisms and more useful (in many cases) than propositional calculus.

Predicate calculus makes heavy use of symbolic notation, using lowercase letters $a, b, c, ..., x, y, z$ to denote the subject (in predicate calculus, often referred to as "individuals"), and uppercase letters $M, N, P, Q, R, ...$ to denote predicates. The simplest of assertions are formed by moving the predicate with the subject.

For example, using the "all" quantifier means that when you have an arbitrary variable you must prove something true about that variable, and then prove that it does not

matter what variable you chose because it will always be true. Thus, from propositional calculus the sentence, "All humans are mortal," becomes, in predicate calculus, "All things x are such that, if x is a human, then x is a mortal." This sentence may also be written symbolically under predicate calculus. (To compare, the sentence "x is a human" is not a statement in propositional calculus [see above] because it involves an unknown entity x; therefore, a truth value cannot be assigned without knowing what x represents.)

What is the **atomic formula** of **predicate calculus**?

The atomic formula of predicate calculus is when a predicate and special case of a subject (individual) are written together. For example, if M is the predicate "to be a human," and b is the subject (individual) "Socrates," then Mb means the assertion "Socrates is a human." This atomic formula is phrased "b is the argument of M." Thus, M, as the predicate, may be applied to any subject, and that subject is then an argument of M. But if c is the subject "Vermont," then Mc is a false assertion, because Vermont is not a human. Some predicates require more than one argument; thus, you can have formulas such as Mxy.

What are **logical operators** in predicate calculus?

Predicate calculus commonly uses seven special symbols—called *logical operators*— to express a formula (in this case, a formula is a meaningful expression built up from atomic formulas by repeated application of the logical operators). The following table lists the symbols and their meanings. (Note: Many of these symbols are also used as logical operators in truth tables; see above):

Logical Operators in Predicate Calculus

Symbol	Name	Usage	Meaning*
&	conjunction	... & ...	"both ... and ..."
∨	disjunction	... ∨ ...	"either ... or ... (or both)"
~	negation	~...	"it is not the case that ..."
⊃	implication	... ⊃ ...	"if ... then ..."
≡	bi-implication	... ≡ ...	"... if and only if ..."
∀	universal quantifier	∀x ...	"for all x, ..."
∃	existential quantifier	∃x ...	"there exists x such that ..."

*Where x is any variable

What is an **algorithm**?

The word "algorithm" is a distortion of Muhammad ibn Musa al-Khuwarizmi's name (783–c. 850; also seen as al-Khowarizmi and al-Khwarizmi), the Persian mathematician who wrote about algebraic methods (for more about al-Khuwarizmi, see "History of Mathematics"). In general, an algorithm is a specific set of instructions that, if fol-

lowed correctly, will lead to a recognizable end result. Simply put, a recipe is an example of an algorithm. For example, if there are two different recipes for making apple pie—one calling for cutting fresh apples for the filling, the other calling for apples from a can—the end results will be the same: an apple pie.

In mathematics, most algorithms include a finite sequence of steps that repeat, or require decisions using logic and comparisons until the final result is found (often called a computation). The best example is the long-division algorithm, in which the remainders of partial divisions are carried to the next digit or digits. For example, in the division of 1,347 by 8, a remainder of 5 in the division of 13 by 8 is placed in front of the 4, and 8 is then divided into "54," and so on. More advanced use of algorithms are found in a type of logic called metamathematics (see below).

How is a **decision problem** connected to algorithms?

A decision problem is also known as an *Entscheidungsproblem,* which stems from the German. Decision problems bring up the question of whether an algorithm represents a specific mathematical assertion or not, as well as whether it has or does not have a proof.

What is **metamathematics**?

Metamathematics is the study of mathematical reasoning in a general and abstract way, usually by trying to understand how theorems are derived from axioms. Thus, it is often called proof theory (for more information about axioms, see below). It does not study the objects of a particular mathematical theory, but examines the mathematical theories themselves with respect to their logical structure. Metamathematics is also used in logic to study the combination and application of mathematical symbols; this is often referred to as metalogic.

What is **Gödel's Incompleteness Theorem**?

Austrian-American mathematician and logician Kurt Gödel (1906–1978) is best known for his studies in mathematical logic—in particular, his "incompleteness theorem," presented in 1931. This theorem shows that an infinite number of propositions that can't be derived from axioms of a system may be proved by metamathematical means external to mathematics. In other words, mathematics abounds with questions that have a "yes or no" nature; the incompleteness theory suggests that such questions will always exist. (For more about Gödel, see "History of Mathematics.")

What are some more recent **philosophies** of **mathematical logic**?

Mathematical knowledge and logic in the late 20th and early 21st centuries has been greatly impacted by the development of predicate calculus and the digital computer. Out of these ideas—not to mention centuries of mathematics and logic groundwork—come three of the latest philosophical doctrines of mathematical thought.

The Greek philosopher Plato's notion that people are born possessing all knowledge inspired the mathematical philosophy of Set-theoretical Platonism, which deals with the concept of infinite sets. *Library of Congress.*

Formalism is the idea that mathematics is truly formal; therefore, it is only concerned with the algorithmic manipulation of symbols. In formalism, predicate calculus does not denote predicates—or anything else—meaning mathematical objects do not exist at all. This definitely fits into today's world of computers, especially in the field of artificial intelligence. But this philosophy does not take into account human mathematical understanding, not to mention mathematical applications in physics and engineering.

Constructivism was a "fringe" movement at the turn of the 21st century. Constructionists believe that mathematical knowledge is obtained by a series of purely mental constructions, with all mathematical objects existing only in the mind of the mathematician. But constructivism does not take into account the external world, and when it is taken to extremes it can mean that there is no possibility of communication from one mind to another. This philosophy also runs the risk of rejecting the basic laws of logic. For example, if you have a mathematical problem with a yes or no nature, and the answer is unknown, then neither "yes" nor "no" is in the mind of the mathematician. This means that a disjunction is not a legitimate mathematical assumption, and, thus, ideas such as Aristotle's law of the excluded middle ("either or") are cast aside. Not many modern mathematicians want to throw out centuries of logic.

Set-theoretical Platonism sounds as if mathematicians are regressing back to Plato's time. In reality, this philosophy is based on a variant of the Platonic doctrine of recollection in which we are born possessing all knowledge, and our realization of that knowledge is contingent on our discovery of it. In the set-theoretical Platonism, infinite sets exist in a non-material, purely mathematical realm. By extending our intuitive understanding of this realm, we can cope with problems such as those encountered by the Gödel Incompleteness Theorem. But this philosophy, like the others, has a seemingly infinite number of gaps, especially the question of how a theory of infinite sets can be applied to a finite world.

What do these philosophies tell about the state of modern mathematics and logic? Like many abstract and complex studies, philosophies come and go; some are good, some seemingly on the mathematical fringe. But they also show us that there is cur-

rently no single philosophy that truly defines our mathematical and logic foundations, especially when it comes to combining both mathematical knowledge and the application of mathematics to physical reality.

AXIOMATIC SYSTEM

What are **axioms** and **postulates**?

These two words are often treated as the same; in fact, some mathematicians consider the word *axiom* a slightly archaic synonym for *postulate*. Although both are considered to be a proposition (statement) that is true without proof, there are subtle differences.

An axiom in mathematics refers to a general statement that is true without proof, and it is often related to equality, such as "two things equal to the same thing are equal to each other," and those related to operations. They should also be consistent—it should not be possible to deduce any contradictory statements from the axiom.

A postulate is also a proposition (statement) that is true without proof, but it deals with a specific subject matter, such as the properties of geometric figures. Thus, it is not as general as an axiom. For example, Euclidean geometry is based on the five postulates known, of course, as Euclid's postulates. (See below; for more about Euclid, see "History of Mathematics" and "Geometry and Trigonometry.")

What is an **axiomatic system**?

An axiomatic system is a logical system that has a definite set of axioms; from these axioms, theorems can be derived. In each system, propositions (statements) are proved on the basis of a limited number of axioms or postulates—all with a few undefined terms. The other terms are defined on the basis of the undefined terms. One of the first axiomatic systems was Euclidean geometry.

Overall, an axiomatic system has several basic components: the undefined terms of the system (primitives); well-formed formulas, or how symbols are put into the system based on certain allowed rules, sometimes called defined terms; axioms, or what is also known as "self-evident truths" of the system; theorems, or statements that are proved based on axioms or other proven theorems; and, finally, the rules of inference, or those that allow moves from certain formulas to other formulas.

How are some parts of an **axiomatic system** further **defined**?

There are several terms that further define an axiomatic system. All of them are slightly intertwined, depending on the system.

The absence of contradiction—or the ability to prove a proposition (statement) and its negative are both true—is known as *consistency*. Independence is not neces-

One of the most well-known axiomatic systems was developed by the Greek mathematician Euclid (c. 325–c. 270 BCE). He presented 13 books of geometry and other mathematics titled *Elements* (or *Stoicheion* in Greek). Included in these books were theorems about geometry and numbers derived from five postulates about points, lines, circles, and angles, four axioms about equality, and one axiom stating "the whole is greater than the part." A more modern axiomatic system is the axiomatic set theory, which is based on eight axioms and three undefined terms.

sary to an axiomatic system, but consistency is definitely necessary. The opposite of consistency in an axiomatic system is inconsistency.

An axiomatic system is called *independent* if no other axioms can be derived (or proved) from other axioms in the system; in other words, the entire axiomatic system will be termed independent if all of its underlying axioms are independent. The independence of a system is usually determined after the consistency. An axiomatic system that is *dependent* has some axioms that are redundant; this is also called *redundancy*.

An axiomatic system is *complete* if no additional axiom can be added to the system without making the new system dependent or inconsistent. In other words, the aim is to prove or disprove any statement about the objects in the system from the axioms alone. In complete systems, every true proposition about the defined and undefined terms can be proved from the axioms. Systems with the logic based on true or false propositions connected by "and," "or," and "not" are complete, as are those that include quantifiers. More complex systems, such as set theory, are not considered complete.

What is an **undefined term**?

In terms of axiomatic systems, undefined terms are also called *primitives*. Although it sounds like "double-speak," these primitives are object names, but the objects they name are left undefined. (The axioms are statements within the system that make assertions about the primitives.) If a meaning is attached to a primitive, it is called an interpretation.

Undefined terms are also found in a mix of axiomatic systems and geometry in which definitions are formed using known words or terms to describe a new word. There are three words in geometry that are not formally defined—point, line, and plane—because they can't be described without using words that are themselves undefined. These terms are fundamentally important in the study of geometry, because they are needed to further describe even more complex objects such as circles and triangles. (For more about geometry, see "Geometry and Trigonometry.")

What are **theorems, corollaries,** and **lemmas**?

In mathematics and logic a *theorem* is a statement demonstrated to be true by accepted mathematical operations and arguments. In general, a theorem is usually based on some general principle that makes it part of a larger theory; it differs from an axiom in that a proof is required for its acceptance. Some of the more well-known theorems are named after their discoverers, such as the Pythagorean Theorem (involving right triangles) and Fermat's Last Theorem. It is interesting to note that American Richard Feynman (1918–1988), one of the most brilliant physicists of the 20th century, stated that any theorem, no matter how difficult to prove in the first place, is viewed as "trivial" by mathematicians once it has been proved. Thus, according to Feynman, there are only two types of mathematical objects: trivial ones and those that have not yet been proved.

A *corollary* is a theorem that has been proved in only a few steps from an established theorem, or one that follows as a direct consequence of another theorem or axiom. Finally, a *lemma* is a theorem proved as a preliminary or intermediate step in the proof of another, more basic theorem; or, it is a brief theorem used to prove a larger theorem.

What are **existence theorems**?

An existence theorem is one that has a statement beginning with "There exist(s) …," or, more generally, "for all x, y, \ldots there exist(s)…." Existence theorems are presented in several ways, including showing the exact formulas for the solution, describing in their proofs' iteration processes how to approach the problem, and by simply deducing the solutions without showing any methods as to how it was determined.

Many mathematicians do not believe in existence theorems, stating that any theorems in which entities cannot be constructed are worthless. Other mathematicians cite the existence of such theorems—but prefer to use tried-and-true theorems that offer specific proven methods.

What is a **proof**?

A proof is simply the process of showing a theorem to be correct, although the process itself might not be simple. These mathematical arguments are often quite rigorous, and they are used to demonstrate the truth of a given proposition. The result of the proved statement is a theorem.

Interestingly enough, there are several computer systems now being developed to automate proofs. But some mathematicians (mostly purists) do not believe these computer-assisted proofs are valid; they believe that only humans can understand the nuances and have the intuition needed to develop a theorem's proof. One good example is called the four-color theorem: Its proof relies on meticulous computer testing of many separate cases, all of which can't be verified by hand.

Are there **different** types of **proofs**?

There are several different types of proofs in mathematical logic. *Direct proofs* are based on rules that result in one true proposition from two propositions. They show that a given statement is true by simply combining existing theorems with or without some mathematical manipulations. For example, if you have two sides of a triangle with the same length, a definition and theorem show that a line bisecting their vertex produces two congruent triangles—a direct proof that the angles at the other two vertices have the same size.

An example of deduction can be illustrated by our friend Spot the dog. If Spot is a dog, and we know that all dogs have four legs, then Spot must have four legs. *The Image Bank/Getty Images.*

In logic, *indirect proofs* are also called "proofs by contradiction," and are known in Latin as *reductio ad absurdum* ("reduced to an absurdity"). This type of proof initially assumes that the opposite of what you are trying to prove is true; from this assumption, certain conclusions can be drawn. One then searches for a conclusion that is false because it contradicts given or known information. Sometimes, a given piece of information is contradicted, which shows that, since the assumption leads to a false conclusion, the assumption must be false. If the assumption is false (the opposite of the conclusion one is trying to prove), then it is known that the goal conclusion must be true. All of this has therefore been shown "indirectly."

Finally, a *disproof* is a single instance that contradicts a proposition. For example, the disproof of "all primes are odd" is the true statement "the number 2 is a prime and not odd." If a disproof exists for a proposition, then the statement is false.

What are **deduction** and **induction** and how are they used in mathematics?

Deduction in logic is when conclusions are drawn from premises and syllogisms (for more information on these terms, see above). In this instance, a deduction is a form of inference or reasoning such that the conclusion is true if the premises are true; or, based on general principles, particular facts and relationships are derived. Deductive logic also means the process of proving true statements (theorems) within an axiomatic system; if the system is valid, all of the derived theorems are considered valid. For example, if it is known that all dogs have four legs and Spot is a dog, we logically deduce that Spot has four legs; other examples of deductive reasoning include Aristotle's syllogisms.

What is *modus ponens*?

The Latin term *modus ponens* means "mode that affirms," or in the case of logic, stands for the rule of detachment. This rule (also known as a rule of inference) pertains to the "if … then" statement and forms the basis of most proofs: "If *p* then *q*," or if *p* is true, then the conclusion *q* is true. It is often seen as the following:

If p, then q.
p. Therefore, q.

To see this another way:

p ⇒ q: "If it is raining, then there are clouds in the sky."
p: "It is raining."
q: "There are clouds in the sky."

There are several ways to break down the *modus ponens*. The argument form has two premises: The "if-then" (or conditional claim), or namely that p implies q; and that p (called the antecedent of the conditional claim) is true. From these two premises it can be logically concluded that q (called the consequent of the conditional claim) must be true as well; in other words, if the antecedent of a conditional is true, then the consequent must be true.

Induction is a term usually used in cases concerning probability, in which the conclusion can be false even when the premises are true. In contrast to deduction, the premise provides grounds for the conclusion, but it does not necessitate it. Inductive logic generates "correct" conclusions based on observation or data. (But note that not all inductive logic leads to correct generalizations, making the validity of many such arguments probabilistic or "iffy" by nature.)

One can see how both these processes work in the scientific world, especially in the scientific method in which general principles are inferred from certain facts. For example, by observation of events (induction) and from principles already developed (deduction), new hypotheses are formulated. Hypotheses are then tested by applications; and as the results satisfy the conditions of the hypotheses, laws are developed by induction. Future laws are then often developed, many of them determined by deduction.

What is a **conclusion** in logic?

A conclusion is a statement (proposition) found by applying a set of logical rules (syllogisms) to a set of premises. In addition, the final statement of a proof is called the proof's conclusion. For example, in a statement that includes "if … then," the result following the "then" in the statement is called the conclusion.

What are some examples of paradoxes throughout history?

The oldest paradoxes may be from the Greek Epimenides the Cretan (lived sometime during the 6th century BCE), who stated, "All Cretans are liars." If this statement is true—and any other culture you would care to put in the Cretans' place—then the implication is that the statement is a lie. This is also called the "Liar's paradox."

The number of paradoxes has continued almost *ad infinitum* since then. Some of the more popular ones include those listed as *Zeno's paradoxes*. They are named after Greek philosopher Zeno of Elea (c. 490–c. 425 BCE), a disciple of the philosopher Parmenides, who believed that reality was an absolute, unchanging whole—and, thus, that many things we take for granted, such as motion, were simply illusions. In order to defend his master's highly debated philosophy, Zeno developed his paradoxes. Most of Zeno's paradoxes are still highly debated by modern mathematicians and philosophers, thus proving another paradox: Nothing truly changes throughout history—or does it?

What is a **fallacy**?

A fallacy is an incorrect result—in this case, one arrived at through misleading reasoning when examining a logical argument. One of the more common fallacies in logic is believing incorrectly that if "p implies q" is true, "then q implies p" is also true. The idea of such invalid arguments was well known in the past: With Greek mathematician Aristotle's syllogisms, an argument was valid if it adhered to all the laws; to be false, it only needed to break one law. Euclid, another Greek mathematician, was known to have written an entire book on fallacies in geometry, but the book has since been lost.

What is a **paradox**?

In logic, paradoxes are statements that seem to be self-contradictory or contrary to one's expectations. These arguments imply both a proposition and its opposite. One of the most famous paradoxes was stated by English logician Bertrand Russell (1872–1970) in 1901 and deals with sets: "If sets that are not members of themselves are normal, is the set of normal sets itself normal?" (For more information about Russell's paradoxes in set theory, see below.)

What are some **paradoxes** that deal with **space and time**?

There are numerous paradoxes that deal with the counterintuitive aspects of continuous space and time. One of the most well known is the *dichotomy (or racetrack) paradox*.

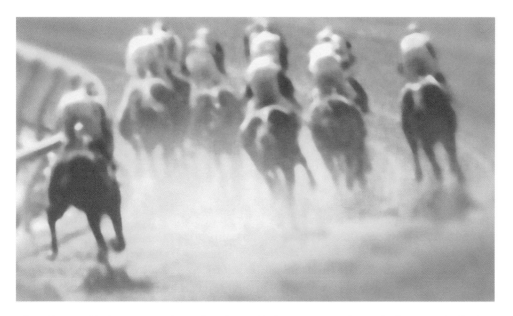

When a horse and jockey race around a track, as they circle they must repeatedly cut the distance to the finish line in half. The dichotomy paradox says that if this is literally the case, the horse will never complete its race; it will just make the distance to the finish line smaller and smaller. *Taxi/Getty Images.*

Before an object can travel a distance *d,* it must keep traveling "in halves": In terms of the racetrack, in order to reach the end of the course, a person would have to first reach the halfway mark, then the halfway mark of the remaining half, then the halfway mark of the final fourth, then of the final eighth, and so on *ad infinitum* (to infinity). Therefore, the distance can never truly be traveled to reach the end of the racetrack.

The *Achilles and the tortoise paradox* is a version of the tortoise and the hare, but with a very different resolution than the well-known fable. In this paradox, Achilles gives the slower tortoise a head start; Achilles starts when the tortoise reaches point *a*. But by the time Achilles reaches *a*, the tortoise has already moved beyond that point, to point *b*; when Achilles reaches *b*, the tortoise is at point *c*, and so on *ad infinitum*. Since this process goes on forever, Achilles can never catch up with the tortoise.

Another paradox is the *arrow paradox*. In this case, an arrow in flight has a certain position at a given instant in time, but that is indistinguishable from a motionless arrow in the same position. So how is the arrow's motion perceived?

Finally, one of the most interesting and insightful paradoxes is attributed to Socrates—thus, it is called the *Socrates' paradox*. It is based on Socrates' statement, "One thing I know is that I know nothing."

SET THEORY

What is **set theory**?

Set theory is the mathematical theory of sets and is associated with logic; it is also considered the study of sets (collections of objects or entities that can be real objects or intellectual concepts) and their properties. (For more about sets, see below.) Under formal set theory, three primitives (undefined terms) are used: S (the set), I (the identity), and E (the element). Thus, the formulas *Sx, Ixy, Exy* mean "x is a set," "x is identical to y," and "x is an element of y," respectively.

Overall, set theory fits in with the aims of logic research: to find a single formula theory that will unify and become the basis for all of mathematics. As it turns out, sets lead directly to a vast amount of data encompassing all of modern mathematics. There are also a number of different set theories, each having its own rules and axioms. No matter what version, set theory is not only important to mathematics and logic but also to other fields, such as computer technology and atomic and nuclear physics.

What are **naive** and **axiomatic set theory**?

The naive set theory is not one that takes everything for granted. It is actually a branch of mathematics that attempts to formalize the nature of the set using the fewest number of independent axioms possible. But it is not the answer to formalizing sets, as it quickly leads to a number of paradoxes. Because of this, mathematicians use a more formal theory called the *axiomatic set theory,* which is a version that uses axioms taken as uninterpreted rather than as formalization of preexisting truths. (For more about axiomatic systems, see elsewhere in this chapter).

What is **Russell's Paradox**?

Russell's Paradox is one of the most famous of the set theory paradoxes. It first appears when studying the naive set theory: In this case, R is the set of all sets that are not members of themselves; from there, R is neither a member of itself nor not a member of itself. The paradox sets becomes evident when one tries to reason how a set appears to be a member of itself if and only if it is not a member of itself.

Discovered by Welsh mathematician and logician Bertrand Arthur William Russell (1872–1970) in 1901, the paradox sparked a great deal of work (and controversy) in logic, set theory, and especially in philosophy and foundations of mathematics. The reason why it became so important was its effect on mathematics: It created problems for those who based mathematics on logic, and it also indicated that something was wrong with Georg Cantor's intuitive set theory. (For more about Russell and the paradox, see "History of Mathematics.")

> ## Who developed set theory?
>
> German mathematician George (Georg) Ferdinand Ludwig Philipp Cantor (1845– 1918) is most well known for his development of set theory (for more information on Cantor, see "History of Mathematics"). His *Mathematische Annalen* is a basic introduction to set theory in which he built a hierarchy of infinite sets according to their cardinal numbers. In particular, using one-to-one pairing, he showed that the set of real numbers has a higher cardinal number than does the set of rational fractions.
>
> Unlike most subjects in mathematics, Cantor's set theory was his creation alone. But like many brilliant, revolutionary thinkers throughout history, his ideas were highly criticized by his contemporaries. This strong opposition contributed to the multiple nervous breakdowns he suffered throughout the last 33 years of his life, which ended tragically in a mental institution.

What is a **set**?

Simply put, a set is a collection of objects or entities; these are called the *elements* of the set. The number of elements in a set can be large, small, finite, or infinite. The informal notation used for sets is sometimes seen as $x = \{y, z, \ldots\}$, with brackets used to contain the elements within the set. It is stated as, "x is a set consisting of the elements y, z, and so on." But more commonly sets are seen as capital letters and elements as lowercase letters, such as *a* is an element of set *A*.

How does one **interpret sets**?

There are several ways to look at sets. Two sets (or more) are considered identical if, and only if, they have the same collection of objects or entities. This is a principle known as *extensionality*. For example, the set $\{a, b, c\}$ is considered to be the same as set $\{a, b, c\}$, of course, because the elements are the same; the set $\{a, b, c\}$ and the set $\{c, b, a\}$ are also the same, even though they are written in a different order.

It becomes more complex when sets are elements of other sets, so it is important to note the position of the brackets. For example, the set $\{\{a, b\}, c\}$ is distinct from the set $\{a, b, c\}$ (note that the brackets differ); in turn, the set $\{a, b\}$ is an element of the set $\{\{a, b\}, c\}$. (It is a set included between the outside brackets.)

Another example that shows how sets are interpreted includes the following: If *B* is the set of real numbers that are solutions of the equation $x^2 = 9$, then the set can be written as $B = \{x : x^2 = 9\}$, or *B* is the set of all *x* such that $x^2 = 9$. Thus *B* is $\{3, -3\}$.

What are **open** and **closed sets**?

Formally, an *open set* is one in which every element in the set has a *neighbor* in the set or does not include its boundary. A *closed set* is one that does include a boundary of a set, or one where some elements have neighbors not in the set.

What is a **null** or **empty set**?

A null or empty set contains no elements; an empty set is considered to be a subset of every other set. The opposite of an empty set is, logically, a *nonempty* set, or one that is not empty. The notations for empty set are {} and \varnothing, but not (), as it is sometimes written in texts. Interestingly enough, an empty set is considered to be both open and closed for any set X.

What is the **set theory approach** to arithmetic?

The set theory approach to arithmetic is defined in terms of the nonnegative whole numbers 0, 1, 2, 3.... These numbers are identified with specific sets based on the placement and number of brackets. For example, mathematicians identify 0 with the empty set { }; 1 is identified with {{ }}; 2 is identified with {{}, {{}}}; 3 is identified with {{ }, {{ }}, {{ }, {{ }}}}, and so on (each bracket is an interpretation of the empty set and 1, or {{}}).

What are the **basic symbols** used to **operate on sets**?

When set theory founder Georg Cantor developed the symbols for sets, he used a single horizontal overbar to denote a set with no structure besides order; thus, it represented the order type of the set. A double bar meant that there was no order from the set, which is also called the cardinal number of the set (see below).

The more common symbols accepted in today's set theory include the following:

Symbols Used to Operate Sets

Symbol	Meaning
∩	"and" or intersection (the intersection of two sets)
∪	"or" or union (the union of two sets)
∈	set membership (such as $a \in S$ in which a is an element of the set S) and subset (such as $A \in B$ or every element of A is also an element of B)
	not a set membership (such as $a \square S$, in which a is not an element of S; this symbol is larger and thicker than set membership
\varnothing	empty or null set (a set with no elements)
{}	braces or brackets (to enclose the listed elements of a set; empty brackets are called an "empty set," or a set with no elements)

Symbol	Meaning
\|	such that, as in {x \| x is a rational number}, is a set "x, such that x is a rational number"
:	also used as "such that"
∉	is not an element of
⊆	is a subset of (as in $A \subseteq B$)
⊂	is a proper subset of
⊄	is not a proper subset of
A'	complement of A

How are some **symbols** used in **operations on sets**?

There are many ways to operate on sets. The following lists some of the more simple operations on sets, where *E, F,* and *G* are sets:

$E \cap F = F \cap E$

$E \cup F = F \cup E$, in which both are commutative operations;

$(E \cap F) \cap G = E \cap (F \cap G)$

$(E \cup F) \cup G = E \cup (F \cup G)$, in which both are associative operations;

$(E \cap F) \cup G = (E \cup G) \cap (F \cup G)$

$(E \cup F) \cap G = (E \cap G) \cup (F \cap G)$, in which both are distributive operations.

What are the **basic set operations**?

There are several basic set operations, the most common being the intersection of sets, union of sets, and the complement of sets. The following lists these operations (note: the first two operations obey the associative and commutative laws, and together they obey the distributive law):

Intersection—The intersection of two sets is the set of elements common to the two sets. For example, the intersection of sets *A* and *B* is the set of elements common to both *A* and *B*. This is usually written as *A* ∩ *B*. Thus, if *A* = {1, 2, 3, 4} and *B* = {3, 4, 5}, then the intersection of *A* and *B* would be {3, 4}.

Union—The union of sets is the combining of members of the sets. For example, the union of two sets *A* and *B* is the set obtained by combining members of sets *A* and *B*. This is usually written as *A* ∪ *B*. Thus, if *A* = {1, 2, 3, 4} and *B* = {3, 4, 5}, then the union of *A* and *B* would be {1, 2, 3, 4, 5}.

Complement or complementation—When the set of all elements under consideration must be specified, it is called the universal set. And if the universal set is *U* = {1, 2, 3, 4, 5} and *A* = {1, 2, 3}, then the complement of A (or A') is the set of all elements in the universal set that are not *A,* or {4, 5}. The intersection between a set and its

complement is the empty or null set (\varnothing); the union of a set and its complement is the universal set.

What are **cardinal** and **ordinal numbers** and **finite sets** in set theory?

Cardinal and ordinal numbers are used in reference to numbers: Ordinal numbers are used to describe the position of objects or entities arranged in a certain sequence, such as first, second, third, and so on; cardinal numbers are natural numbers, or 0, 1, 2, 3, and so on. (For more information about cardinal and ordinal numbers, see "Math Basics.")

But cardinal numbers used in set theory describe the number of members in a set. Both ordinal and cardinal numbers are further used to describe infinite sets and are prefaced with "first ordinal" or "first cardinal" infinities. The first ordinal infinity applies to the smallest number greater than any finite ordered set of natural numbers. The first cardinal infinity applies to the number of all the natural numbers. (For more about infinite sets, see below).

A *finite set* is one that is not infinite. It can be numbered from 1 to *n,* for some positive integer *n*. This number *n* is also called the set's cardinal number; thus, for a certain set *A,* the cardinality is denoted by *card(A)*. There are a number of rules to cardinal numbers and finite sets. For example, if two sets bisect, then they are said to have the same cardinality (or power). The empty set is considered to be a finite set, with its set's cardinal number being 0.

What is a **universal set**?

A universal set actually applies to sets that are not universal but are chosen from a specific type of entity, such as sets of numbers or letters. Thus, the set of all the elements in a set theory problem are collectively called the universal set. In reality, however, the "set of all things" does not exist because there is no largest or all-inclusive set, and so the true universal set is not recognized in standard set theory.

What is a **subset** and **proper subset** in set theory?

Simply put, a *subset* is a portion of a set. If set *B* is a subset of set *A,* then all elements of set *B* are also elements in set *A*. If *A* and *B* are equal, then both sets are subsets of themselves; the empty set is also considered a subset of every other set. A *proper subset* is a subset other than the set itself.

When is a set a **superset**?

A superset is one that contains all the elements of a smaller set. For example, if *B* is a subset of *A,* then *A* is a superset of *B*; in other words, *A* is a superset of set *B* if every element in *B* is in *A*. Like a proper subset, there are also proper supersets (or a superset that is not the entire set).

What is Zermelo's axiom of choice?

Although it sounds like something on a Greek restaurant menu, Zermelo's axiom of choice is actually a fundamental axiom in set theory. It states that given any set of mutually exclusive nonempty sets, there is at least one set that contains exactly one element in common with each of the nonempty sets.

This was actually one of David Hilbert's problems that needed to be solved by mathematicians of his day (for more about David Hilbert, see earlier in this chapter, and in "History of Mathematics"). German mathematician Ernst Friedrich Ferdinand Zermelo (1871–1953) took on the task, and in 1904 he developed what is called the well-ordering theorem, which says every set can be well ordered based on the axiom of choice.

This brought fame to Zermelo, but it was not accepted by all mathematicians who balked at the lack of axiomatization of set theory (for more about axiomatic set theory, see above). Although he finally did axiomatize set theory and improve on his theorem, there were still gaps in his logic, especially since he failed to prove the consistency in his axiomatic system. By 1923, German mathematician Adolf Abraham Halevi Fraenkel (1891–1965) and Norwegian mathematician Albert Thoralf Skolem (1887–1963) independently improved Zermelo's axiomatic system, resulting in the system now called *Zermelo-Fraenkel axioms* (Skolem's name was not included, although another theorem is named after him). This is now the most commonly used system for axiomatic set theory.

What does it mean if **a set is countable**?

If a set is countable (or denumerable), it means that it is finite. This also means that the set's members can be matched in a *one-to-one correspondence*—in which each element in one set is matched exactly with one element in the second, and vice versa—with all the natural numbers, or with a subset of the natural numbers. Mathematicians often say, "*A* and *B* are in one-to-one correspondence," or "*A* and *B* are bijective." (For more about one-to-one correspondence, see "Math Basics.") In set theory, all finite sets are considered to be countable, as are all subsets of the natural numbers and integers. But sets such as real numbers, points on a line, and complex numbers are not countable.

What is **combinatorics**?

Combinatorics is a branch of mathematics—overall, called combinatorial mathematics—that studies the enumeration, combination, and permutation of sets and the mathematical relations that involve these properties, defined as:

Enumeration—Sets can be identified by the enumeration of their elements; in other words, determining (or counting) the set of all solutions to a given problem.

Combination—Combination is how to count the many different ways elements from a given set can be combined. For example, the 2-combinations of the 4-set {A, B, C, D} are {A, B}, {A, C}, {A, D}, {B, C}, {B, D}, {C, D}.

Permutation—Permutation is the rearrangement of elements of a set into a particular order, often in a one-to-one correspondence. The number of permutations of a particularly sized set with n members is written as the factorial $n!$. For example, a set with 4 members would have 4 in first place to 1 in the last place. This would equal $4 \times 3 \times 2 \times 1 = 4!$, or 24, permutations of 4 members. (For more information about factorials, see "Algebra.")

What is an **ordered pair**?

An ordered pair is two quantities—usually written as (a, b)—that have a significant order; thus, (a, b) does not equal (b, a). Ordered pairs are used in set theory to define members in a function.

Ordered pairs are also valuable in linear equations and graphing, in which the x coordinate is the first number and the y coordinate is the second number, or (x, y). They are used on a grid to locate a point. (For more information about ordered pairs and graphs, see "Geometry and Trigonometry.")

How do **functions** pertain to **sets**?

A function in sets pertains to a correspondence between two sets called the *domain* and *range;* each member of the domain has exactly one member of the range. It is often called a many-to-one (or sometimes one-to-one) relation. For example, f = {(1, 2), (3, 6), (4, −2), (8, 0), (9, 6)} is a function, with each set of numbers being an ordered pair. This is because it assigns each member of the set {1, 3, 4, 8, 9} exactly one value in the set {2, 6, −2, 0, 6}. It never has two ordered pairs with the same x and different y values. In this case, the domain is {1, 3, 4, 8, 9} and the range is {2, 6, −2, 0, 6}.

To show an example that is *not* a function, f = {(1, 8), (4, 2), (3, 5), (1, 3), (6, 11)} is *not* a function because it does not assign each member of the set exactly one value: It assigns $x = 1$ to both $y = 8$ and $y = 3$, or it has two ordered pairs that have the same x values to two different y values, (1, 8) and (1, 3). (For more information about functions, see "Algebra.")

How else is the **term function used**?

Unfortunately (as with many mathematical terms), there is often more than one function for the word "function." For example, contrary to the definition above, function can also mean the relationships that map single points in the domain to multiple points in the range—called *multivalued functions*—which is mainly used in the theo-

ry of complex functions. To further confuse matters, there are also functions called non-multivalued functions.

Where did the **symbol** for **infinity originate**?

Infinity is represented by the symbol ∞, a sign introduced by John Wallis in 1655 in his treatise "De sectionibus conicus." Historians believe that Wallis, a classical scholar, adopted the sign from the late-Roman symbol for "1,000." Whether it was from there or another source, the result was (and remains) the same: "a figure-eight on its side," as many people describe the infinity symbol.

The expanses of the universe seem infinite to us, but in mathematics the concept of infinity reaches even beyond the edges of the universe toward numbers that are inconceivably large. *The Image Bank/Getty Images.*

Are there different **types of infinity** in mathematics?

To most of us, the universe represents infinity, but in mathematics it is the unbounded quantity that is greater than every real number. Called *potential infinity* in mathematics, it is the potential for infinity that exists with natural numbers because one can always conceive of a number greater than any given number. Another type of infinity in mathematics is *completed infinity,* which refers to the size of an infinite set (such as all the points on a line). At the end of the 19th century, German mathematician George (Georg) Ferdinand Ludwig Philipp Cantor (1845–1918) showed that different orders of infinity existed and that the infinity of points on a line was of a greater order than that of prime numbers.

ALGEBRA

THE BASICS OF ALGEBRA

What is the **origin** of the word "algebra"?

The word "algebra" comes from the title of the book *Al jabr w'al muqābalah* by Persian mathematician Muhammad ibn Musa al-Khuwarizmi (783–c. 850; also seen as al-Khowarizmi or al-Khwarizmi). The book is roughly translated as *Transposition and Reduction,* in which he explains the basics of algebraic methods. (For more information about the history of algebra, see "History of Mathematics.")

What early **mathematicians** are thought to be responsible for **originating** the use of **algebraic methods and ideas**?

To some scholars, Greek (Hellenic) mathematician Diophantus (c. 210–c. 290) is considered the "father of algebra," as he developed his own algebraic notation. His words were noted and preserved by the Arabs; the translation of his words into Latin in the 16th century led to many algebraic advances. In more "modern" times, French mathematician François Viète (1540–1603; also known by the Latin name Franciscus Vieta) is often credited as the "founder of modern algebra." (For more information about Diophantus and Viète, see below and "History of Mathematics.")

What is **algebra**?

Depending on whether one is a student or professional mathematician, the word can mean either of the following (both of which are further described elsewhere in this chapter). *School algebra* is what mathematicians refer to as the algebra we learn in middle and high school and call "arithmetic." But for most people, algebra means

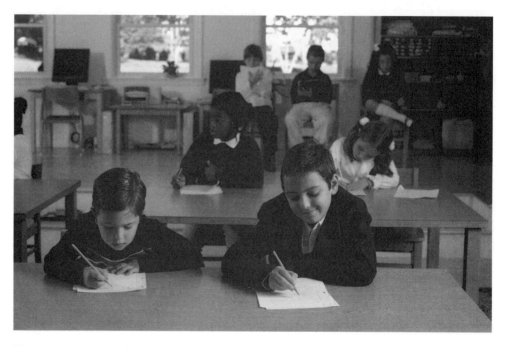

When most people think of "algebra," their thoughts turn to the polynomial equations they learned in school. But mathematicians have a much broader definition of the term, which includes concepts of "abstract algebra." *Stone/Getty Images.*

solving polynomial equations with one or more variables; the solutions to such equations are often obtained by the operations of addition, subtraction, multiplication, division, raising to a power, and extracting a root. (For more information about all these operations, see "Math Basics.") This also includes determining the properties of functions and graphs.

But mathematicians use the word "algebra" most often in reference to the abstract study of number systems and operations within them, such as groups, rings, and invariant theory. This is called *abstract algebra.*

How else is the word **algebra used**?

Algebra may be defined as the subjects of arithmetic and abstract algebra, but there are other meanings. These algebras involve vectors and matrices, the algebra of real numbers, complex numbers, and quaternions (an operator or factor that changes one vector into another). There are also those exotic algebras "invented" by mathematicians—and usually named after the inventor—with the majority not truly understood except, perhaps, by their creators.

ALGEBRA EXPLAINED

What is an **expression** in mathematics?

An expression in mathematics is a statement that uses either numbers, variables, or both. For example, the following are all mathematical expressions:

y

4

$6 - 4$

$5 \times x - 7$

$4 + 5 \times (3 - 2)$

$x + 4 \times (7 - x)$

In order to write an expression from a written mathematical problem, one has to interpret the text. For example, if one person weighs 100 pounds and another weighs y pounds; the expression for their combined weights would be $100 + y$.

What are **equations**?

In its simplest form, an equation is represented by expressions written with an equal sign in between; the two entities on either side are equal to each other. They are among the simplest mathematical problems most people deal with (most people have solved equations in their daily lives without realizing it). For example, when students first learn addition in school, they typically work on equations such as: ___ $+ 5 = 7$, in which the blank needs to be filled. This problem could also be expressed as $x + 5 = 7$, a simple equation. In this case, the equation is solved when x equals 2. The following are also equations:

$6 = 6$

$x = 8$

$y + 8 = 14$

$x - 4 = 15 - x$

$5xy = 8xy^2 + 4$

There are also some fundamental properties of equations that are good to know. They include symmetric properties (if $a = b$, then $b = a$); substitution (if $a = b$, then a may be replaced by b); addition (if $a = b$, and c is a number, then $a + c = b + c$); and multiplication (if $a = b$, and c is a number, then $a \times c = b \times c$).

What are **algebraic equations**?

An *algebraic equation,* as with the equations defined above, is a statement in which two numbers, letters, or expressions are equal. But algebraic equations take this idea even further: Most of the time, the objective is to try to simplify the numbers and the one (or

more) variables in the equation. These can further be defined as any combination of variables or constants linked together by any operation—addition, subtraction, multiplication, division (except division by zero). This type of algebraic equation is often referred to as a polynomial (see elsewhere in this chapter for more about polynomials).

Who was the **first mathematician** to write and **solve general algebraic equations**?

French mathematician François Viète (1540–1603; also known as Franciscus Vieta) is often referred to as the "founder of modern algebra." He was not a professional mathematician, but he contributed a great deal to the understanding and spread of modern symbolic algebra. Although some of his work paid tribute to ancient mathematical traditions, Viète created a kind of "new math": It was not one based on the traditional geometric visualizations, but rather expressed as abstract formulas and general rules. But Viète still divided algebra into distinct branches partially derived from Greek mathematics: *zetetics* (translating a problem into an equation), *poristics* (proving theorems through equations), and *exegetics* (solving equations). He also was the first to combine algebra with geometry and trigonometry. (For more about Viète, see "History of Mathematics.")

What are **variables** in algebraic equations?

Variables are the symbols (usually a letter such as x or y) used in algebraic equations that represent an unknown number—and on whose value a function, polynomial, and so on, depends. Variables remain unknown until the equation is solved; thus, they are sometimes referred to as *unknowns* in an algebraic equation.

It is not always easy to work with variables, as there are so many letters used throughout various equations. But in many mathematical and scientific texts, there are some variables that are customary to use. They are listed as follows:

n indicates natural numbers or integers
x represents real numbers
z stands for complex numbers

What are some **other terms** used in dealing with **algebraic equations**?

There are many terms in algebra, including those dealing with equations. The following lists some of the more common ones:

Equality and *inequality*—An equality is a mathematical statement that shows the equivalence of two quantities. For example, if a is equal to b, it is written as the equality $a = b$. An inequality is just the opposite: a does not equal b, or $a \neq b$.

Formula—A formula is a rule, fact, or principle expressed in terms of mathematical symbols, including equations, equalities, identities, or inequalities. (Note: The

plural of formula in Latin is "formulae," but it has become more readily accepted as "formulas.")

Identity—An identity is a mathematical relationship equating one quantity to another that initially may appear to differ; it also means an equation that is always true, such as the Pythagorean theorem (for more about identities, see below).

How can **word problems** be expressed as **equations**?

There seem to be a gazillion word problems out there—just ask anyone taking such tests in grade or high school! The one thing most have in common is that they can be expressed as an equation—many as algebraic equations—in which there are known and unknown quantities. In almost all instances, there are key words that lead the reader to determine not only the numbers and variables, but also what operations to use in the equation to determine the answer. The following lists some simple key words and their corresponding operation:

Common Key Words Used in Word Problems

Key Words	Operation	Examples
sum, total, more than	addition	The sum of my weight and 10 equals 130 ($y + 10 = 130$); the groceries at one store were \$3.00 while the total for two stores was \$4.00 ($y + 3 = 4$); seven more than the price is \$126 ($y + 7 = 126$)
difference, discrepancy	subtraction	The difference (or discrepancy) between her age and her sister's age, who is 30, is 10 ($y - 30 = 10$)
times, multiplied by	multiplication	Three times his brother's age is 6 ($3 \times y = 6$); eight multiplied by her weight is 96 ($y \times 8 = 96$); the product of his weight and 6 is 36 ($y \times 6 = 36$)

What are **independent** and **dependent variables** in algebra?

Variables can be broken down into independent and dependent variables. An independent variable is a quantity that increases or decreases (is variable), or that has an infinite number of values in the same expression. For example, in the expression $x^2 + y^2 = r^2$, x and y are variables. A dependent variable is a quantity that varies but is produced by changes in the independent variable. In other words, the dependent variable's value is dependent on the independent variable. For example, in the expression $f(x) = y$, x is the independent variable and y is a dependent variable (because y is dependent on the value of x).

135

Are there **differences** between **independent** and **dependent variables** in mathematics and statistics?

Yes, there are subtle differences between these two types of variables in mathematics and statistics. In mathematics, independent variables are those whose value determines the value of other variables; in statistics, they are a manipulated variable in an experiment or study whose presence or degree determines the change in the dependent variable.

Dependent variables in mathematics are those variables whose value is determined by an independent variable; in statistics, they are the observed variables in an experiment or study whose changes are determined by the presence or degree of one or more independent variables. (For more information about variables in statistics, see "Applied Mathematics.")

What is a **solution**?

When an equation has a variable, the number that replaces the unknown and makes the equation true is called a *solution*. For example, for the equation $(5 \times y) + 2 = 12$, y would be 2, or "2" is the solution that makes the equation true. To work this out, the process would look like this: $(5 \times 2) + 2 = 12$; $10 + 2 = 12$; and, finally, $12 = 12$. Remember, when finding a solution for an equation, one must follow any parentheses, exponents, multiplications, divisions, additions, and subtractions in the correct order of operation.

How do you **simplify** an **algebraic equation**?

The best way to *simplify* an equation is to combine like terms, which makes the equation simpler to solve. Numbers may be combined, as well as any terms with the same variable.

Terms can be combined either by adding or subtracting variables of the same kind. One can also use multiplication and division to simplify an equation by multiplying or dividing each side by the same number (except 0). The following are some examples:

- Adding like terms $4x + 3x = 14$, simplifies to $7x = 14$.

- The equation $4 + 8x + 10 - 4x - 2 = 20$ can be simplified by combining the like terms, which gives the simplified result $12 + 4x = 20$.

- If necessary, do a combination of addition, subtraction, multiplication, or division. For example, the equation $2x - 2 = 4x + 3$, simplifies to $2x = 4x + 5$ (by adding 2 to both sides); then subtract $2x$ from both sides, simplifying to $0 = 2x + 5$. (Note: Since subtracting any

René Descartes, who is more often remembered for developing the concept of Cartestian coordinates, also originated the idea of using letters when writing equations that include unknown values. *Library of Congress.*

number is the same as adding its negative, it is often more helpful to replace subtractions with additions of a negative number.) Finally, subtract 5 from both sides (or $2x = -5$), divide both sides by 2, and the result is $x = -5/2$ (or -2.5).

- To simplify expressions raised to a power, certain rules should be followed. For example, for $(x + 3)^2 - 4x$, square $x + 3$, or $(x + 3)(x + 3)$, first squaring the first term (x squared equals x^2), then the second (3 squared equals 9), then multiply and add the inner and outer terms together ($3x + 3x = 6x$). By combining like terms, the entire equation results in the simplified expression $x^2 + 6x + 9 - 4x$, which finally equals $x^2 + 2x + 9$.

What are some **examples** of **algebraic equation solutions**?

The following lists some simple solutions to selected algebraic equations:

- To solve for the equation $4x - 4 = 12$
 add 4 to each side: $4x = 16$
 then divide both sides by 4: $x = 16/4$
 then solve for x: $x = 4$

- To solve for the equation $(x + 3)^2 - 4x = (x - 1)^2 + 3$
 expand each side by first doing the operations within the parentheses: $x^2 + 2x + 9 = x^2 - 2x + 4$

There is a certain order to working out an algebraic equation. The following lists the operations in their correct order:

1. Do what is inside the parentheses
2. Do the exponents
3. Do all the multiplications and divisions from left to right
4. Do all the additions and subtractions from left to right

There are also rules when it comes to grouping in algebraic equations, especially when working with parentheses. When an expression has parentheses within parentheses, work from the inside out, removing the innermost parentheses first. And remember, if there is a positive sign in front of the parentheses, it does not change any sign inside the parentheses; if there is a negative sign in front of the parentheses, it *will* change all the signs within the parentheses.

then add $-x^2$ to each side (another way of subtracting): $2x + 9 = -2x + 4$
then add $2x$ to each side: $4x + 9 = 4$
then add -9 to each side: $4x = -5$
then multiply each side by 1/4 (divide by 4): $x = -5/4$

What are **functions**?

Functions are mathematical expressions describing the relationship between variables and involve only algebraic operations. If there is one independent variable, the dependent variable, for example y, can be determined through the function. This is often seen written as $y = f(x)$, spoken as "y equals f of x" (functions also use notation with x; for example, $f(x) = 2x + 1$ is a function). Functions within equations are also common; for example, in the equation $-x^2 + y = 3$, the y represents a function of x. This equation can also be written as $y = 3 + x^2$.

Functions do not always have to be in terms of $f(x)$. They can just as easily be termed $g(x)$, depending on the equation. But note: The equation $x^2 + y^2 = 9$ is *not* a function, as x and y are both independent variables.

What is a **coefficient**?

In an algebraic equation, a coefficient is simply a multiplicative factor. In the majority of cases, the coefficient is the numerical part (most often a constant) of the equation. Thus, it is called a numerical coefficient. For example, in $3x = 6$, the coefficient is 3;

Are there different types of functions?

Yes, there are different types of functions—so many, in fact, that the topic of "functions" is a book in itself. In particular, algebraic equations include polynomial and rational expression functions. For example, polynomial equations include linear (first degree) functions, such as $f(x) = 2x$; a quadratic (second degree) function example is $f(x) = x^2$ (for more about polynomials and degrees, see below).

But "algebraic and polynomial functions" are not the only use of the term "function"—so don't get confused. There are also non-algebraic functions called exponential functions, and the inverses of exponential functions, which are called logarithmic functions. Set theory emphasizes the use of functions (for more about functions and sets, see "Foundations of Mathematics"); and there are trigonomic functions that include the relationships of sine, cosine, and tangent functions (for more information about trigonometry, see "Geometry and Trigonometry"). There also are continuous or discontinuous functions, transcendental functions, and even real and complex functions (all this may or may not be connected to algebra). The list goes on, but it is easy to see that mathematicians love the word "function."

in $-3x = 6$, the coefficient is -3, as the coefficient takes on the sign of the operation. Terms such as xy may not appear to have a numerical coefficient, but it is 1—a number that is not written, but assumed.

Coefficients do not have to be just numbers: In the equation $5x^3y$, the coefficient of x^3y is 5. But in addition, the coefficient of x is $5x^2y$, and the coefficient of y is $5x^3$. Coefficients are also seen in functions; for example, in the function $f(x) = 2x$, the 2 is the coefficient.

How can **functions** be **defined** based on **variables**?

A function having a single variable is said to be *univariate;* with two variables, it is *bivariate;* and with more than two variables it is *multivariate* (although two variables are considered multivariate by some people).

What is a **linear equation**?

As the term suggests, linear equations have to do with lines; and in algebra, a linear equation means certain equations (or functions) whose graph is a line (for an extensive explanation of graphs, see "Geometry and Trigonometry"). More specifically, in

The first mention of diophantine equations was by Greek (Hellenic) mathematician Diophantus (c. 210–c. 290 CE). In his treatise *Arithmetica,* he solved equations with several variables for integral solutions—what we call diophantine equations today. (For more about Diophantus in history, see "History of Mathematics.") These are represented by one equation with at least two variables, such as x and y, and whose solutions have to be whole numbers (or integers). These equations either have no solutions, or an infinite or finite number of solutions. *Diophantine analysis* is the mathematical term for how to determine integer solutions for such algebraic equations.

algebra, a linear equation is one that contains simply the variable, which makes them one of the simplest types of equations. For example, a linear equation in one variable has one unknown (the variable) represented by a letter; this letter, usually x, is always to the power of 1, meaning there is no x^2 or x^3 in the equation.

For instance, $x + 3 = 9$ is a simple linear equation. To solve such an equation, one must either add, subtract, multiply, and/or divide both sides of the equation by numbers and variables—and do this in the correct order—to end up with a solution: a single variable and single number on opposite sides of the equals sign. In this case, the solution to the linear equation is $x = 6$.

Finally, linear equations can be further broken down. For example, in the linear equation $ax + by + cz + dw = h$, in which $a, b, c,$ and d are known numbers and $x, y, z,$ and w are unknown numbers, if $h = 0$, the linear equation is said to be *homogeneous*.

What is the **absolute value** of a number?

The absolute value of a real number is the number stripped of any negative value. Therefore, the absolute value of a number will always be greater than or equal to zero. (Formally, the absolute value is considered the distance of a number from zero on a number line.) The symbol for "absolute value" is the number inside two parallel vertical lines ($|\ |$). For example, the absolute value of x is given as $|x|$. If the number is negative within the absolute value sign, it will automatically become positive. In numerical form, $|3|$ equals 3 and $|-3|$ equals 3.

When discussing complex numbers, the absolute value often means squaring the numbers, then taking the square root of those numbers. For example, the common way of writing complex equations is $z = a + bi$; the absolute value of z becomes $|z| = \sqrt{a2 + b2}$. For instance, if $z = 3 - 4i$, then $|z| = \sqrt{3^2 + 4^2}$, then $|z| = 5$.

What is a system of equations?

A system of equations is any set of simultaneous equations—two or more—that are intertwined and have to be determined together. They are usually a finite set of equations with the same unknowns, all of which have common solutions. A set of linear equations is said to be a *linear system,* while a set of homogeneous linear equations is called a *homogeneous linear system*. The number of equations is finite, meaning the solutions don't go on forever like some with billions of answers. But a word of caution: Some problems use systems with hundreds of equations and just about the same number of variables.

ALGEBRAIC OPERATIONS

What are the **common** types of **mathematical operations**?

Operations are ways to obtain a single entity from one or more entities by manipulating numbers (and letters) in certain ways. They include the *elementary operations* most of us are familiar with: addition, subtraction, multiplication, division, cubing, squaring, and integer root extraction. There are numerous types of other operations, too, including *binary operations,* in which two quantities or expressions x and y interact in set theory (for more about set theory, see "Foundations of Mathematics").

What is the algebraic concept of **inverse**?

"Inverses" in algebra are operations (or numbers) that "undo" each other. For example, if one multiplies 4 by its inverse, or 1/4, the solution is 1; thus, the rule for multiplicative inverse for x (with $x \neq 0$) is $1/x$, as in $x(1/x) = 1$. In addition, if you add -4 to 4, you get zero (0); thus, the additive inverse of x is $-x$, as in $x + (-x) = 0$.

What are **identity** and **conditional equations**?

Identity and conditional equations are ways in which numbers associate with each other. When an equation is true for every value of the variable, then the equation is called an identity equation. It is often denoted as I or E (the E is from the German *Einheit,* or "unity"). For example, $3x = 3x$ is an identity equation, because x will always be the same number. Zero is the identity element for addition, because any number added to 0 does not change the value of any of the other numbers in the operation (or $x + 0 = x$). The number 1 is the identity element of multiplication, as any number in an operation multiplied by 1 does not change the value of that number. Multiplication identity is often written as $x \times 1 = x$.

When an equation is false for at least one value, it is called a conditional equation. For example, $6x = 12$ is conditional because it is false when $x = 3$ (and any number other than 2). In other words, if at least one value can be found in which the equation is false (or the right side is not equal to the left side) then the equation is called a conditional equation.

How do **numbers associate** with each other?

Generally in mathematics, there are certain properties of operations that determine how numbers associate with each other. *Closure* is a property of an operation that reveals how numbers associate with each other; in particular, when two whole numbers are added, their sum will be a whole number. Closure as a property of multiplication occurs when two whole numbers are multiplied and their resulting product is a whole number.

An *associative property* means that for a given operation that combines three quantities (two at a time), the initial pairing of the quantities is arbitrary. For example, when doing an addition operation, the numbers can be combined in two ways: $(a + b) + c = a + (b + c)$. Thus, when adding the numbers 3, 4, and 5, this means that they may be combined as $(3 + 4) + 5 = 12$ or $3 + (4 + 5) = 12$. Following the same logic for multiplication, the associative law states that $(a \times b) \times c = a \times (b \times c)$. In fact, in an associative operation, the parentheses that indicate what quantities are to be first combined can be omitted; an example of the associative law for addition is $3 + 4 + 5 = 12$, and for multiplication, $2 \times 3 \times 4 = 24$. But not all operations are associative. One good example is division: You can't divide in the same way as you added or multiplied above. For example, the result of dividing three numbers differs. The operation $(96 \div 12) \div 4 = 2$ is not the same as $96 \div (12 \div 4) = 32$.

Like the associative property, the *commutative property* is another way of looking at how numbers associate with each other in operations. In particular, this law holds that for a given operation that combines two quantities, the order of the quantities is arbitrary. For example, in addition, adding $4 + 5$ can be written either as $4 + 5 = 9$ or $5 + 4 = 9$, or expressed as $a + b = b + a$. When working on a multiplication operation, the same rule applies, as in $a \times b = b \times a$. Again, not all operations are commutative. For example, subtraction is not, as in the equation $6 - 3 = 3$, which is not the same as $3 - 6 = -3$. Division also is not commutative. For instance $6 \div 3 = 2$ is not the same as $3 \div 6 = 1/2$.

The final property of an operation is the *distributive property.* In this rule, for any two operations the first is distributive over the second. For example, multiplication is distributive over addition; for any numbers a, b, and c, $a \times (b + c) = (a \times b) + (a \times c)$. For the numbers 2, 3, and 4, you would have $2 \times (3 + 4) = 14$ or $(2 \times 3) + (2 \times 4) = 14$. Formally, there is a right and left distribution—left is listed above; right is $(a + b) \times c = (a \times c) + (b \times c)$. In most cases, both are commonly referred to as distribu-

> ## What does iteration mean?
>
> In most English texts, iteration means to repeat, and this holds true for mathematics. In the case of numbers, iteration means a procedure in which the result is fed back and the procedure repeated; in some instances, it is repeated over and over. For example, to find the square root of 39, you can use iteration—or repeat a procedure to find the solution. Knowing that the solution must be close to 6 (or the square root of 36, a number close to 39), one can divide 39 by 6 (39/6) and get 6.5. Next, average 6 and 6.5 to get 6.25; then iterate again, dividing 39 by 6.25 (39/6.25) = 6.24 (the actual square root of 39 is 6.244997 …).
>
> One of the most obvious instances in which iterations take place is in a calculator or computer. For example, in order to get the square root of 39, as in the example above, a calculator (or computer) automatically uses iteration to calculate the answer to a certain decimal place. The more numbers in a procedure, the more iterations are needed, which is why supercomputing has become such a great asset not only to mathematics but many other sciences as well.

tivity. Again, not all operations are distributive. For example, addition is not distributive over multiplication, as in $a + (b \times c) \neq (a + b) \times (a + c)$.

What is the **factorial** of a number?

A factorial is the product of consecutive natural numbers for all integers greater than or equal to 0. Factorials usually start with 1; the symbol for factorial is an exclamation point (!). For example, 4 factorial (4!) is $1 \times 2 \times 3 \times 4$, or 24. In the numerical system, consecutive factorials are 1 (1! = 1), 2 (2! = 1×2), 6 (3! = $1 \times 2 \times 3$), 24 (4! = $1 \times 2 \times 3 \times 4$), 120 (5! = $1 \times 2 \times 3 \times 4 \times 5$), 720 (6! = $1 \times 2 \times 3 \times 4 \times 5 \times 6$), 5040 (7! = $1 \times 2 \times 3 \times 4 \times 5 \times 6 \times 7$), 40,320 (8! = $1 \times 2 \times 3 \times 4 \times 5 \times 6 \times 7 \times 8$), 362,880 (9! = $1 \times 2 \times 3 \times 4 \times 5 \times 6 \times 7 \times 8 \times 9$), and so on.

There are two additional rules: The number 0 factorial (0!) = 1, and the factorial values for negative integers are not defined. Factorials are most often used in reference to counting numbers, statistics (especially in probability calculations), calculus, and physics.

EXPONENTS AND LOGARITHMS

What is an **exponent** in terms of **algebra**?

An exponent is actually raising a number to a certain power; this is written as a superscript to the right of a real number, such as 3^4 (expressed as "three raised to the fourth

power," or "three with an exponent of four.") (For more information on exponents, see "Math Basics.") The exponent represents the number of times a number is being multiplied. The above example, for instance, actually means "$3 \times 3 \times 3 \times 3$," which is equal to 81. The power can be an integer (negative or positive numbers), real number, or even a complex number. This can also be thought of as taking the quantity b, the base number, to the power of another quantity often called e, the exponent. (In many computer-oriented texts, this is written as $b \wedge e$.)

Exponents are important to algebra as they are often included in algebraic equations. The process of performing the operation of raising to a power is known as *exponentiation*. Exponents are also often associated with functions. For example, in the function $f(x) = x^2$, the 2 is the exponent.

What is a **base** in **algebra**?

The base is used in algebra in connection with powers. In fact, it is called the base of a power—or the number that is used as a factor a given number of times. In the example 3^4, 3 is the base. The base can either be the number used with an exponent to create a power, such as the 3 in 3^4, or a number written as a subscript, such as with a logarithm (for example, $\log_a x$, in which a is the base number). (See below for more information about logarithms; for more information about bases, see "Math Basics.")

What are some **simple rules** of **exponents**?

There are several simple rules when it comes to exponents. These include the following:

- The equation $x^1 = x$ (or a number raised to the 1 power is the number itself; this is also called the "rules of 1").
- The equation $x^0 = 1$ (unless $x = 0$, which is considered undefined; this is also called the "zero rule").
- A number without an exponent has an exponent of 1, as in $20 = 20^1$.
- A negative exponent indicates that the number is to be divided by the exponent instead of multiplied. For example, 3^{-3} is equal to $1/(3^3)$, or 1/27. But there is a restriction to this rule: $x^{-n} = 1/x^n$ only when x is not zero; if x is 0, then x^{-n} is undefined.

What are some **rules** for **combining exponents**?

There are also rules for combining exponents (called the *laws of indices*):

- To multiply exponents with identical bases, add the exponents, such as $3^2 \times 3^3 = 3^5$ (3 is the base).
- To multiply like exponents, combine terms, such as $10^2 \times 2^2 = (10 \times 2)^2 = 400$.
- To divide identical bases, subtract the exponents, such as $10^3/10 = 10^{3-1} = 10^2 = 100$ (the denominator 10 has an assumed exponent of 1).

What are the **connections** between **logarithms and algebra**?

Logarithms are the numbers of the power to which a base must be raised in order to get a given positive number. For example, the logarithm of 100 to the base 10 is 2, or $\log_{10} 100 = 2$. This is because $10^2 = 100$. *Common logarithms* are positive numbers that use the number 10 as the base; they are written as $\log x$. Those using the number symbolized by e as the base are called *natural logarithms* (also phrased as logarithms with a base e); the natural logarithm of a number x is written as $\ln x$.

What's the connection? Because logarithms are really exponents, they satisfy all the usual rules of exponents. Consequently, tedious and long algebraic calculations such as those involving multiplication and division can be replaced by the

Even the simple act of cranking up your sound system involves math. The decibel scale used by your loudspeakers and amplifiers employs the concept of logarithms. *The Image Bank/Getty Images.*

simpler processes of adding or subtracting the corresponding logarithms. In general, logarithmic tables are usually used for this purpose—although calculators, computers, and the Internet often replace the need for such tables.

What was the **progression** of **logarithm development**?

The invention of logarithms was a long process, starting with Scottish mathematician John Napier (1550–1617; also known as Laird of Merchiston), who first came up with the idea of logarithms in 1594. But the actual invention and announcement of logarithms would take another 20 years: In 1614, Napier would publish *Mirifici logarithmorum canonis descripto* (*Description of the Wonderful Canon of Logarithms*), which offered tables and rules for their use.

Not long afterward, in 1617, English mathematician Henry Briggs (1561–1630) published *Logarithmorum chilias prima* (*Logarithms of Numbers from 1 to 1,000*), introducing the concept of common logarithms—or logarithms based on the powers of ten. And finally, independently from Briggs and Napier, came Swiss mathematician Joost Bürgi (1552–1632), who in 1620 presented *Arithmetische und geometrische Progress-tabulen,* a German work presenting the discovery of logarithms.

These discoveries differed in several ways: Napier's approach was algebraic; Bürgi's was geometric. There were differences from the common and natural logarithms we

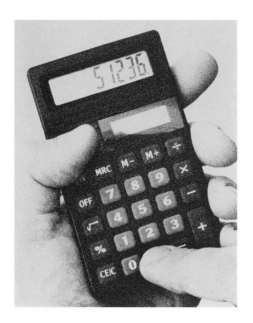

Slide rulers and calculating systems based on Napier's bones were used for centuries before the invention of today's handy battery-operated and solar calculators. *Stone/Getty Images.*

use today. And neither Napier nor Bürgi mentioned the concept of a logarithmic base—something that Briggs presented.

By 1624 Briggs would write *Arithmetica logarithmica* (*The Arithmetic of Logarithms*), extending his common log tables from 1 to 20,000 and from 90,000 to 100,000. But the work on logarithms did not end with Napier, Briggs, or Bürgi. Natural logarithms eventually evolved out of Napier's original work. Defining logarithms as exponents was finally recognized by English mathematician John Wallis (1616–1703), who presented them in his 1685 publication, *De algebra tractatus* (*Treatise of Algebra*).

For what other **invention** was **John Napier** known?

Scottish mathematician John Napier may have been known for his contributions to logarithms, but he was also the inventor of a tool called Napier's Bones (also known as Napier's Rods). These were multiplication tables inscribed on strips of animal bone or wood. Wilhelm Schickard would eventually build the first calculating machine based on Napier's bones, a device that could add, subtract, and—with help—multiply or divide.

Napier was also the "instigator" in another discovery: In 1621 English mathematician and clergyman William Oughtred (1575–1660) used Napier's logarithms as the basis for the slide rule (a ruler-like instrument used long before hand-held calculators came into vogue). Oughtred not only invented the standard rectilinear slide rule, but also the circular slide rule, which was an extremely useful tool that remained in common usage for more than three hundred years. (For more about Oughtred, see "Math Basics.")

What are the **properties of logarithms**?

Logarithms have certain properties, depending on the interpretations of an equation. The following lists some of the most common properties (these rules are the same for all positive bases):

- $\log_a 1 = 0$, because $a^0 = 1$. For example, in the equation $14^0 = 1$, the base is 14 and the exponent is 0. Because a logarithm is an exponent, this would mean the equation can be written as a logarithmic equation, or $\log_{14} 1 = 0$ (zero is the exponent).

- $\log_a a = 1$, because $a^1 = a$. For example, in the equation $3^1 = 3$, the base is 3 and the exponent is 1; the result is 3, with the corresponding logarithmic equation being $\log_3 3 = 1$.

- $\log_a a^x = x$, because $a^x = a^x$. For example, $3^4 = 3^4$, with the base as 3. The logarithmic equation becomes $\log_3 3^4 = 4$.

What is an **exponential function**?

Along with exponents come exponential functions, or the relationship between values of a variable and the numbers formed by raising some positive number to the power of those values. In functional notation, an exponential function is written $f(x) = a^x$, in which a is a positive number; for example, the function $f(x) = 2^x$ is an exponential function.

In logarithmic terms, an exponential function is most commonly written as exp (x) or e^x, in which e is called the base of the natural logarithm. These types of functions are usually shown on a graph. (For more information about graphs, see "Geometry and Trigonometry.") As a function of the real variable x, the resulting graph of e^x is always positive, or above the x axis and increasing from left to right. Although the line of such a function never touches the x axis, it gets very close to it.

What does *e* **represent** in terms of **logarithms**?

No, e is not the code name in a James Bond movie. When talking about logarithms (or logs), in the majority of mathematical circles, it means the base of the natural logarithm. It is yet another irrational, transcendental number (such as pi, or π) that has a plethora of names: It has been called everything from the logarithmic constant and Napier's number to Euler's constant and the natural logarithmic base. One of the best ways to define e is to use the expression $(1 + x)^{(1/x)}$; e is the number that this expression approaches as x gets smaller and smaller. Substituting in values for x gives one a better idea: if $x = 1$, the result is 2; if $x = 0.5$, the result is 2.25; when $x = 0.25$, the result is 2.4414 ...; if $x = 0.125$, the result is 2.56578 ...; if $x = 0.0625$, the result is 2.63792 ...; and so on. This is why approximations are often used in solving equations using e.

147

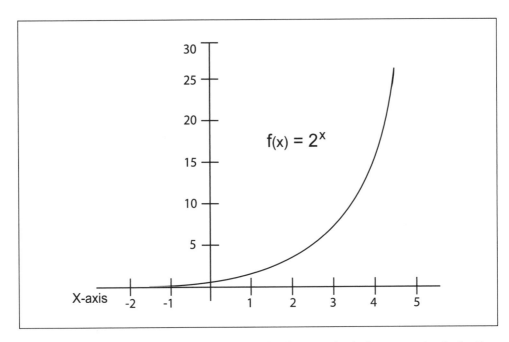

In this example of an exponential function, when 2 is raised to the power of x, the line representing the function always lies above the x axis.

What are the **rules** for **combining logarithms**?

There are certain rules for combining logarithms. In the following cases, let a be a positive number that does not equal 0; n is a real number; and u and v are positive real numbers:

> Logarithmic Rule 1: $\log_a(uv) = \log_a(u) + \log_a(v)$
> Logarithmic Rule 2: $\log_a(u/v) = \log_a(u) - \log_a(v)$
> Logarithmic Rule 3: $\log_a(u)^n = n\log_a(u)$

This can be expressed as follows: In rule one, multiplication inside the log is turned into addition outside the log (and vice versa); in rule two, division inside the log is turned into subtraction outside the log (and vice versa); and in rule three, an exponent on anything inside the log can be moved to the front of the log as a multiplier (and vice versa). But remember, these rules only apply if the bases are the same. For example, because the bases are not the same in $log_a(u) + log_b(v)$, this expression can't be simplified.

How do you **expand logarithms**?

Like an algebraic expression, it is possible to expand logarithms, which is a way of "picking apart" an expression. The following lists two examples of expanding a logarithmic expression:

- To expand $\log_2(3x) = \log_2(3) + \log_2(x)$
- To expand $\log_2(12/x) = \log_2(12) - \log_2(x)$

Is it possible to **simplify logarithms**?

Yes, as in algebraic equations, it is possible to simplify logarithms, but in different ways. The following lists some examples:

- To simplify $\log_3(x) + \log_3(y) = \log_3(xy)$
- To simplify $\log_3(6) - \log_3(4) = \log_3(6/4)$
- To simplify $2\log_3(x) = \log_3(x^2)$

How can the **base of logarithms be changed**?

The base of logarithms can be changed from one that is not 10 or e to an equivalent logarithm with base 10 or e. The following gives the formula for such a transition, in which a, b, and x are real positive numbers (but neither a nor b are equal to 1, and x is greater than 0):

Convert $\log_a x$ to the base b by using the formula
$(\log_b x)/(\log_b a)$

What is an example of a **log table**?

An example of a log table can be seen in Appendix 2 at the back of this book.

How are **equations** with **exponents and logarithms** solved?

The way to solve an exponential equation is relatively easy: Take the log of both sides of the equation, then solve for the variable. For example, to solve for x in the equation $e^x = 60$;

1. First, take the natural log (ln) of both sides: $\ln(e^x) = \ln(60)$
2. Simplify using the logarithmic rule #3 (see above) for the left side: $x \ln(e) = \ln(60)$
3. Then simplify again, since $\ln(e) = 1$ to: $x = \ln(60) = 4.094344562$
4. And finally, check your answer (using log tables or your calculator) in the original equation $e^x = 60$: $e^{4.094344562} = 60$ is definitely true.

The way to solve a logarithmic equation is equally easy: Just rewrite the equation in exponential form and solve for the variable. For example, to solve for x in the equation $ln(x) = 11$:

1. First, change both sides so they are exponents of the base e: $e^{ln(x)} = e^{11}$
2. When the bases of the exponent and logarithm are the same, the left part of the equation becomes x, thus, it can be written: $x = e^{11}$

3. To obtain x, determine the solution for e^{11}, or x is approximately 59,874.14172.

4. And finally, check your answer (using tables or your calculator) in the original equation $ln(x) = 11$: $ln(59{,}874.14172) = 11$ is definitely true.

POLYNOMIAL EQUATIONS

What is a **polynomial**?

In its simplest form, a polynomial is a mathematical equation that involves a sum of powers in one or more variable, all of which is multiplied by coefficients. In such equations, variables and numbers on both sides of the equal sign are considered polynomials. For example, by expanding out the expression $(x-2)^3$ (or multiplying out the equation) we discover that $(x-2)^3 = x^3 - 6x^2 + 12x - 8$, which is a polynomial equation.

Are there **other ways** of **describing polynomials**?

Yes, there are other ways to describe polynomials. In particular, a polynomial with only one variable is called a *univeriate polynomial*. A *multivariate polynomial* is one with more than one variable.

There are other terms to further define polynomials, including the following:

Monomial—A monomial is a one-term polynomial (*mono* means "one"). For example, $3x$ is monomial.

Binomial—A binomial is a two-term polynomial (*bi* means "two"). For example, $3x^2 - 10$ is a binomial.

Trinomial—A trinomial is a three-term polynomial (*tri* means "three"). For example, $4x^3 + 3x + 6$ is a trinomial.

What is the **degree** of a **polynomial equation**?

The highest order power (or exponent) used in a univariate polynomial is called its *polynomial degree* or *order;* it is also defined as the largest (or maximum) sum of exponents that appear on the variables in the equation's terms. Commonly, the word "degree" is used more frequently because "order" has another meaning when discussing polynomials.

You can also talk about degrees in reference to monomials, binomials, and trinomials. The degree of a monomial is the sum of the degrees of the variable in the equation; the degrees of the other polynomial equations are the greatest of the degrees in terms after the equation has been simplified. For example, a third degree trinomial equation is $x^3 - 3x - 2 = 0$.

What are **quartic equations**?

Quartic equations are polynomial equations whose highest power of the unknown variable is four. Put another way, quartic equations are algebraic equations whose highest exponent (or degree or order) is four. (For more about the history of solving quartic [and cubic] equations, see "History of Mathematics.") But take note: Quartic equations are *not* the same as quadratic equations—or second degree equations in one variable—so don't mix them up.

What are the **names of polynomial equations** with different degrees?

The polynomial equations with different degrees (or orders)—especially the lowest degrees—are as follows:

- *first degree* (1) polynomial = linear
- *second degree* (2) polynomial = quadratic
- *third degree* (3) polynomial = cubic
- *fourth degree* (4) polynomial = quartic
- *fifth degree* (5) polynomial = quintic
- *sixth degree* (6) polynomial = sextic

How do you **multiply polynomial equations**?

To multiply two monomials, multiply the coefficients and then multiply the variables (and when multiplying the variables, keep the variables and add the exponents). The following are several examples of multiplying various polynomial equations:

Multiplying monomials:

$$(5x)(6x^2)$$
$$= (5 \times 6)(x^{2+1}) \text{ (multiply the numbers and add the exponents)}$$
$$= 30x^3$$

Multiplying a monomial and polynomial (binomial):

$$4y(2y - 8)$$
$$= 4y(2y) - 4y(8) \text{ (multiply } 4y \text{ times both terms)}$$
$$= 8y^2 - 32y \text{ (depending on what is on the other side of the equation, you can further simplify by dividing by 8, or } y^2 - 4y)$$

Multiplying polynomials:

$$6y^3(8y^6 + 5y^4 - 3y^3)$$
$$= 6y^3(8y^6) + 6y^3(5y^4) - 6y^3(3y^3)$$
$$= 48y^9 + 30y^7 - 18y^6$$

How do you **divide polynomials**?

To divide a polynomial by a monomial, divide each term in the polynomial by the monomial (to divide monomials, divide the coefficients and then subtract the exponents) or:

$(A + B + C)/M = A/M + B/M + C/M$

For example, to solve the equation

$10x^5/2x^3 = (10/2) (x^{5-3}) = 5x^2$

What does **factoring polynomials** mean?

When a polynomial is written as the product of two or more polynomial equations, the polynomial has been factored. This allows a complicated polynomial to be broken up into easier, lower-degree pieces; and it makes the equation easier to solve. One way to look at it is that factoring a polynomial is the opposite process from multiplying polynomials.

One of the most basic ways to factor a polynomial is similar to factoring a number. When a number is factored, the result will be the prime factors that multiply together to give the number (for example, $6 = 2 \times 3$, or $12 = 2 \times 2 \times 3$; see "Math Basics" to learn more about prime factors). With polynomials, this is often called "taking out a common factor": If every term in a polynomial expression has several factors, and if every term has at least one factor that is the same, then that factor is called a *common factor*. If this is the case, then the common factor can be removed from every term and multiplied by the whole remaining expression.

For example, for the equation $2x^2 + 8x$, the first term has factors of 2 and x, while the second term has factors of 2, 4, and x. The common factors are 2 and x, making $2x$ the overall common factor. This makes the expression equal to $2x(x + 4)$. Thus, it is easy to see that when a polynomial is factored, it results in simpler polynomials that can be multiplied together to give the initial polynomial.

What is **differences of squares**?

Differences of squares is another way to factor an expression into a form that is essentially [something]2 subtracted from [something]2. Mathematically, an equation that resembles $(a^2x^2 - b^2)$ is a difference of squares and can be factored into $(ax + b) (ax - b)$, with the factors being identical, except for the sign. This, in turn, equals $(ax)(ax) - abx + abx - b^2$. The two middle terms ($- abx$ and $+ abx$) further cancel each other out, resulting in $(a^2x^2 - b^2)$, which results in a way of factoring called the difference of squares. For example, take the expression $16 - s^2$. From the above, we know that the form $(a^2x^2 - b^2)$ equals $(ax + b)(ax - b)$. Thus, by "substitution," and if $a = 1, x^2 = 16$, and $b = s$, we can say that $16 - s^2 = (4 + s)(4 - s)$, with the resulting factoring of the equation using difference of squares.

What is a perfect square?

There are many equations that can be factored into a perfect square. Any expression written in the form $x^2 + 2ax + a^2$ is a perfect square—an expression written as [something]2. To determine if an expression is a perfect square, first see if the constant term is a square number—in other words, can the square root of the number be taken to get an integer for an answer. If so, determine if the square root of the constant, multiplied by 2, gives the coefficient of the linear term (or the x term). If it does, the original expression may be factored into a perfect square. (Note: The above procedure only works when the coefficient of x^2 is 1.)

For example, in the equation $x^2 + 8x + 16$, the constant term (16) is already a perfect square (the square root of 16 is 4). Since $2(4) = 8$, the original expression can be written as a perfect square. Because we know $x^2 + 2ax + a^2$ is a perfect square, and equals $(x + a)^2$, by substituting the common factor 4 into the equation, we find that $x^2 + 8x + 16 = (x + 4)^2$.

What is the **difference** and **sum of cubes**?

Similar to squares, there is also the difference and sum of cubes that deals with the factoring of polynomials. The *difference of cubes* takes the form $a^3 - b^3$, and can be factored into $(a - b)(a^2 + ab + b^2)$. Thus, if an expression resembles $(a^3 - b^3)$, then $(a - b)$ is a factor; use long division to find the remaining factor(s).

The *sum of cubes* takes the form $a^3 + b^3$, and can be factored into $(a + b)(a^2 - ab + b^2)$. Thus, if an expression resembles $(a^3 + b^3)$, then $(a + b)$ is a factor. Again, use long division to find the remaining factor(s).

How do you find the **roots** of a **polynomial**?

Finding the root, also called a zero, of a polynomial is one way to solve for the equation. In other words, the root of an equation is simply a number (or numbers) that solves the equation.

For example, for second-degree polynomials we can find the roots by *completing the square*. Picking apart an equation is the best way to see this:

1. $3x^2 - 4x + 1 = 0$
2. $(1/3)(3x^2 - 4x + 1) = (1/3)0$ (making the coefficient of the x^2 term into a 1)
3. $x^2 - (4/3)x + 1/3 = 0$
4. $(x^2 - (4/3)x) + 1/3 = 0$ (group the x and x^2 terms together)
5. $(x^2 - (4/3)x + (-2/3)^2) - (-2/3)^2 + 1/3 = 0$ (determine the coefficient of the x term, divide it by 2 and then square; add and subtract that term)

153

6. $(x - 2/3)^2 - 4/9 + 1/3 = 0$

7. $(x - 2/3)^2 - 1/9 = 0$ (add together the $4/9 + 1/3$ by converting the denominator to 9, in which $1/3$ becomes $3/9$)

8. $(x - 2/3)^2 = 1/9$ (move the $1/9$ to the other side of the equation by subtracting it from both sides)

9. $x - 2/3 = 1/3$ or $x - 2/3 = -1/3$

That means that $x = 1$ or $x = 1/3$ are the two roots that make the equation true (just substitute either number into the initial equation to see that they are both true).

What are examples of **polynomials** with **one root and no roots?**

The following is an example of a polynomial with only one root:

$x^2 + 6x + 9 = 0$

$(x^2 + 6x + (6/2)^2) - (6/2)^2 + 9 = 0$

$(x + 3)^2 - 9 + 9 = 0$

$(x + 3)^2 = 0$

$x + 3 = 0$

$x = -3$, or the polynomial has only one root $x = -3$

But not all polynomials have roots. The following is an example of a polynomial with no root:

$2x^2 - 6x + 8 = 0$

$(1/2)(2x^2 - 6x + 8) = (1/2)0$

$x^2 - 3x + 4 = 0$

$(x^2 - 3x + (-3/2)^2) - (-3/2)^2 + 4 = 0$

$(x - 3/2)^2 - 9/4 + 4 = 0$

$(x - 3/2)^2 + 7/4 = 0$

$(x - 3/2)^2 = -7/4$

Because a real number squared is greater than or equal to 0, that means $(x - 3/2)^2$ will always be greater than or equal to 0. Thus, the answer can't be $-7/4$, a negative number, and there are no real roots for this polynomial.

What is a **quadratic equation?**

A quadratic equation is a second-degree (order) polynomial equation, thus it is guaranteed to have two solutions, both of which may be real or complex. This is seen in the standard form $ax^2 + bx + c = 0$.

The roots of x can be found in the equation by factoring and completing the square (a method of transforming a quadratic equation so that it is in the form of a

What is the Fundamental Theorem of Algebra?

The Fundamental Theorem of Algebra (FTA) is nothing new; it was first proved by mathematician Carl Friedrich Gauss (1777–1855) in 1799. The equation was as follows:

$a_n x^n + a_{n-1} x^{n-1} + \ldots + a_1 x^1 + a_0 = 0$ (as long as n is greater than or equal to 1 and a_n is not zero, and has at least one root in the complex numbers).

The proof of this theorem goes on for pages—far beyond the scope of this book. What all those proofs, numbers, and letters boil down to is that a polynomial equation must have at least one number in its solution. It also tells us when we have factored a polynomial completely. Simple enough, but like much of mathematics, someone had to prove it.

But that is not all: The FTA is not constructive, and therefore it does not tell us how to completely factor a polynomial. In other words, in reality, no one really knows how to factor a polynomial; we only know how to apply techniques to certain kinds of polynomials. In fact, French mathematician Evariste Galois (1811–1832), who died tragically in a duel, proved that there will never be a general formula that will solve fifth degree or higher polynomials.

perfect square). For example, to solve the equation $x^2 - 3x = 4$ by factoring, write the equation in standard quadratic equation form: $x^2 - 3x - 4 = 0$. Then, factor the form: $(x - 4)(x + 1) = 0$.

In order for these numbers to equal zero, we determine that for $(x - 4)$, x would be 4; and for $(x + 1)$, x would equal -1. Thus, the solutions are 4 and -1.

Can **all quadratic equations** be solved by **factoring**?

Don't be fooled: Not all quadratic equations can be solved by factoring. For example, $x^2 - 3x = 3$ is not solvable with this method. One way to solve quadratic equations is by completing the square; still another method is to graph the solution (a quadratic graph forms a parabola—a U-shaped line seen on the graph). But one of the most well-known ways is by using the quadratic formula.

$$\frac{-b + \sqrt{b^2 - 4ac}}{2a} \text{ and } \frac{-b - \sqrt{b^2 - 4ac}}{2a}$$

For example, if we want to find the roots of the polynomial $x^2 + 2x - 7$, we can replace the "corresponding" numbers from the initial equation into the quadratic equation $ax^2 + bx + c = 0$. Thus, $a = 1$, $b = 2$, and $c = -7$. Substituting these numbers into the quadratic formula, we solve for:

$$\frac{-2 + \sqrt{2^2 - 4(1)(-7)}}{2(1)}$$

$$\frac{-2 - \sqrt{2^2 - 4(1)(-7)}}{2(1)}$$

That equals

$$\frac{-2 \pm \sqrt{32}}{2}$$

What is the **discriminant** of a **quadratic equation**?

In the quadratic equation $ax^2 + bx + c = 0$, the value of $b^2 - 4ac$ is the discriminant—the same numbers and letters that are under the square root sign of the quadratic formula. This is actually the products of the squares of the polynomial root differences. In other words, this quantity characterizes certain properties of the quantity's roots. The discriminant is often used for such mathematical concepts as metrics, modules, quadratic fields, and polynomials.

MORE ALGEBRA

What is an **array**?

A mathematical array is a list of lists, or a rectangular arrangement of objects; it is a useful way of keeping a collection of things. Arrays are orderly arrangements of objects in columns and rows, with the objects most often being numbers. For example, the most common arrays are two-dimensional with a certain amount of rows and columns:

$$\begin{array}{ccc} 3 & 0 & 1 \\ -2 & 4 & 0 \end{array}$$

In this case, this is a 2-by-3 array (rows are always mentioned before columns). The order of numbers and/or elements in an array may sometimes, but not always, be significant (as it is in a matrix; see below).

What is a **matrix**?

A matrix is a concise way of representing and working with linear transformations; also, it is a rectangular array or grid of numbers or variables that allows the user to perform certain mathematical operations. They are usually symbolized as large parentheses or two large pair of parallel double lines surrounding the array of numbers or variables. These numbers can be manipulated to solve systems of equations or problems with many differ-

Who invented matrices?

Although a simple form of matrices may have been used by the Mayans (and maybe other cultures; see below), the true mathematical use of a matrix was first formulated around 1850 by English mathematician, poet, and musician James Sylvester (1814–1897). In his 1850 paper, Sylvester wrote, "For this purpose we must commence, not with a square, but with an oblong arrangement of terms consisting, suppose, of m lines and n columns. This will not in itself represent a determinant, but is, as it were, a Matrix out of which we may form various systems of determinants by fixing upon a number p, and selecting at will p lines and p columns, the squares corresponding of pth order." In this case, Sylvester used the term matrix to describe its conventional use, or "the place from which something else originates."

But the matrix story was not all about Sylvester. In 1845 Sylvester's collaborator, English mathematician Arthur Cayley (1821–1895), used a form of matrices in his work *On the Theory of Linear Transformations;* by 1855 and 1858, Cayley began to use the term "matrix" in its modern mathematical sense. Although he was an avid mountaineer and a lawyer for close to a decade and a half (which is how he met Sylvester), during his free time Cayley published more than 200 mathematical papers. He also contributed a great deal to the field of algebra, initiated analytic geometry of n-dimensional spaces, and developed the theory of invariants, among other mathematical feats.

Sylvester also remained brilliant throughout his life. He founded the *American Journal of Mathematics* in 1878; and at the ripe age of 71, he invented the theory of reciprocants (differential invariants).

ent variables or numbers by addition, subtraction, multiplication, or other methods. Each row and column of a given matrix must have the same number of elements.

Any time one has a list of numbers, or a table of numbers in a specific order, concerning anything at all (prices, grades, populations, coordinates of points, production tables …), it can be considered to be a matrix. When the idea of the matrix was first conceived, its development dealt with transformation of geometric objects and solution of systems of linear equations. Historically, the early emphasis was on the determinant (see below), not the matrix; today, especially in linear algebra, matrices are considered first.

What are some **examples** of **matrices**?

The dimensions of a matrix are the number of rows (horizontal numbers) and columns (vertical numbers); it is written as the rows first, columns second. The following are some simple examples of matrices, all of which differ in their dimensions:

A 3 by 2 (3 × 2) matrix:

$$\begin{bmatrix} 1 & 2 \\ 3 & 4 \\ 5 & 6 \end{bmatrix}$$

A 2 by 3 (2 × 3) matrix:

$$\begin{bmatrix} 1 & 2 & 3 \\ 4 & 5 & 6 \end{bmatrix}$$

A 4 by 4 (4 × 4) matrix (when the row and column dimensions are the same, it is called a "square matrix"):

$$\begin{bmatrix} 1 & 2 & 3 & 4 \\ 5 & 6 & 7 & 8 \\ 9 & 1 & 2 & 3 \\ 4 & 5 & 6 & 7 \end{bmatrix}$$

How do you **add matrices**?

It is pretty easy to add matrices. For example, for (1 2) + (−1 −2), you add the matrices (or in this case, row numbers) simply by adding the corresponding terms in each matrix. Thus, the matrix of (1 2) + (−1 −2) = (1−1 2−2) = (0 0).

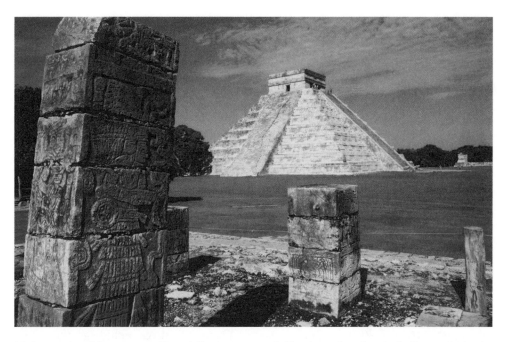

All that remains of the once-great Maya civilization are remarkable ruins such as these in the Yucatan Peninsula of Mexico. They are reminders of an advanced culture that had made great strides in mathematics, including the use of matrices. *The Image Bank/Getty Images.*

What is the **identity matrix**?

The identity matrix is the *n*-by-*n* matrix that has all ones down the main diagonal and zeroes everywhere else; it must also be a square. The following is the identity matrix.

$$\begin{bmatrix} 1 & 0 & 0 \\ 0 & 1 & 0 \\ 0 & 0 & 1 \end{bmatrix}$$

Or in more formal and explicit terms:

$$\begin{bmatrix} 1 & 0 & \cdots & 0 \\ 0 & 1 & \cdots & 0 \\ \vdots & \vdots & \ddots & \vdots \\ 0 & 0 & \cdots & 1 \end{bmatrix}$$

What happens when you **multiply** a **matrix** by the **identity matrix**?

When you multiply any *n*-by-*n* matrix by the identity matrix, you get that same matrix back again. Therefore, let the letter *I* represent the *n*-by-*n* identity matrix, and *A* rep-

159

resent any other n-by-n matrix. We then have $A \times I = A$ and $I \times A = A$. This is much like the situation when using the real numbers: $x \times 1 = x$ and $1 \times x = x$.

How are **matrices used**?

Matrices are used in a multitude of fields, from mathematics and science to certain humanities fields. For example, they are used in physics to determine the equilibrium of rigid bodies; in graph theory, fractals, and solutions of systems of linear equations in mathematics; and in forest management, computer graphics, cryptology—even electrical networks.

ABSTRACT ALGEBRA

What is **abstract algebra**?

Abstract algebra is a collection of mathematical topics that deal with algebraic structures rather than the usual number systems. These structures include groups, rings, and fields; branches of these topics include commutative and homological algebras. In addition, linear algebra and even elementary number theory (see "Math Basics") are often included under abstract algebra.

What is an **algebraic structure**?

An algebraic structure is made up of a set (collection of objects called elements; for more information about sets, see "Foundations of Mathematics") together with one or more operations on the set that satisfy certain axioms. The algebraic structures get their names depending on the operations and axioms. For example, algebraic structures include fields, groups, and rings, as well as many other structures with strange names such as loops, monoids, groupoids, semigroups, and quasigroups.

What is a **field**?

A field is an algebraic structure that shares the common rules for operations (addition, subtraction, multiplication, and division, except division by zero) of the rational, real, and complex numbers (but not integers, see below under "ring"). A field must have two operations, must have at least two elements, and must be commutative, distributive, and associative (see above for definitions). Formerly called "rational domain," a field in both French (*corps*) and German (*Körper*) appropriately means "body." A field with a finite number of members is called a *Galois* or *finite field*. Fields are useful to define such concepts as vectors and matrices.

What is a **group** in **abstract algebra**?

A group, usually referred to as *G,* is a finite or infinite set of elements together with a binary operation (often called the group operation) that together satisfy the four fundamental properties—closure, associativity, and the identity and inverse properties (for more information about these properties, see elsewhere in this chapter). A great many of the objects investigated in mathematics turn out to be groups, including familiar number systems—such as the integers, rational, real, and complex numbers under addition; non-zero rational, real, and complex numbers under multiplication; non-singular matrices under multiplication; and so on. The branch of mathematics that studies groups is called group theory, an important area of mathematics that has many applications to mathematical physics (such as particle theory).

What is a **ring**?

A ring is an algebraic structure (some definitions say a set) in which two binary operators (addition and multiplication) in various combinations must satisfy either the additive associative, commutative, identity, and inverse properties, the multiplicative associative property, or the left and right distributivity properties. For example, the elements of one operation, such as addition, must form a group that is commutative, also known as an abelian group. The multiplicative operation must produce unique answers that have the associative property. These two operations are further connected by requiring the multiplication to have a distributive property with respect to the addition. This can be written as follows, with *a, b,* and *c* elements of the ring:

$$a \times (b + c) = (a \times b) + (a \times c) \text{ and } (b + c) \times a = (b \times a) + (c \times a)$$

161

Transcendental numbers are those that are not the root of any integer polynomial, or that are not an algebraic number of any degree. Thus, all transcendental numbers are irrational (rational numbers are algebraic numbers of degree one). The importance of such numbers translates through more than two millennia of history: For example, they provided the first proof that circle squaring was insoluble, which is one of the geometric problems that has baffled mathematicians throughout antiquity.

Rings are usually named after one or more of their investigators. But such a practice usually makes understanding the properties of the various associated rings difficult for anyone other than the mathematician working on the ring.

What is **linear algebra**?

Linear algebra is the study of linear sets of equations. It encompasses their transformation properties and includes the analysis of rotations in space, least squares fitting, and numerous other problems in mathematics, physics, and engineering.

Though Charles Dodgson is more commonly known as the author of *Alice's Adventures in Wonderland* under the pen name of Lewis Carroll, he was also a brilliant mathematician. Among other accomplishments, he devised ways to refine Boolean algebra notations. *Library of Congress.*

What is **Boolean algebra**?

Boolean algebra is an abstract mathematical system used to express the relationship between sets (groups of objects or concepts). It is important in the study of information theory, the theory of probability, and the geometry of sets. The use of Boolean notation in electrical networks aided the development of switching theory and the eventual design of computers.

It was English mathematician George Boole (1815–1864) who first developed this type of logic by demonstrating the algebraic manipulation of logical statements, showing whether or not a statement is true, and showing how a statement can be made into a simpler, more

convenient form without changing its overall meaning. Today, this way of looking at logic is called Boolean algebra. (For more information about Boole, see "History of Mathematics.")

Boolean algebra did not end there: In 1881 the English logician and mathematician John Venn (1834–1923) interpreted Boole's work and introduced a new way of diagramming Boole's notation in his treatise *Symbolic Logic*. This was later refined by the English mathematician Charles Dodgson (1832–1898), who was better known as the writer of *Alice's Adventures in Wonderland* (under the pseudonym Lewis Carroll). Today, when studying sets, we call this method not the Boole, Carroll, or Dodgson diagram, but the Venn diagram. Thus, Boolean notation demonstrates the relationship between groups, indicating what is in each set alone, what is jointly contained in both, and what is present in neither.

GEOMETRY AND TRIGONOMETRY

GEOMETRY BEGINNINGS

What is **geometry**?

Geometry is the study of figures or objects in space—of a certain number of dimensions and types—and focuses on the properties and measurements of points, lines, angles, surfaces, and solids of those objects (or sometimes even the space around them). The word *geometry* is from the Greek words for "earth" and "to measure" (*geometria,* broken down into *gë* and *metreein,* respectively). A person who studies geometry is called a geometer or geometrician.

What are the **divisions** found within the **field of geometry**?

The geometry field has several distinct divisions, including:

Plane geometry—This includes common features such as circles, lines, triangles, and polygons.

Solid geometry—This also includes such figures as circles and lines, as well as polyhedrons.

Spherical geometry—This includes shapes such as spherical triangles (see below) and polygons.

Analytic geometry—Also called coordinate geometry, this includes the study of figures in terms of their positions, configurations, and separations.

There are other types of geometry, too, including projective geometry and non-Euclidean geometry. Most of these are more complex forms of geometry that each have their own special reasons for use.

When did **geometry originate**?

The field of geometry was probably developed by several cultures over millennia, but only in crude, elementary forms. Some of the first to actually work with geometry were the cultures of the Mesopotamian region around 3500 BCE (especially the Babylonians). They were the earliest peoples to know about what is now called the Pythagorean theorem (in fact, the Greek mathematician and philosopher Pythagoras of Samos [c. 582–c. 507 BCE] may have actually learned about this theorem in his travels to the east), and they possessed all the theorems of plane geometry that the Greeks attributed to Thales. The Egyptians came next, using geometric methods mainly for construction of huge monuments. This included the sundry pyramids and monuments of the region, some of which still dot the landscape today—a tribute to their builders who used geometric techniques.

Were the **Greeks** involved in **geometry**?

The Greeks were known to have extensive knowledge of geometry, producing many great geometers. With this and other contributions in mathematics, the Greeks profoundly changed the approach and character of the entire mathematical field. It is thought that Thales of Miletus (c. 625–c. 550 BCE; Ionian) first introduced geometry to the Greeks. As a merchant traveler, he was exposed to the Babylonian concept of measurement, from which practices sprang geometry. Thales used his geometric knowledge to solve problems such as the height of the pyramids and the distance of ships from the shoreline.

Greek geometer Hippocrates of Chios (470–410 BCE) was one of the first to present an axiomatic approach to geometry, as well as the first to work on the elements almost a century before Euclid (see below). Hippocrates may have worked on geometry and such problems as squaring the circle, but he lacked common sense and was duped by many people.

Zeno of Elea (c. 490–c. 425 BCE) raised questions about lines, points, and numbers—all part of geometry—with his many paradoxes (for more information about Zeno and his paradoxes, see "Foundations of Mathematics"). Another important figure is Eudoxus of Cnidus (408–355 BCE), who worked on geometric proportions and theories for determining areas and volumes.

Others followed these geometers, including Archimedes (c. 287–212 BCE; Hellenic), who worked on mechanics and took the first steps toward integral calculus. Apollonius of Perga (262–190 BCE), or the "great geometer," first named and presented theories on conic sections in his book *Conics,* and he introduced the terms "parabola," "ellipse," and "hyperbola." There was also Pappus of Alexandria (290–350), who presented the basis for modern projective geometry (the geometry that deals with incidences, or whether elements such as lines, planes, and points coincide or not).

What are the five postulates of Euclid?

Euclid was also famous for his postulates, propositions (statements) that are true without proof and deal with specific subject matter, such as the properties of geometric objects (for more information about postulates, see "Foundations of Mathematics"). Along with definitions, Euclid began his text *Elements* with five postulates. These postulates are as follows (some of which may seem obvious to us now, but in Euclid's time they had yet to be formally recorded):

- It is possible to draw a straight line from any point to another point.

- It is possible to produce a finite straight line continuously in a straight line.

- It is possible to describe a circle with any center and radius.

- All right angles are equal to one another.

- Given any straight line and a point not on it, there "exists one and only one straight line which passes" through that point and never intersects the first line, no matter how far the lines are extended. Another way to say this is: One and only one line can be drawn through a point parallel to a given line. This is also called the parallel postulate.

Mathematicians first believed this last postulate could be derived from the first four, but they now consider it to be independent of the others. In fact, this postulate leads to Euclidean geometry, and eventually to many non-Euclidean geometries that are made possible by changing the assumption of this fifth postulate.

Like many early attempts at explaining mathematics, not all these postulates tell the entire geometric story. There were still a large number of gaps, many of which were gradually filled in over time.

What **Greek mathematician** wrote the book *Elements*?

The Greek mathematician and geometrician Euclid (c. 325–c. 270 BCE) made some of the most significant improvements to geometry in his time. (For more about Euclid, see "History of Mathematics" and "Foundations of Mathematics.") One contribution was his collection of 13 books on geometry and other mathematics, titled *Elements* (or *Stoicheion* in Greek). This work has been called the world's most definitive text on geometry. The first six books offer elementary plane geometry, with sections on triangles, rectangles, circles, polygons, proportions, and similarities; the rest of the books present other mathematics of his day, including the theory of numbers (books 7 to 10), solid geometry, pyramids, and Platonic solids. These books were used for centuries in western Europe; in fact, the elementary geometry many students learn in high school today is still largely based on Euclid's ideas on the subject.

What is **Euclidean geometry**?

Euclidean geometry is named after Euclid, the famous Greek mathematician. It is geometry mostly based on Euclid's fifth postulate—the parallel postulate—and is sometimes called parabolic geometry. Plane geometry is described as two-dimensional Euclidean geometry, while three-dimensional Euclidean geometry is known as solid geometry.

What was **François Viète's contribution** to geometry?

French mathematician François Viète (or Franciscus Vieta, in Latin; 1540–1603), although thought of as the "founder of modern algebra," also introduced a connection between algebra, geometry, and

Modern graphic designers and engineers use computer animation programs to create three-dimensional imagery on two-dimensional screens, thus combining concepts of Euclidean geometry in modern-day technology. *Taxi/Getty Images.*

trigonometry. He also included trigonometric tables in his *Canon Mathematicus* (1571), along with the theory behind their construction. (For more about Viète, see "History of Mathematics" and "Algebra.")

What was **Gaspard Monge's** connection to geometry?

French mathematician, physicist, and public official Gaspard Monge (also known as Comte de Péluse; 1746–1818) was the first to lay down ideas about modern descriptive geometry, a field that is essential to mechanical and architectural drawing. He is also called the founder of differential geometry. As one of the founders of the École Polytechnique, he served as professor of descriptive geometry, and around 1800 published the first textbook on the subject based on his lectures, aptly called *Géométrie descriptive*. Today, the system once called "géométrie descriptive" is now known as orthographic projection, a graphical method used in modern mechanical drawing.

BASICS OF GEOMETRY

What is a **mathematical space**?

Outer space may be the "final frontier" to some people, but back on Earth there are also numerous types of space in mathematics. For the most part, mathematical space consists of points, sets, or vectors. Each space and the members of that space obey certain mathematical properties. Most spaces are named after their principal investigator,

How do we interpret dimensions in everyday life?

We are all familiar with dimensions around us, although we may not be aware of them. Most people are familiar with the ideas of two- (such as a drawing on paper) and three-dimensional objects (ordinary objects, including an apple or a car, exist in three-dimensional space), but there are others as well.

Zero dimension can be thought of as a point in space. One dimension can be visualized by a line or a curve in space. Another way of understanding one dimension is with time, something we think of as consisting of only "now," "before," and "after." Because the "before" and "after"—regardless of whether they are long or short—are actually extensions, time becomes similar to a line (as in "timeline")—or a one-dimensional object.

Two dimensions are defined by two coordinates in space, such as a rectangle. One of the most obvious two-dimensional objects we see all around us are paintings and photographs—although they represent a three-dimensional object. Even this page you are reading can be considered a two-dimensional object, though strictly speaking, the thickness of the paper gives it a third dimension. Three dimensions are considered the space we occupy, as three dimensions gives everything around us depth. Our binocular vision allows us to see depth (things in three dimensions), which is why everything becomes "flat" or two-dimensional when we view the world through just one eye.

One can also conceive of four or more dimensions, but there are few common examples. Most hyper-dimensional aspects are used by mathematicians, various scientists, and even economists. They need such dimensional analysis for their complex mathematics, such as for modeling weather patterns or the ups and downs of the stock market.

including Euclidean and Minkowski space. One of the most general types of mathematical spaces is called the topological space.

How is a **dimension described** in mathematics?

In mathematics, a dimension is the number of coordinates (or parameters) required to describe points of—or even points on—a mathematical object (usually geometric in nature). The dimension of an object is often referred to as its dimensionality. (For more about coordinates, see elsewhere in this chapter; for more about dimensions and science, see "Math in the Natural Sciences.")

Each dimension represents points in space—from a single to multiple points. The concept of dimension is important in mathematics, as it defines a geometric object

The ups and downs of stock markets, such as this one in Tokyo, Japan, are often illustrated using two-dimensional line graphs. Sometimes, though, economists employ even more complex, four-dimensional models. *Stone/Getty Images.*

conceptually and/or visually. In fact, the idea of dimensions can even be applied to abstract objects that can't be directly visualized. Mathematicians most often display such dimensions on graphs using a single point (for example, x) to represent one dimension; two points (usually x and y, or an ordered pair) to represent two dimensions; and three points (usually x, y, z) for three dimensions. The four- (and higher) dimensional analogs of three-dimensional objects often retain the prefix "hyper-" such as hypercube and hyperplane. The basic geometric structures of higher-dimensional geometry—the line, plane, space, and hyperspace—all consist of an infinite number of points arranged in specific ways.

What is **Euclidean space**?

Euclidean (also seen as Euclidian) space is often called Cartesian space, or more simply, *n-space*. It is made up of n dimensions and is a set of points, with each point represented by a coordinate of n components. Space with two to three dimensions—and that does not use Einstein's and others' ideas of relativistic physics—are considered Euclidean space.

In Euclidean space, distance is defined by certain "rules": The distance between two points is positive, unless they are the same points; the distance from points a to b

How is a curve defined in geometry?

A curve is a continuous collection of points drawn from one-dimensional space to *n*-dimensional space; it is also considered an object that can be created by moving a point. But note: Our usual use of the word "curve" does not mean a straight line, but in mathematics, a line or triangle is often referred to as a curve.

Different forms of geometry define curves in various ways. Analytic geometry uses plane curves—such as circles, ellipses, hyperbolas, and parabolas—which are usually considered as the graph of an equation or function. The properties of these curves are largely dependent on the degree of the equation in the case of algebraic curves (curves with algebraic equations) or on the particular function, as in the case of transcendental curves (curves whose equations are not algebraic). Even more complex are space curves, all of which require special techniques used only in differential geometry.

will be the same as from *b* to *a*; and the distance between the points does not change if they are totally shifted over in one direction (such a sliding over of the points is called *translation*; for more on this, see elsewhere in this chapter). In addition, the Pythagorean theorem is valid for three points that are the vertices (the intersection points of the sides of an angle) of a right triangle.

What are some of the **basic "building blocks"** of geometry?

There are several basic "building blocks" of geometry, all of which have to do with the objects we often see in geometry. A zero-dimensional object that is specifically located in *n*-dimensional space using *n* coordinates is called a *point*. The idea of a point may be obvious to most people, but for mathematicians, describing and dealing with points is not straightforward. For example, Euclid once gave a vague definition of a point as "that which has no part."

Euclid also called the *line* a "breadthless length," and further called a straight line one that "lies evenly with the points on itself." Modern mathematicians define lines as one-dimensional objects (although they may be part of a higher-dimensional space). They are mathematically defined as a theoretical course of a moving point that is thought to have length but no other dimension. They are often called straight lines, or by the archaic term, a right line, to emphasize the fact that there are no curves anywhere along the entire length. It is interesting to note that when geometry is used in an axiomatic system, a line is usually considered an undefined term (for more about axiomatic systems, see "Foundations of Mathematics"). In analytic geometry, a line is

Parallel lines are found all around us in life, such as in the way we farm land to grow crops in neat rows. *National Geographic/Getty Images.*

defined by the basic equation $ax + by = c$, in which a, b, and c are any number, but a and b can't be zero at the same time.

A *line segment* is the shortest curve—which is actually straight—to connect two different points. It is a finite portion of an infinite line. Line segments are most often labeled with two letters corresponding to the endpoints of the line segment, such as a and b, and written as ab.

Distance is the length of the path between two points, or the length of a segment, for example, between points a and b. When talking about the distance between any two points associated with a and b (as real numbers) on a number line, distance becomes the absolute value of $b - a$ ($|b - a|$).

Another basic building block is the *ray*. Think of a ray as a laser beam: It originates at one point and continues in one direction toward infinity. A ray is defined as part of a line on one side of a point, and includes that point. Two letters are needed to name a ray. For example, Ray *AB* is defined by point *A*, where it begins, and point *B*, the point the ray goes through. (But note: Ray *AB* is not the same as ray *BA*.) If the initial point is not included, the resulting figure is called a *half line*. In geometry, a ray is usually considered a half-infinite line with one of the two points *A* and *B* taken to be at infinity.

What does the word **parallel** mean in geometry?

In geometry, parallel means two lines in two-dimensional Euclidean space that do not intersect; or two lines in the same plane that never meet and that maintain the same

distance from each other at every point. In three-dimensional Euclidean space, these lines also do not intersect, maintaining a constant separation between points closest to each other on the two lines. In analytic geometry, parallel means those lines with the same slope, as well as other curves with the same slope for every x value. The symbol for parallel lines is ||. For example, $A \parallel B$ means that line A is parallel to line B.

Angle ABC is formed by rays BA and BC linked at point B.

What is an **angle**?

An angle is a major concept in both geometry and trigonometry. An angle is formed by two rays that begin at the same point; a *straight angle* is one in which the two rays lie on the same line. It can also be described as two planes coming from a common line.

Angles are named in several ways: a capital letter at its vertex (the common points for both rays, see illustration, ∠B), a small letter within the angle, a number within the angle, or by three capital letters (ABC or CBA, with the middle letter representing the vertex and the other two points the rays). The two rays that form the angle are called the sides of the angle, such as side BA and side BC.

How are **angles measured**?

Angles are measured in two major ways: degrees and radians. Similar to degrees on a temperature scale, *degrees* in mathematics—especially when discussing angles—are usually denoted by the symbol °; they are divided into 60 arc minutes, and arc minutes are divided into 60 arc seconds. The multiples of 60 are thought to be connected to the Babylonian's sexagesimal number system (one based on the number 60), in which the year was composed of 360 days (or 12 months at 30 days each). If the vertex of an angle and one side are fixed and the other side is rotated about the vertex, it sweeps out a circle of 360° with each complete rotation. Said another way, a full rotation is 360 degrees.

Radians (denoted as rad) are real numbers represented as an angle; they are the central angle of a circle determined by two radii and an arc joining them. In degrees, a radian is about 57.29578 degrees or $180/\pi$ degrees, because a semicircle contains π radians. For example, $\pi/6$ is equal to a 30 degree angle; because a straight angle is π radians (or 180°), if you divide 180 by 6, (or $\pi/6$), it is equal to 30 degrees. Radians are most often used in probability and statistics, or calculus, especially to obtain the derivative of trigonometric functions.

173

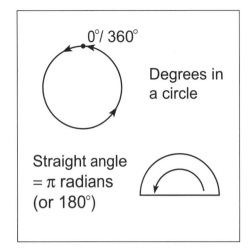

The angles of a circle are typically measured either in degrees or radians (rad). There are 360 degrees in a circle, or $2\pi \times 1$ rad.

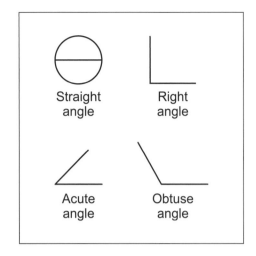

The four basic types of angles are straight, acute, obtuse, and right.

How are the **simplest angles** classified?

Angles are usually expressed in terms of "rotation," in which a full rotation is equal—in most cases—to a 360 degree circle. Half of a full rotation is called a *straight angle;* a quarter of a full rotation is called a *right angle* (90 degrees). An angle less than a right angle is called an *acute angle;* an angle greater than a right angle is called an *obtuse angle.*

What are **other types of angles**?

There are numerous other types of angles, some of which are examined in the following list:

Adjacent angle—Adjacent angles are two angles side-by-side that have a common vertex (endpoint) and ray; angles are adjacent if a common line is between the other sides.

Vertical angles—Vertical angles are two angles whose sides form two pairs of opposite rays; because of this, vertical angles have no sides in common and are directly across from one another. They are also considered a pair of opposite congruent angles where two lines intersect. Vertical angles include intersecting lines that define a plane, with the vertical angles always in the same plane.

Congruent angles—Congruent angles have the same size and shape; vertical angles are considered to be congruent.

Exterior and interior angles—Interior angles are those inside a polygon, such as a triangle; exterior angles are outside the polygon formed by extending one ray (side) outside the polygon from a vertex.

Alternate and corresponding angles—A pair of angles that lie on opposite sides and opposite ends of a transversal (a line that cuts two or more lines in the same plane) are called alternate angles; both these angles are equal if the lines cut by the transversal are parallel. They are also broken into alternate interior angles (the "inside" angles) and alternate exterior angles (the "outside" angles). For two parallel lines, the alternate interior and exterior angles are equal—or congruent. Corresponding angles are pairs of angles that lie on the same sides and ends of a transversal.

Dihedral angle—A dihedral (meaning "two bases") angle is one of the four angles formed where two non-parallel planes meet.

Complementary angles—Two angles that add up to a right angle are complementary. For example, the two acute angles in a right triangle are always considered complementary, as their sum adds up to 90 degrees of the right triangle.

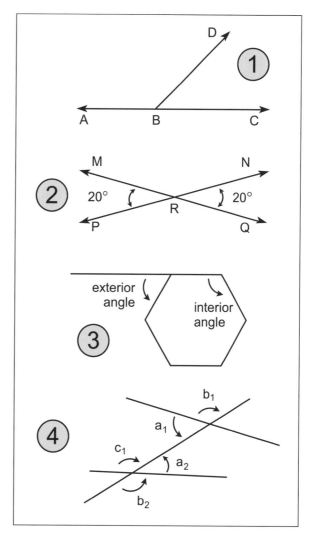

Among the types of angles are: 1. adjacent angles; 2. vertical (congruent) angles; and 3. exterior and interior angles. In illustration 4 we see alternate interior angles (a_1 and a_2), alternate exterior angles (b_1 and b_2), and corresponding angles (b_1 and c_1).

Supplementary angles—Two angles that add up to a straight angle (180 degrees) are supplementary. If the angles are both supplementary to the same angle, then the two angles are congruent.

What are **perpendicular, orthogonal, normal,** and **tangent lines**?

Lines are also classified by their relationship to other angles and lines. *Perpendicular lines* are two lines, segments, or rays that intersect to form a right (90 degree) angle. **175**

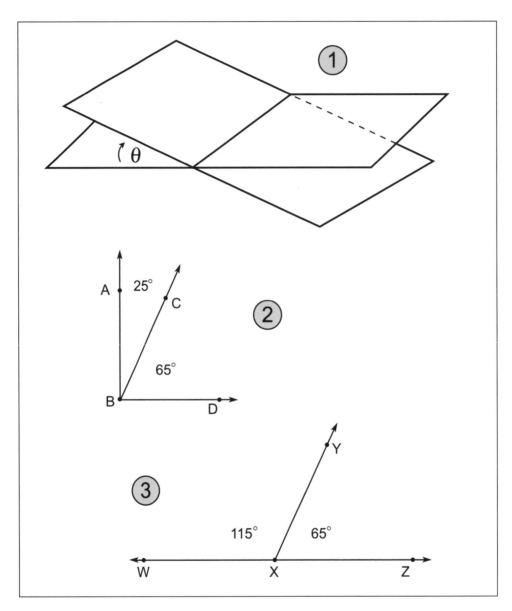

In the above, illustration 1 shows dihedral angle θ; illustration 2 shows complementary angles; and illustration 3 shows supplementary angles.

Orthogonal lines are another way of saying lines are perpendicular—but it is mostly used in terms of functions, transformations, and vectors in other mathematical fields. *Normal lines* are perpendicular lines where each line is perpendicular to a curve (including a line) or a surface (including a plane). And finally, a line is considered to be *tangent* to a circle if it intersects the circle at exactly one point—also called the *point of tangency*. Such lines are also called tangential lines.

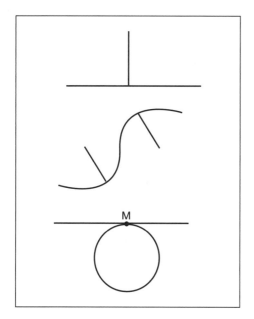

Examples of a perpendicular line (top), normal lines (middle), and tangential lines (bottom), with point M being the point of tangency.

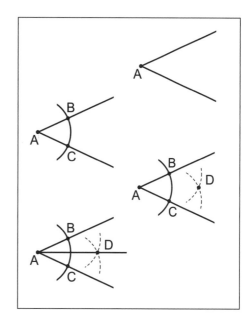

To determine a bisector, follow these steps from top to bottom: Draw an angle; draw an arc from the angle's vertex that touches the two angle lines; draw two more arcs from the points where the first arc meets the angle lines; draw a line from point A to point D.

What does the term **bisect** mean?

The term "bisect" is important in geometry; it means to cut into two (or divide into half) mainly lines (or two-dimensional figures) and angles. To compare, a bisected line segment means finding the plane, line, or point that is the midpoint of the line segment. This is also called a segment bisector. An angle bisector is a ray in the interior of an angle that forms two equal angles. First locate the point on each ray that is equally distant from the vertex; then draw a third point equally distant from each of the first two rays. A line that extends through the third point and the vertex is the angle bisector.

To draw a bisector, follow this sequence (as per the illustration above): First, draw an angle; then draw an arc centered at the vertex (endpoint), in which B and C are the intersections of the arc and angle lines at equal distances from the vertex; next draw an arc centered at B and one centered at C—both with the same radii—inside the angle; finally, extend a line from the vertex to the point D where the arcs of B and C intersect—making AD the bisector of the angle at A.

What are **geometric postulates**?

Similar to other parts of mathematics, there are many geometric postulates, or statements that are assumed to be true without proof. From these postulates, theo-

rems—another type of mathematical statement—can be proven (in addition, theorems are proven by definitions or previously proven theorems). An example of a postulate in geometry is, "Through any two points there is exactly one line." Another is, "If two points lie in a plane, then the entire line containing those two points lies in the plane."

What are **proofs** and **theorems** in geometry?

Proofs are extremely important to geometry. Similar to other divisions of mathematics, proofs are defined as sequences of justified conclusions used to prove the validity of an "if-then" statement. (For more information about postulates, theorems, and undefined terms, see "Foundations of Mathematics.")

There are essentially five steps in showing that any proof is a good proof: state the theorem to be proved; list what information is available; draw an illustration (if possible) to represent the information; state what is to be proved. Finally, develop a system of deductive reasoning, especially concentrating on statements that are accepted to be true; along with the true statements, add any necessary undefined terms.

In geometry, in order to prove a theorem, you need to use definitions, properties, rules, undefined terms, postulates, and (possibly) other theorems. And like hyperlinking to other text with Internet links, such theorems can be used throughout geometry (and other mathematics) in the proofs of other new, more difficult theorems.

PLANE GEOMETRY

What is **plane geometry**?

Plane geometry is simply the study of two-dimensional figures in a plane. Most mathematicians further define the plane as Euclidean (for more information about Euclidean geometry, see above). It examines such objects as circles, lines, and polygons.

How is the term "surface" used in mathematics?

When most of us think of the word "surface" we often envision our own world—the thin outer crust of soil and rock we walk and live on. In engineering, the term means the outer part (or the skin with a thickness of zero) of a body. In science it can apply to a plethora of objects, from geologic structures to micrometer-sized particles.

In mathematics, "surface" also has numerous meanings. The most common denotes a two-dimensional topological space or three-dimensional Euclidean space. A surface in mathematics can be complicated and complex, such as in certain fractals, or extremely simple, such as in a plane.

What does a **"plane"** mean in geometry?

A plane—in geometry or any other field of mathematics—means a surface such that a straight line joining any two of its points lies totally in that surface. A plane is considered to be two-dimensional; when a plane is discussed with higher dimensions, it is called a hyperplane. Thus, in the majority of mathematical discussions, a plane can be thought of as a two-dimensional group of points that reach out to infinity in all directions (for more information about dimensions, see above).

What is a **polygon**?

The word polygon means "many angles" (Greek *poly* for "many"; *gonis* for "angle"). It is a closed figure in a plane made up of line segments (straight with no curves) that intersect only at their vertices (endpoints). In other words, no sides (lines) can touch each other except at endpoints. A polygon has the same number of sides as it has vertices.

What are some **divisions of polygons**?

There are two major divisions of polygons: *Regular polygons* are convex polygons with equal sides and length; thus, all sides and angles are congruent (equal). For example, one of the most famous regular octagons is the stop sign used along roads in the United States: a closed polygon with eight equal sides. The naming of the various polygons can be challenging, though. For example, a polygon called a regular triangle is also called an equilateral triangle; another name for the polygon called a regular quadrilateral is a square. *Irregular polygons* are those with sides of differing lengths and variable angles. Therefore, unless all the sides of the polygon are of the same length and all the angles are of the same measure, the polygon is said to be irregular.

But don't be fooled: The names for the various polygons—such as hexagon, nonagon, and pentagon, depending on number of sides—don't just apply to the regu-

179

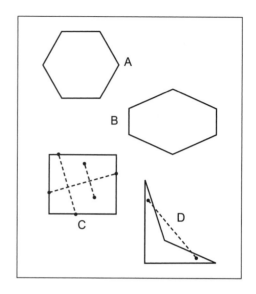

Examples A, B, and C above are polygons, while examples D and E are not.

Types of polygons include regular hexagons (A), irregular hexagons (B), convex polygons (C), and concave polygons (D).

lar polygons, but rather to *any* two-dimensional closed figure with the number of sides as described by its name. For example, the two figures shown above are both pentagons—A is a regular hexagon and B is an irregular hexagon.

Polygons are described in other ways, too. *Convex polygons* are those in which every line drawn between any two points inside the polygon lie entirely within the figure. Opposite from the convex polygons are the *concave polygons*—those that are essentially "caved in," with some of the sides bent inward. If a line is drawn between two points inside a concave polygon, the line often passes outside the figure. Another type of polygon is a *star polygon,* in which a star figure is drawn based on equidistant points connected on a circle.

What are the **names** given to **regular polygons**?

Polygons are classified according to the number of sides they have. A polygon with n sides is called an n-gon. The following lists some of the names of polygons, depending on the number of sides. It is also possible to substitute "n-gon," when the name is not known (for example 14-gon or 20-gon).

Names for Regular Polygons

Sides	Polygon Name
3	trigon or triangle
4	quadrilateral or tetragon

Sides	Polygon Name
5	pentagon
6	hexagon
7	heptagon
8	octagon
9	nonagon or enneagon
10	decagon
11	hendecagon or undecagon (even less frequently as unidecagon)
12	dodecagon
13	tridecagon or triskaidecagon
14	tetradecagon or tetrakaidecagon
15	pentadecagon or pentakaidecagon
16	hexadecagon or hexakaidecagon
17	heptadecagon or heptakaidecagon
18	octadecagon or octakaidecagon
19	enneadecagon or enneakaidecagon
20	icosagon
30	triacontagon
40	tetracontagon
50	pentacontagon
60	hexacontagon
70	heptacontagon
80	octacontagon
90	enneacontagon
100	hectogon
10,000	myriagon

Some texts list a two-sided polygon as a "digon," but this is only meant for theoretical mathematics.

What are **triangles** and how are they **classified**?

Triangles (or "three angles") are polygons with three sides. A triangle's three line segments (or sides) are joined together at three vertices (endpoints). For all triangles, the sum of a triangle's three interior angles is equal to a straight angle, or 180 degrees.

Triangles are classified by either the lengths of their sides, or, more commonly, the measurement of their angles. All triangles have at least two acute angles, but the third angle, which can be used to classify the triangle, can be acute, right, or obtuse (for more about these angles, see above). The types of triangles based on angles are:

Right triangle—There is one angle of 90 degrees.

181

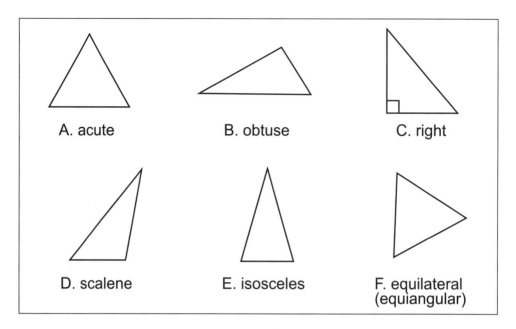

The basic types of triangles are acute (A), obtuse (B), right (C), scalene (D), isosceles (E), and equilateral (F).

Acute triangle—All three angles are less than 90 degrees, or the triangle has three acute angles.

Obtuse triangle—There is one angle greater than 90 degrees, or the triangle has one obtuse angle.

Equiangular triangle—An acute triangle in which all angles are congruent, or when all three angles are equal.

Triangles can be further classified by their sides, as in the following:

Scalene triangle—There are no sides—and therefore no angles—that are equal; in other words, no sides are congruent.

Isosceles triangle—Two of the sides are equal (congruent), and thus the base angles are equal.

Equilateral triangle—When all three sides are equal (congruent).

What are the **parts** of **right and isosceles triangles**?

There are special names for the parts of a right triangle. The *hypotenuse* is the side opposite the 90 degree angle—the side that will always be the triangle's longest. The shorter sides are called the *legs*.

Similar to the right triangle, the isosceles triangle has specific names for its angles and sides: Congruent (equal) sides are, as with the right triangle, called the

legs; the angle formed by these two legs is called the *vertex angle;* and the *base* is the side opposite the vertex angle. The two angles that are formed by the base and the legs are called the triangle's *base angles.*

How is the **Pythagorean Theorem** connected to **right triangles**?

The Pythagorean Theorem deals with right triangles: Simply stated, the sum of the squares of the legs equals the square of the hypotenuse (longest side). The converse of the Pythagorean Theorem is also true: If the sum of the squares of a triangle's two sides equals the square of the longest side, then the triangle is a right triangle. (For an illustration of the Pythagorean Theorem, see "History of Mathematics.")

What are the various **types of quadrilaterals**?

There are several types of quadrilaterals, which are polygons with four sides. Interestingly enough, some definitions can be "combined"; for example, if a quadrilateral is both a rhombus and a rectangle, it is truly a square. The following lists the common quadrilaterals:

Square—The most obvious quadrilateral is the square. It is an equiangular quadrilateral with four right angles (it is also defined as having four congruent sides).

Rectangle—The second most well-known quadrilateral is the rectangle, a quadrilateral with four right angles, with the opposite sides parallel and congruent, and opposite angles congruent.

Parallelogram—A parallelogram is a quadrilateral with both pairs of opposite sides parallel; thus, opposite sides and angles are congruent.

Rhombus—A rhombus is a parallelogram with four equilateral (or congruent) sides.

Trapezoid—A trapezoid is a quadrilateral with exactly one pair of parallel sides. This is also seen in books as "a quadrilateral with at least one pair of parallel sides," but this latter definition is often debated among mathematicians, as the meaning is not the same as the first statement. With trapezoids, the parallel sides are called the *bases;* the nonparallel sides are called the *legs.*

Isosceles trapezoid—An isosceles trapezoid is one with nonparallel sides that are equal in length, or a trapezoid with a pair of equiangular base angles. The legs of an isosceles trapezoid are congruent. (For more about these figures, see elsewhere in this chapter.)

What is a **circle**?

A circle is one of the most fundamental shapes in geometry—and one shape we commonly see every day. For mathematicians, a circle is defined as a set of points on a plane at a certain distance from a center point. In reality, a circle is a polygon with an infinite number of sides.

The parts of a circle include the center (o), radius (r), diameter (d), and circumference (c).

The distance of a line segment from the center to the points on the circle is called the *radius* (or a line segment whose endpoints are on any point on the circle and its center). A line segment that travels from one endpoint on the circle, through the radius (center of the circle), and to another endpoint directly opposite is called the *diameter;* two times the radius of a circle is the diameter. The outer perimeter of the circle is called the *circumference*. The *chord of a circle* is a line segment whose two endpoints are on the circle. *Concentric circles* are two or more circles that lie in the same plane and have the same center, but with different radii. Circles with the same radius are called *congruent circles*. (For more about circle measurements, see elsewhere in this chapter.)

Why are **arcs** and **angles** important to **circles**?

Everything starts with the angle whose vertex is at the center of the circle (logically called the *central angle of a circle*.) All the central angles of a circle add up to 360 degrees. Every central angle cuts the circle into two *arcs:* the *minor arc* (always less than 180 degrees) and the *major arc* (always more than 180 degrees. Thus, the measure of the minor arc is actually the measurement of the central angle, while the measurement of the major arc is 360 degrees minus the measure of the central angle. An *arc length* is the distance between an arc's endpoints along the path of the circle. *Congruent arcs* are arcs with the same measurements. When the diameter of a circle separates the circle into two congruent arcs it is called a *semicircle*.

SOLID GEOMETRY

What is **solid geometry**?

Solid geometry is the study of objects in three-dimensional Euclidean space. It deals with solids, as opposed to plane geometry, which deals with two dimensions. This part of geometry is concerned with entities such as polyhedra, spheres, cones, cylinders, and so on. (For more about Euclidean space and dimensions, see elsewhere in this chapter.)

In geometry, solids are defined as closed three-dimensional figures, or any limited portion of space bounded by surfaces. They differ in subtle ways from what we perceive

as solids: We see solids in terms of what surrounds us—three-dimensional figures with their surfaces the actual objects we perceive. Geometric solids are actually the union of the surface and regions of space; in a way, this adds another dimension to two-dimensional space.

What is a **polyhedron**?

The word polyhedron comes from the Greek *poly* (meaning "many") and the Indo-European word *hedron* (meaning "seat"). In geometry, a polyhedron is considered to be a three-dimensional solid that represents the union of polygonal regions, usually joined at their edges and having no gaps.

Polyhedrons are classified as *convex,* or if you extend any side, the figure lies on only one side of the plane (for example, a pyramid and cube); a *concave* polyhedron is one that can extend on both sides of such a plane (for example, a form that exhibits concavity on one or all sides of the polyhedron). The plural of polyhedron is often seen as "polyhedrons," but it is more correct to use "polyhedra."

What are the various **types of polyhedra**?

As with most forms, polyhedra are divided into many names, depending on the number of faces. The following lists some of them, based on their number of sides:

Types of Polyhedra

Sides	Polyhedra Name
4	tetrahedron
5	pentahedron
6	hexahedron
7	heptahedron
8	octahedron
9	nonahedron
10	decahedron
11	undecahedron
12	dodecahedron
14	tetradecahedron
20	icosahedron
24	icositetrahedron
30	triacontahedron
32	icosidodecahedron
60	hexecontahedron
90	enneacontahedron

What are the Platonic solids?

Platonic solids are also called regular solids, regular figures (a term also used in reference to polygons), regular polyhedra, or "cosmic figures." These solids are convex polyhedra that have equal faces made up of congruent convex regular polygons. (To compare in terms of a two-dimensional polygon, a regular figure means that both the sides and the angles between them are equal.)

There are considered to only be five of these solids: the cube, dodecahedron, icosahedron, octahedron, and tetrahedron. These solids were described by Greek philosopher Plato (c. 428–348 BCE) in his work *Timaeus*—thus the name Platonic solids. His definitions were certainly more whimsical than today's, as he believed the major "elements" were made up of atoms shaped like certain polyhedra. He associated the tetrahedron with the "element" fire, the cube with the earth, the icosahedron with water, the octahedron with air, and the dodecahedron with the material that makes up the constellations and heavens. The mathematical proofs of these solids were worked out long ago by Greek mathematician and geometrician Euclid (c. 325–c. 270 BCE) in the last part of his *Elements* (for more information about Euclid, see elsewhere in this chapter, as well as "Foundations of Mathematics").

How are some **common solids** defined?

There are numerous objects defined in solid geometry. The following lists the most common (and some interesting) ones:

Cone—A cone can be both a surface and a solid. A solid cone is bounded by a region enclosed in a closed curve on a plane and a surface formed by segments joining each point of the closed curve to a point that is not in the plane. (Note: Two solid cones seen with their pointed ends together help define conic sections; for more about conic sections, see elsewhere in this chapter.)

Pyramid—A pyramid is a polyhedron (see above) with one face a polygon and all other faces as triangles with a common endpoint (vertex). They are named based on the polygon's base, such as the triangular pyramid, square pyramid, and so on. Some of the most famous "solid pyramids" are the sandstone pyramids of Egypt. These are actually called right square pyramids because the base of the polygon is a square and the vertical line from the vertex meets the center of the base.

Cylinder—A cylinder can be both a surface and a solid. A solid cylinder is one that forms by rotating a circle about an axis through the midpoints of the opposite side; it is also called a right circular cylinder. One of the most well-known cylinders is possibly sitting right next to you: A coffee cup, with its cylindrical shape, and the bases (in this case, the base and rim) considered to be congruent circles.

Prism—A prism is a polyhedron with two parallel, congruent faces that make up the bases of the shape; the other, lateral faces are considered to be parallelograms. If the lateral faces are rectangles, the prism is called a right prism.

Parallelepiped—This strange-sounding polyhedron is one that has all its faces as parallelograms, or a prism with parallelogram bases. The most familiar parallelepiped is a simple box—also called a rectangular parallelepiped—that has rectangles for all the six faces. (For more about these figures, including how to calculate their areas, see elsewhere in this chapter.)

How are **spheres described**?

The Greek philosopher Plato first described the five regular solids known as Platonic solids in his book *Timaeus. Library of Congress.*

A sphere in solid geometry is considered to be the set of points (or the "skin" of the sphere) in three-dimensional Euclidean space that are equidistant from the sphere's singular, central point. Or, more simply, a sphere is a perfectly round, three-dimensional object. The term "sphere" also extends into other dimensions; for example, a sphere in two dimensions is also called a circle.

What is **spherical geometry**?

Spherical geometry is the study of objects on the surface of a sphere; this differs from the type of geometry studied in plane or solid geometry. In spherical geometry, there are no parallel lines, and straight lines are actually great circles, so any two lines meet in two points. In addition, the angle between two lines is the angle between the planes of the corresponding great circles. There are also entities called *spherical triangles* (or Euler triangles), when three planes pass through the surface of a sphere and through the sphere's center of volume; they have three surface angles and three central angles. There are also *spherical polygons,* in which a closed geometric object on the surface of a sphere is formed by arcs of great circles.

MEASUREMENTS AND TRANSFORMATIONS

What is the **perimeter** of a **two-dimensional geometric figure**?

The perimeter of a two-dimensional figure is what we perceive as the "outside" of the object. Mathematicians describe a perimeter more precisely as the sum of the lengths of a polygon's sides; the perimeter of a simple closed curve is measured as its length.

What is the **area** of a **two-dimensional geometric figure**?

The area of two-dimensional figures varies depending on the object. For the simpler two-dimensional figures (square, rectangle, parallelogram), the area can actually be found by calculating the number of square units in the interior of the object. Such a task can be difficult and time consuming, so mathematicians simply multiply the height (h) times the base (b), sometimes said as length times height or length times width (with rectangles). (For more information about the area of geometric figures, see the Appendix 3: Common Formulas for Calculating Areas and Volumes of Shapes.)

How is the **area determined** for some common **two-dimensional polygons**?

Finding the areas of polygons is not as simple as determining the areas of a rectangle and square. In order to find the area of certain polygons, one essentially "breaks the shape down" into smaller shapes with simpler formulas; these types of shapes are called *composites*. The formulas for many of these polygons are all adaptations from the rectangle area formula—or height (h) times base (b). For example, a triangle is actually exactly half of a rectangle. Thus, the formula to find the area of a triangle is half the base times the height ($1/2bh$). In the case of the trapezoid, the figure can be divided into triangular sections, with the area equal to one half times the two bases times the height, or $1/2(B + b)h$.

The triangle formula also is used to find other regular polygon areas, but they are less obvious. For a regular polygon, a feature called the *apothem* is necessary for finding the area. This is the height of one of the congruent triangles inside the regular polygon. In general, to find the polygon's area, you need to find the area of one triangle and multiply it times the number of sides. For example, to find the area of a hexagon, divide the figure into six triangles, each with l equal to the height of each interior triangle or apothem; l is also half of the smallest interior dimension of the hexagon, called w. Thus, the area of a hexagon is the square root of 3 divided by 2 times w squared (or $\sqrt{3}/2 \times w^2$).

There is still another way to determine the area of a polygon by using the Pythagorean Theorem, a method that uses length and height, with the resulting formula for the area of a polygon looking much different. Both methods give the same

How is the term "surface area" used in geometry?

Logically, the surface area is the area of a given surface. There are several ways to interpret this in geometry: Area can mean the extent of the surface region on a two-dimensional plane. Surface area (often called the lateral surface area, although there is a difference) formulas for three-dimensional objects are more complex—all the surface areas are added around the outside of the object, from a cube to a sphere.

Surface area is commonly denoted as S for a surface in three dimensions, and A for the surface area of a two-dimensional plane (commonly, it is simply called "the area"). But be careful: The surface area of a three-dimensional object should not be confused with "volume"—or the total amount of space an object occupies. (For more about volume, see below.)

results in numerical terms. For example, the area of a hexagon using the smallest interior dimension would be 0.866 times the square of the smallest width, w; the area of a regular hexagon using the Pythagorean Theorem is 2.598 times the square of its side length. These may be different ways of presenting the solution, but both give the same correct area measurements.

What are the **measurements** of a **circle**?

There are many measurements of a circle. The perimeter of a circle is actually called the *circumference;* to calculate this, multiply pi (π) times the diameter, d (πd), or pi (π) times twice the radius, r ($2\pi r$). The area of a circle is calculated by multiplying pi (π) times the radius squared, or πr^2.

How was the **area of a circle** first determined?

When it comes to determining the area of a circle, there are many historical claims to this solution. One of the earliest techniques may have been the Chinese "comb" method, in which a circle is cut into n wedges, each $360/n$ degrees and each piece identical (with the same area). To see how this works, take the bottom half of a unit circle and cut it into wedges like slices of a pie; place all the wedges next to each other, with the points up (like the teeth of a comb or animal). Then split the top half of the circle in the same way, putting them next to each other above the other wedges, but this time point the tips of the wedges down. Close the "teeth"; as n goes to infinity, the shape of the combined wedges approaches a rectangle. Taking the formula for determining the area of a rectangle (width times height), the width is πr (or half the circumference) and the height is r. Thus, the area is πr^2.

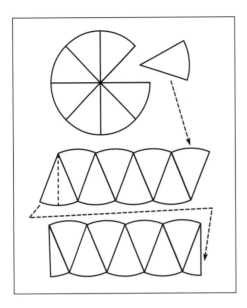

To calculate the area of a circle, the Chinese devised the Chinese comb method in which a circle is divided into pie slices and rearranged into a rectangle to greatly simplify the math involved.

What other way did **Archimedes** use to find the **area of a circle**?

The Greek mathematician Archimedes (c. 287–212 BCE; Hellenic) also found a way to determine the area of a circle similar to the Chinese comb: a method he first recorded in his work, *Measurement of a Circle* (c. 225 BCE). He also used a sequence of wedges to determine the area of a circle; as the number of wedges (or triangles) inside the circle increased toward infinity, they became infinitely thin. By giving each small triangle a base (*b,* a line connecting the points where the wedges touched the circle's circumference), he determined that the area was 1/2 times the radius (*r*) times the base, summed over all the infinitesimal triangles (or sum (1/2) *rb*). Because they all had the same height, that was factored out. Thus, the area became (1/2)*r* (sum (*b*)) = 1/2*rc,* with *c* being the circumference, or the sum of the bases (*b*) of all the triangles (since the bases make up what is perceived as the circle's circumference). This is interpreted as one half times the radius times the circumference ($c = 2\pi r$), which is the same as saying πr^2. (For more about Archimedes and his wedges, see "Mathematical Analysis.")

How are the **surface area** and **volume** of a **three-dimensional geometric figure** calculated?

The *surface area* (often abbreviated S.A.) of a three-dimensional geometric figure is the total surfaces of the solid; it actually has units of distance or length squared. For example, the surface area of a cube is $6a^2$, in which *a* is the length of the sides. To translate, a cube has sides of equal lengths (*a*); the area of a cube is the sum of the areas of the six squares (a^2) that cover it.

For more "diverse" figures, the surface area is actually equal to the *lateral area* plus the area of each base. For example, the surface area of a prism or cylinder is the lateral area plus the area of each base. (Because the bases for a prism or cylinder are congruent, this is often expressed as twice the area of the base.) The surface area of a pyramid or cone is the lateral area plus the area of the single base.

The *volume* of a three-dimensional geometric figure is the total amount of space the object occupies; volumes of such objects have units of length cubed. For example,

the volume of a box (also called a rectangular parallelepiped) is length times width times height, or $l \times w \times h$; the volume of a cube is all the sides a cubed, or a^3.

What are **lateral surface areas**?

In many mathematical texts, the *lateral surface area* (L.A.) is given along with— or instead of—the surface area (S.A.). This type of area is the surface area of a three-dimensional figure, excluding the area of any bases. In other words, the lateral surfaces are the side faces (or surfaces) of a solid, or any face or surface that is not a base (or bottom of a figure).

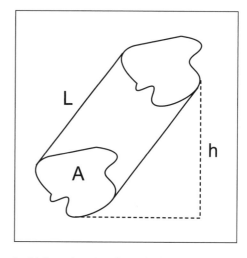

In this irregular prism shape, the lateral area equals the perimeter of shape A multiplied by L, and the surface area equals the lateral area plus two times the area of shape A.

What are the **surface area, lateral area,** and **volume** of a **sphere**?

A sphere is a set of points equidistant from a central point (a center); a sphere in two dimensions is a circle. In the case of a sphere, the lateral and surface areas are the same; therefore, a sphere with a radius r has a surface/lateral area of $4\pi r^2$; the volume of a sphere is $4/3\pi r^3$.

What are **geometric transformations**?

Geometric transformations are rules that mean there is a one-to-one correspondence between two sets of points in two figures. A geometric transformation changes the plane or three-dimensional space. Each transformation can be defined by thinking about where each point is "taken," with a point said to be "taken to its image."

One way to visualize a transformation is to think of expanding, shrinking, or moving one figure to get a second figure. Depending on the transformation, the image will either be congruent (equal) to the original image (called the *identity transformation*), similar but not necessarily equal, or will not resemble the original at all.

What are some **transformations** in geometry?

A transformation that keeps a figure's same shape and size, but moves it to a new location, is called *isometry*. There are several common types of these transformations. *Dilatation* is the only transformation that does not create equal figures. It means to take a shape and make it larger or smaller, but keep the same proportions. In terms of a circle, a dilatation creates another circle with the same center, also known as a con- **191**

centric circles. *Reflection* is similar to what we called a "flip" in elementary school mathematics—like the flip side of an object. One easy way to see this is by noting one's reflection in a mirror—the "figure" is on one side of a line and the mirror image on the other. Reflection twice about two parallel lines is synonymous with translation; reflection twice about two intersecting lines is called rotation.

Another type of transformation is *rotation*. This is simple to understand: in elementary school mathematics it is sometimes called a "turn" or "spin." In this case, one point on a plane remains unchanged while keeping all the distances between the other points the same. Finally, a *translation* (or glide transformation) is similar to what is called a "slide" in elementary school mathematics. All the points in the plane move in the same direction and the same distance; or, the figure slides in a single direction. Translation is also considered to be reflection twice across two parallel lines.

ANALYTIC GEOMETRY

What is **analytic geometry**?

Analytic geometry is actually another name for coordinate (or Cartesian) geometry, a way to geometrically represent ordered pairs of numbers (or coordinates). The objects are described as n-tuples of points, in which $n = 2$ in a plane and $n = 3$ in a space (or two and three dimensions) in some coordinate system. Overall, analytic geometry allows mathematicians to determine the position, configurations, and separations of objects using algebraic representation and manipulation of equations.

What is a **graph**?

The term graph has several different meanings in mathematics. It can mean the interpretations of numbers, including bar graphs, pie charts, and line graphs. For example, a pie chart (graph) is often used to represent percents, such as a breakdown of where a taxpayer's money is spent in the various government agencies for a certain year.

In analytic geometry, a graph is simply a way of plotting—thus, visually representing—points, lines, curves, and solids in order to understand and interpret certain geometric figures and to solve equations. For example, solving an equation with one to two variables (usually written as x and y, or two dimensions) results in a curve on a graph (note: a line is considered a curve in geometry). Equations that contain three variables (usually as x, y, and z, or three dimensions) result in a surface.

What is a **coordinate system**?

A coordinate system is one that uses *coordinates*—a number or numbers that identify a point on a number line, plane, or in space. These points are most often seen on a

Who developed analytic geometry?

Analytic geometry began when French philosopher, mathematician, and scientist René Descartes (1596–1650; in Latin, Renatus Cartesius) published a work explaining how to use coordinates for finding points in space. He was the first to make a graph and presented a geometric interpretation of a mathematical function; this marked a step toward what is now known as Cartesian coordinates, a term derived from Descartes's Latin name. Around the same time—and independently—French mathematician Pierre de Fermat (1601–1665) also did much to establish the ideas of coordinate geometry. But, unlike Descartes, Fermat did not publish his work. Both Descartes's and Fermat's ideas would lead to modern Cartesian coordinates. (For more about Descartes and Fermat, see "History of Mathematics.")

graph and can be a combination of two numbers for a two-dimensional figure or three numbers for three dimensions.

How are **two-dimensional Cartesian coordinates** determined?

With two-dimensional Cartesian coordinates (or rectangular coordinates), an *ordered pair* of two numbers is determined using two axes oriented perpendicular to each other (for more about ordered pairs, see "Foundations of Mathematics"). The ordered pair coordinates are found by specifying points along the two axes: First, along the x-axis, the positive or negative amount on either side of the y, or vertical, axis; this is called the *abscissa* and indicated as x. The other number is the positive or negative amount on either side of the x, or horizontal, axis; this is called the *ordinate* and indicated as y. The coordinates are usually written in terms of x and y at the point as an ordered pair; for example, in the illustration of a two-dimensional coordinate system on p. 194, the ordered pair is (7, 8), in which x is 7 and y is 8.

What are **quadrants**?

Graphs used in the Cartesian coordinate system are broken down into quadrants, or four segments. For example, in a two-dimensional system, the x and y axis—with the origin as zero and the two axes perpendicular to each other—form the four quadrants. In the top left quadrant (quadrant II), the x is negative and the y is positive; in the top right quadrant (quadrant I), both coordinates are positive; in the bottom left quadrant (quadrant III), Cartesian coordinates are both negative; and in the bottom right quadrant (quadrant IV), the x is positive and the y is negative.

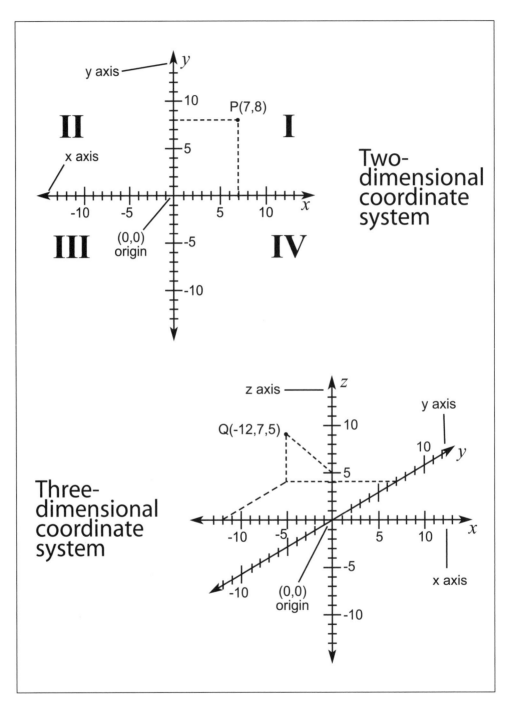

The four quadrants of a two-dimensional coordinate system are indicated above by the Roman numerals I, II, III, and IV. A three-dimensional coordinate system (bottom) adds a z axis so points can be indicated in real space, such as with point Q(-12, 7, 5) in this illustration.

How are **three-dimensional Cartesian coordinates** determined?

The three-dimensional coordinates are those that describe a solid or three-dimensional object with three axes (or three planes): the usual two axes (as in the two-dimensional system), and an additional line. The coordinates are usually written in terms of *x, y,* and *z,* and are often called *ordered triples* or just triples.

What are some **terms** used in the **Cartesian coordinate system** and **graphs**?

There are numerous terms used in the Cartesian coordinate system. Besides the ones already mentioned, the following are some of the most common.

An *intercept* is a point's distance on a coordinate system axis from the origin to where a curve or surface intersects the axis. On a graph, the x-intercept and y-intercept are two important features that show where a line cuts through the x and y axes, respectively. The *origin* is the fixed point from which measurements are taken. In most cases—especially in a standard, simple two-dimensional Cartesian coordinate system—this means the point that represents zero. This is often seen as (0, 0), or the point in which the x and y axes intersect on a graph. In a three-dimensional system, the coordinates are often seen as (0, 0, 0).

A *Cartesian plane* (or coordinate plane) is described as a two-dimensional space made up of points that are identified by their relation to the origin (zero), and the x and y axes. An *axis* (the plural is axes) is a reference line used in a graph or a coordinate system, such as the Cartesian system. For example, the x-axis and y-axis are perpendicular lines on a graph in a two-dimensional system; in a three-dimensional system, they are the x-axis, y-axis, and z-axis.

Collinear points are those that lie on a straight line. Any two points are considered collinear because a straight line passes through both. Many procedures in analytic geometry involve determining the collinear points that represent coordinates that solve an equation. Logically, those points that do not lie on the same line—or, in other words, that do not solve the equation—are called non-collinear points.

How is the **slope measure** and **y-intercept** of a line determined?

On a graph, the slope is the line in a plane completely specified by a point and a number. Mathematicians know that when an equation is written as $y = mx + b$, they can determine the slope measure and y-intercept (or slope intercept) of the line. In this case, the m (or x coefficient) is called the *slope measure* and the y-intercept are the coordinates that solve the equation.

For example, for $y = -2x + 4$, the slope measure is -2 and y-intercept is the coordinates (0, 4). To test this result, replace the x with the first coordinate (0); $-2 \times 0 = 0$; therefore, 4 is left, resulting in y (or the y-intercept) being 4.

What does **point-slope equation of a line** refer to in analytic geometry?

A point-slope equation of a line refers to Cartesian coordinates: If a point has coordinates (x_1, y_1) on a line and a slope equal to m, then the equation $y - y_1 = m(x - x_1)$ is the point-slope equation. Point-slope is most often used when finding an equation of a line.

Why are **functions** important to **analytic geometry**?

The intricacies of functions are important mainly because the function (and the graph of all the points in a function) is one of the basic foundations of analytic geometry. This becomes even more evident when trying to solve complex equations or equations with more than one variable.

For example, when determining the solution to an equation, such as $3x + 4y = 8$, the two resulting numbers—called a *set of ordered pairs of numbers*—is called a *relation*. In turn, a function then becomes a relation in which each first element, such as x, is matched exactly with a second element, such as y. In other words, a function can take on a definite value (or values) when certain values are assigned to other quantities or independent variables of the function. (For more information about functions, see "Foundations of Mathematics" and "Algebra.")

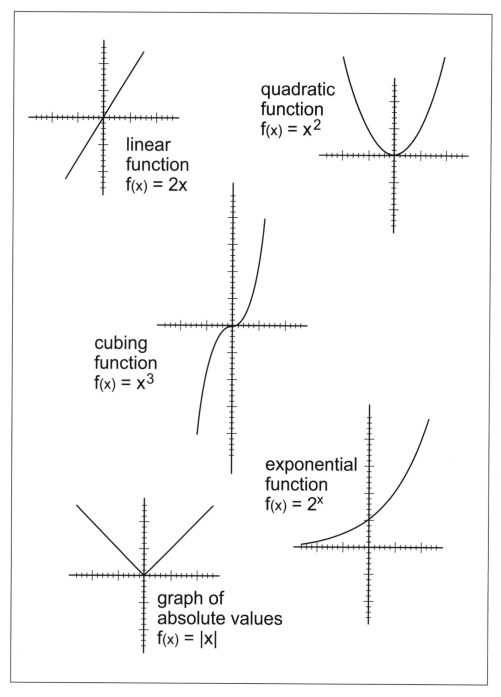

Some common types of graphs illustrate the linear function, quadratic function, exponential function, and absolute values.

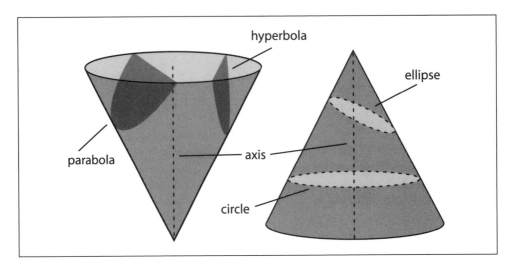

Depending on the way a plane intersects a cone shape, different types of curves will be the result.

What is a **one-to-one function**?

A one-to-one function is one in which each input (number that replaces the variable) has exactly one output (result of the equation). In such cases, the function needs to pass the "horizontal line test," showing that a horizontal line intersects the graph once and only once. For example, for the equation $f(x) = x^2$, if one restricts the answer to $x \geq 0$, the result is a one-to-one function; but the equation with no constraints is not a one-to-one function, as the output value of 4 has the two input values of 2 and -2.

What is the definition of a **conic section**?

This family of curves—first discovered (as far as we know) by the ancient Greeks—is generated by planes intersecting ("cutting into") a cone. The resulting feature depends on the angle between the plane and the axis of the cone, with none of the planes passing through the endpoint (vertex) of the cone. Each plane that cuts a cone creates a two dimensional figure called a *section,* thus the resulting figures are called *conic sections* or *conics*. In general, the possible resulting surfaces are a sphere (the cut perpendicular to the axis), ellipse (or the circle as a special case; a cut moderately inclined to the axis), parabola (a cut parallel to one of the straight lines that generate the cone), and hyperbola (using two cones tip to tip, the cut even more steeply inclined than all the others).

What are **polar coordinates**?

Polar coordinates are actually an alternative system to the Cartesian coordinates. In two dimensions, they mark a point on a plane by its radial distance (r) from an "ori-

gin" and a polar angle (θ). This method also uses trigonometric functions such as sin and cos (sine and cosine; for more about such functions, see trigonometry in this chapter).

Polar coordinates in three-dimensional space—also called spherical coordinates—use r and two polar angles (θ, ϕ) to give the direction from the origin to the point. To compare, a three-dimensional polar coordinate system overlaps the Cartesian system in several ways: For example, θ is the angle between the line to the origin and the z-axis of the Cartesian (x, y, z) system; ϕ is the angle (counterclockwise when viewed from positive z) between the projection of that line onto the (x, y) plane and the x-axis.

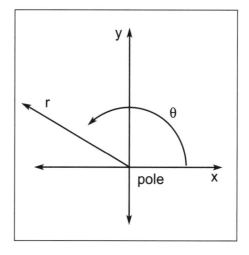

In the above illustration, $x = r \cos \theta$; $y = r \sin \theta$; $r^2 = x^2 + y^2$; and $\theta = \arctan y/x$ ($x=0$).

What is an **Argand diagram**?

An Argand diagram is a graphical way of representing a function of a complex variable, often written as $z = x + iy$, in which x, y, and z are coordinates in three-dimensional space and i is an imaginary number. Its true discoverer is not actually known, but Swiss mathematician Jean Robert Argand (1768–1822) is given credit for the diagram. It is thought that this was also independently discovered by Danish mathematician Casper Wessel (1745–1818), and later by German mathematician and physicist Karl Friedrich Gauss (1777–1855) in 1832 (but he probably determined it much earlier); thus, its other name is the Gaussian plane.

What is an **asymptotic curve**?

On a graph, a line that approaches close to a curve (or even an axis) but never quite reaches it is an asymptotic curve. In an example similar to one of Zeno's paradoxes (see "Foundations of Mathematics"), if a kitten standing a yard from a box walks half the distance to the box each hour, it will technically never reach the box, because the distance it travels each hour is never more than half the remaining distance to the box. If this problem was illustrated as an equation, the answer would never quite reach its solution. A more mathematical example is the exponential function $y = 2^x$, which results in a line that approaches but will never quite reach the x-axis.

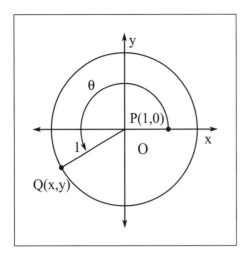

In this example of an angle measured using trigonometry, $x^2 + y^2 = 1$.

TRIGONOMETRY

What is **trigonometry**?

Trigonometry is the study of how the sides and angles of a triangle are related to each other. Interestingly, the angles are usually measured in terms of a circle around the x and y axes; from there, certain formulas are calculated, much as they are in algebra, to determine all the angles and units. Because trigonometry is such a mix of algebra and geometry, it is often considered "the art of doing algebra over a circle." Although "trig" (as it is nicknamed) is a small part of geometry, it has numerous applications in fields such as astronomy, surveying, maritime and aerial navigation, and engineering.

How are **angles measured** in trigonometry?

Angles in trigonometry are measured using a "circle" on x and y axes—often called circle trig definitions. The radian measure of an angle is any real number θ (theta; see illustration). Take an instance in which θ is greater than or equal to zero (θ ≥ 0): Picture taking a length of string and positioning one end at zero; then stretch the other end to 1 on the x axis, to point P(1, 0); this is also considered the radius of the circle. Then, in a counterclockwise direction, swing the string to another position, Q (x, y). This results in θ being an angle (associated with the central angle) with a vertex O or (0, 0) and passing through points P and Q; and because the string is "1 units" in length all the way around, the point from Q to the vertex O will also be 1. The resulting angle θ is measured in degrees—and defined as a part of the circle's total number of degrees (a circle has 360 degrees); it can also be translated into radians.

How are **degrees and radians** translated?

When measuring the number of degrees in a circle, another unit called radians is often used in trigonometry. It is known that a circle is 360 degrees, with 1 degree equal to 60 minutes (60') and 1 minute equal to 60 seconds (60"); this is also called DMS (Degree-Minute-Second) notation. One revolution around the circle also measures 2π radians. Thus, $360° = 2\pi$ radians; or $180° = \pi$ radians. Simply put, to convert radians to degrees, multiply by $180/\pi$; to convert degrees to radians, multiply by $\pi/180$. The following are examples of how to convert degrees and radians:

Convert the angle 236.345° to DMS notation (by breaking down the decimals into minutes and seconds):

236° + 0.345° × 60' /1°
236° + 20.7' = 236° + 20' + 0.7' × 60' /1'
= 236° 20' 42"

Convert the angle 236.345° to radians (convert the entire amount into radians by multiplying π radians by 180°, which is actually equal to "1" since, from above, 180° = π radians):

236.345° × π radians/180° = (236.345 × 3.141592) radians/180°
= 4.124998 radians

And, conversely, convert 4.124998 radians to degrees:

4.124998 radians × 180°/3.141592
= 236.345°

What are the **six trigonometric functions**?

There are six basic trigonometric functions that can be used to interpret the measurement of angles and triangles—most often defined as circle trig definitions. In terms of the illustration below, the following lists the six trig functions of θ, all defined in terms of the coordinates of Q (x, y):

- $\cos \theta = x$
- $\sin \theta = y$
- $\tan \theta = y/x$ if $x \neq 0$
- $\sec \theta = 1/x$ if $x \neq 0$
- $\csc \theta = 1/y$ if $y \neq 0$
- $\cot \theta = x/y$ if $y \neq 0$

The full names for these functions are cos (cosine); sin (also seen as "sine"); tan (tangent); sec (secant); csc (cosecant); and cot (cotangent).

How are **trigonometric functions** used to **describe a right triangle**?

The trigonometric functions can be interpreted using the illustration on p. 202 of a right triangle. If θ is the angle (see illustration), a is one leg (called the *adjacent* because it is adjacent to the angle), b is another leg (called the *opposite,* because it is opposite the angle), and c is the hypotenuse:

- $\cos \theta = a/c$ = adjacent/hypotenuse
- $\sin \theta = b/c$ = opposite/hypotenuse
- $\tan \theta = b/a$ = opposite/adjacent
- $\sec \theta = c/a$ = hypotenuse/adjacent
- $\csc \theta = c/b$ = hypotenuse/opposite
- $\cot \theta = a/b$ = adjacent/opposite

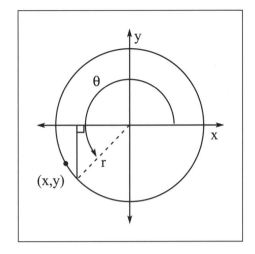

Trigonometric functions can be used to describe a right triangle.

Trigonometric functions of a circle.

What are the same **functions using a circle**?

The following shows the same way of looking at the above triangle using a circle (also see illustration):

- $\cos \theta = x/r$
- $\sin \theta = y/r$
- $\tan \theta = y/x$
- $\sec \theta = r/x$
- $\csc \theta = r/y$
- $\cot \theta = x/y$

What is an **example** of finding a **trig function**?

One example of finding a trig function—including an illustration—follows: Find the trig functions of the angle $\theta = 60°$. In this case, make a circle with a string extending two units; then measure 60° from the vertex (O). Drop a line perpendicular from the point on the circle to the initial line, creating a 90° angle; that makes the last angle 30°. Because the result is a right triangle, with the values of the hypotenuse (2) and the adjacent leg (1), the Pythagorean Theorem (see elsewhere in this chapter) is used to determine the length of the other leg, or $a^2 + b^2 = c^2$:

$$1^2 + b^2 = 2^2$$
$$1 + b^2 = 4$$
$$b^2 = 3$$
$$b = \sqrt{3}$$

Thus, by figuring out the cos (adjacent/hypotenuse) and sin (opposite/hypotenuse) for the angle 60°, Q is at point (1/2, $\sqrt{3}$/2) on the circle. The other trig functions can be determined using the methods in the questions above:

$\cos \theta = 1/2\ (x)$
$\sin \theta = \sqrt{3}/2\ (y)$
$\tan \theta = \sqrt{3}\ (\sqrt{3}/1,\ \text{or}\ y/x)$
$\sec \theta = 2\ (1\ /\ 1/2,\ \text{or}\ 1/x)$
$\csc \theta = 2/\sqrt{3}\ (1/y)$
$\cot \theta = 1/\sqrt{3}\ (x/y)$

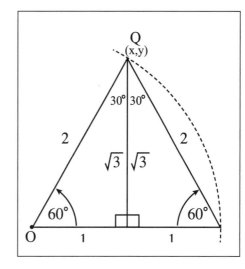

In this example, to find the trig functions of an angle of 60 degrees, draw a triangle with point O at the center of a circle with radius = 2; use the Pythagorean Theorem to determine the length of the line dropping down at a right angle from point Q; and then calculate the trig functions using these measurements.

What are some **identities in trigonometry**?

There are many fundamental identities—an equation that is true regardless of what values are substituted for any of the variables—based on trigonometric functions. Because there are relationships between the trig functions, identities can be used to rewrite equations, allowing the user to simplify or get more information out of an equation. The identities include reciprocal identities, ratio identities, periodicity identities, Pythagorean identities, odd-even identities, sum-difference identities, double angle identities, half angle identities, and several more. The following lists a few such identities:

Pythagorean Identities
- $\cos^2\theta + \sin^2\theta = 1$
- $\tan^2\theta + 1 = \sec^2\theta$
- $\cot^2\theta + 1 = \csc^2\theta$

Reciprocal Identities
- $\sin\theta = 1/\csc\theta$
- $\cos\theta = 1/\sec\theta$
- $\tan\theta = 1/\cot\theta$
- $\sec\theta = 1/\cos\theta$
- $\csc\theta = 1/\sin\theta$
- $\cot\theta = 1/\tan\theta$

Ratio (or Quotient) Identities
- $\tan\theta = \sin\theta/\cos\theta$
- $\cos\theta = \cos\theta/\sin\theta$

203

What is one of the most fundamental identities?

If one examines some of the functions above carefully, the sin and cosine functions are actually the coordinates of a point on the unit circle. This implies that the most important fundamental formula in trigonometry is as follows—one that some people call the "magic identity" but is more commonly known as one of the Pythagorean identities:

$\cos^2\theta + \sin^2\theta = 1$, in which θ is any real number.

This identity can be used in the following step-by-step example:

Show that:

$\sec^2\theta = \tan^2\theta + 1$

Because $\sec\theta = 1/\cos\theta$ (from the reciprocal identities), and $\tan\theta = \sin\theta/\cos\theta$ (from the ratio or quotient identities) then:

$\tan^2 + 1 = (\sin\theta/\cos\theta)^2 + 1 =$
$(\sin^2\theta/\cos^2\theta) + 1 =$
$(\sin^2\theta + \cos^2\theta)/\cos^2\theta$ (where $\cos^2\theta$ divided by itself equals 1).

Then (using the Pythagorean identity):

$1/\cos^2\theta = \sec^2\theta$ (from the reciprocal identity).

What is a **periodic function**?

Periodic functions are those that repeat the same values at regular intervals. For trigonometric functions, the *period* (or the interval of repeated values) is 360 degrees or 2π for the sin, cosine, secant, and cosecant; and 180 degrees or π radians for the cotangent and tangent. (For example, this is often said as "the tangent function has period π.")

What are the **hyperbolic functions (or identities)**?

Similar to the other types of functions and identities above, hyperbolic functions (also called hyperbolic identities) are used to make it easier to find solutions to equations. In fact, there are corresponding trigonometric and hyperbolic identities. For example, the most commonly used trig identity, $\cos^2\theta + \sin^2\theta = 1$, has a corresponding hyperbolic identity:

$\cosh^2 x - \sinh^2 x = 1$

In this case, the minus sign is used instead of a plus and the "cos" and "sin" changed to "cosh" and "sinh," respectively (also changed is the "θ" symbol to x; this is

because many trig functions use the angle as the argument, while hyperbolic functions generally do not).

But these equations are not exactly the same. In particular, when one has a product of two sines it is replaced by minus a product of the two sinh's. For example, for the trig term \sin^2, the hyperbolic identity uses $-\sinh^2$ (this is called Osborn's rule). This is not always straightforward because the minus sign is often hidden.

OTHER GEOMETRIES

What is **projective geometry**?

Projective geometry is a branch of geometry that deals with the properties of geometric objects under projection; it was formerly called "higher" or "descriptive" geometry. Projection includes the transformation of points and lines in one plane onto another plane. This is done by connecting the corresponding points on the planes using parallel lines—similar to shining a light on a person's profile and producing a silhouette on a nearby wall. Each point on the wall is a "projection" of the person's head.

What is **non-Euclidean geometry**?

Non-Euclidean geometry is a branch of geometry that deals with Euclid's fifth postulate—the "parallel postulate" that only one line is parallel to a given line through a given external point. This postulate is replaced by one of two alternative postulates. The results of these two alternative types of non-Euclidean geometry are similar to those in Euclidean geometry, except for those propositions involving explicit or implicit parallel lines. (For more information about non-Euclidean geometry, see "History of Mathematics.")

What are **hyperbolic** and **elliptical geometry**?

Hyperbolic and elliptical geometry are the non-Euclidean alternative geometries mentioned above. The first alternative is to allow two parallels through any particular external point—or *hyperbolic geometry*. This studies two rays extending out in either direction from a point P, and not meeting a line L; thus, the rays are considered to be parallel to L. This also helps prove the theorem that the sum of the angles of a triangle is less than 180 degrees. It is called hyperbolic because a line in the hyperbolic plane has two points at infinity; this is similar to drawing a hyperbola that has two asymptotes.

The second alternative, called elliptical geometry, has no parallels to a given line L through an external point P. In addition, the sum of a triangle's angles is greater than 180 degrees. Sometimes called Riemann's geometry (who developed the idea even further; see below), it is called elliptic in general because a line in the plane of this geometry has no point at infinity (where parallels may intersect it), which is similar to an

ellipse that has no asymptotes. (For more information about hyperbolas, ellipses, and asymptotes, see elsewhere in this chapter).

Who developed the two alternatives to Euclidean geometry?

Hyperbolic geometry was first announced by Russian mathematician Nikolai Ivanovich Lobachevski (1792–1856; also seen as Lobatchevsky) in 1826. He challenged Euclid's fifth postulate that one and only one line parallel to a given line can be drawn through a fixed point external to the line. Instead, he developed a self-consistent system of geometry in which the "flawed" postulate was replaced by one allowing more than one parallel through the fixed point.

This idea was already developed independently by Hungarian János (or Johann) Bolyai (1802–1860) in 1823 (after several attempts to prove the Euclidean parallel postulate, he developed his system by assuming that a geometry could be constructed without the parallel postulate) and German mathematician, physicist, and astronomer Karl Friedrich Gauss (1777–1855; also seen as Johann Carl [or Karl] Friedrich Gauss) in 1816. But, as often happens in science and mathematics, the person who publishes an idea first gets most of the credit; this time, Lobachevski was the first to publish.

What is topology?

In general, topology is a branch of mathematics that examines patterns of geometric figures based on position and relative position without regard to size. It has also been

Who further developed the ideas in non-Euclidean geometry?

In 1854 German mathematician Georg Friedrich Bernhard Riemann (1826–1866) presented several new general geometric principles, laying the foundations of a non-Euclidean system of geometry called elliptical, or Riemann geometry. In this he represented elliptic space and generalized the work of German mathematician Karl Friedrich Gauss in differential geometry. This would eventually provide the basic tools for the general theory of relativity's mathematical expression. (For more about Riemann, see "History of Mathematics.")

called "rubber-sheet" geometry because objects examined in topology are under continuous transformation (also called topological transformation or homeomorphism). The shapes can be stretched, bent, and twisted in various ways, as long as they are not torn or cut as they are changed. For example, a circle is considered to be topologically equivalent to an ellipse (it can be deformed by stretching) and a sphere is equivalent to an ellipsoid. This includes changing an object based on a one-to-one correspondence. (For more about changing such shapes and one-to-one correspondence, see elsewhere in this chapter.)

MATHEMATICAL ANALYSIS

ANALYSIS BASICS

What is **mathematical analysis**?

Mathematical analysis is a branch of mathematics that uses the concepts and methods common to the field of calculus. The key to mathematical analysis is the use of infinite processes; in turn, that involves passage to a limit, or, in other words, the basic branch of *calculus*. For example, a circle's area can be thought of as the limiting value of the areas of regular polygons within the circle, as the number of the polygons' sides increase indefinitely. The reason for calculus is simple: It is one of the most powerful and flexible tools not only in mathematics but in virtually every scientific field.

What are the various **forms of calculus**?

As mentioned above, calculus uses infinite processes that involve passage to a limit. In order to solve its functions, it includes a formal set of mathematical rules applied to changing quantities. Many mathematicians break down calculus into two branches. The first is *integral calculus,* or the general problems of measuring length, area, volume, and other quantities as limits by approximating polygonal shapes; it finds the quantity when the rate of change is known. The second branch deals with discovering the tangent line to a curve at a specific point—or *differential calculus*. In this case, the calculus determines the quantity's rate of change. There also are several other forms of mathematical analysis that depend on calculus methods, such as vector analysis, tensor analysis, differential geometry, and complex variable analysis.

What culture took the first steps in the **development of mathematical analysis**?

Mathematical analysis—and, thus, the ideas of calculus—took centuries to develop. Probably the first to present some solid concepts in the field were the Greeks whose

most important contribution was the method of exhaustion (expanding the measurements of an area to take in more and more of the required area).

For example, Zeno of Elea (c. 490–c. 425 BCE) based many problems on the infinite; Leucippus of Miletus (fl. c. 435–c. 420 BCE), Democritus of Abdera (460–370 BCE; a student of Leucippus who also proposed an early theory about how the universe was formed), and Antiphon (c. 479–411 BCE; who some historians believe tried to square the circle) would all contribute to the method of exhaustion. Eudoxus of Cnidus (c. 400–347 BCE) would be the first to use the method on a scientific basis. Archimedes (c. 287–212 BCE; Hellenic)—considered one of the greatest Greek mathematicians—took mathematical analysis one step further: He more fully developed the theory presented by Eudoxus that would eventually lead to integral calculus.

What is considered one of **Archimedes's** most **significant contributions** to mathematics?

Archimedes made many significant contributions to mathematics, though not all mathematicians would agree with the label "most significant." But one of his contri-

butions *did* advance the field of calculus by showing that the area of a segment of a parabola is 4/3 the area of a triangle with the same base and vertex (endpoint), and 2/3 the area of the circumscribed parallelogram.

To figure this out, he constructed an "infinite" sequence of triangles (or wedges), finding the area of segments composing the parabola. He began with the first area, *A,* then added more triangles between the existing ones and the parabola to get areas of:

$A, A + A/4, A + A/4 + A/16, A + A/4 + A/16 + A/64$, and so on.

Based on his iterations, he determined the following (the first time anyone had determined the summation of an infinite series; for more about infinite series, see below):

$$A(1 + 1/4 + 1/4^2 + 1/4^3 + \ldots) = A(4/3)$$

English mathematician Isaac Barrow's work on tangents laid important groundwork for Isaac Newton's later work on calculus. *Library of Congress.*

Archimedes also applied this *method of exhaustion* (not literally becoming tired, but close to it) to approximate the area of a circle, which, in turn, led to a better approximation of pi (π). Using such integrations, he also determined the volume and surface area of a sphere and cone, the surface area of an ellipse, and many others. His work is considered the first steps toward integration that would eventually lead to integral calculus. (For more information about Archimedes and his wedges, see "Geometry and Trigonometry.")

What did **Isaac Newton** contribute to **mathematical analysis**?

English mathematician and natural philosopher (otherwise called a physicist) Isaac Newton (1642–1727) was one of the greatest scientists who ever lived. Overall, he contributed to physics (such as the discovery of his three famous laws of motion); fluid dynamics (fluid motion); the union of terrestrial and celestial mechanics using the principle of gravitation—thus explaining Kepler's laws of planetary motion; and he even explained the principle of universal gravitation.

By 1665 Newton had not only begun his work on differential calculus, but he also had published one of his greatest scientific works—*Philosophiae naturalis principia mathematica* (*The Mathematical Principles of Natural Philosophy*), often shortened to

The Principia or just *Principia*. In it he presents his theories of motion, gravity, and mechanics, thus explaining the bizarre orbits of comets, tides and tidal variations, the movement of the Earth's axis (called precession), and motion of the Moon. And although he used calculus to find many of his scientific results, Newton also explained them using older geometric methods in the book. After all, calculus was very new. Perhaps he was the first scientist-writer to make sure everyone understood what he was proposing.

Sir Isaac Newton, best known for his laws of motion, also made important contributions in areas such as fluid dynamics and celestial mechanics. *Library of Congress.*

Why was **Gottfried Wilhelm Leibniz** important to calculus?

German philosopher and mathematician Baron Gottfried Wilhelm Leibniz (1646–1716) was a contemporary of Isaac Newton. He is considered by some to be a forgotten mathematician, being overshadowed by Newton, but his contributions to mathematics were just as important. In addition to many other contributions, he (independently of Newton) developed infinitesimal calculus and was first to describe it in print. Because his work on calculus was published three years before Isaac Newton's, Leibniz's system of notation was universally adopted, including the notation for an integral. In 1684, he published *Nova methodus pro maximis et minimis, itemque tangentibus,* a work detailing differential calculus and containing the familiar *d* (or *d/dx*) notation, along with the rules for calculating the derivatives of powers, products, and quotients.

What are the different **categories** of **modern calculus**?

Modern calculus is divided into numerous types. The following lists just a few of these categories:

Basic calculus—Basic calculus is the branch of mathematics concerned with limits and with the differentiation and integration of functions. There is also *advanced calculus,* which takes an even more complex view of calculus, with an emphasis on proofs.

Differential calculus—Differential calculus deals with the variation of a function with respect to changes in the independent variable(s). It does this by determining derivatives and differentials.

Integral calculus—Integral calculus (logically) deals with integration and its application to solve differential equations; it is also used to determine areas and volumes.

Other various analyses—Other parts of calculus entail various types of analyses, such as vector, tensor, and complex analyses, and differential geometry.

Also remember that the term "calculus" is a generic name for any area of mathematics dealing with calculation; thus, arithmetic could be called the "calculus of numbers." It is also why there are such terms as *imaginary calculus*—or a method of looking at the relationships between real or imaginary quantities using imaginary symbols and quantities in algebra—that *do not* mean the type of calculus discussed elsewhere in this chapter.

SEQUENCES AND SERIES

What is a **sequence**?

A sequence is defined as a set of real numbers with a natural order. A sequence is usually included in brackets ({}), with the *terms,* or parts of a sequence, separated by commas. For example, if a scientist collects weather data every day for many days, the first day of collecting can be written as x_1 data; then x_2 for the second day, and so on until x_n, in which n is the eventual number of days. This can be written as $\{x_1, x_2, \ldots x_n\}_{n \geq 1}$. In general, the sequence of numbers in which x_n is the *nth* number is written using the following notation: $\{x_n\}_{n \geq 1}$.

A sequence can get larger or smaller. For example, in the sequence for $\{2^n\}_{n \geq 1}$, the solution is $2 \leq 4 \leq 8 \leq 16 \leq 32$, and so on, with the numbers getting larger. Whereas, for $\{1/n\}_{n \geq 1}$, the sequence becomes $1 \geq 1/2 \geq 1/3 \geq 1/4 \geq 1/5$, and so on, with the numbers getting progressively smaller. This does not mean that sequences only get progressively larger and smaller; certain solutions for sequences include a mix of the two.

What is the **range of a sequence**?

The range of a sequence is merely a set that defines the sequence. The range is usually represented by the set $\{x_1\}$, $\{x_2\}$, $\{x_3\}$, and so on; it is also written as $\{x_n; n = 1, 2, 3, \ldots\}$. For example, in the question above, the data for each day that the scientist collects from his weather experiment is the range. Another example is the range of the sequence $\{(-1)^n\}_{n \geq 1}$: It is the two-element set $\{-1, 1\}$.

When is a **sequence monotonic**?

A sequence is called monotonic if one of the following properties hold: In the sequence $\{x_n\}_{n \geq 1}$, it is increasing if and only if $x_n < x_{n+1}$ for any $n \geq 1$, or it is decreasing if and only if $x_n > x_{n+1}$ for any $n \geq 1$.

For example, in order to check that the sequence $\{2^n\}_{n \geq 1}$ is increasing: Let $n \geq 1$; that gives $2^{n+1} = 2^n \, 2$. Because 2 is greater than 1, which means that $1 \times 2^n < 2 \times 2^n$; thus $2^n < 2^{n+1}$, which shows the sequence is increasing.

What are the **bounds of a sequence**?

Once again, take the sequence $\{x_n\}_{n \geq 1}$. This sequence is *bounded above* if and only if there is a number M such that $x_n \leq M$ (the M is called an *upper-bound*). In addition, the sequence is *bounded below* if and only if there is a number m such that $x_n \geq m$ (the m is called a *lower-bound*). For example, the sequence $\{2^n\}_{n \geq 1}$, is bounded below by 0 because it is positive, but not bounded above.

The sequence is usually said to be merely *bounded* (or "bd" for short) if both of the properties (upper- and lower-bound) hold. For example, the harmonic sequence $\{1, 1/2, 1/3, 1/4 \dots\}$ is considered bounded because no term is greater than 1 or less than 0; thus, the upper- and lower-bounds, respectively, apply.

What is a **limit of a sequence**?

The limit of a sequence is simply the number that represents a kind of equilibrium reached in the sequence. It is also phrased "approaches as closely as possible." (Limit is also a term used in calculus in relation to a function; see elsewhere in this chapter.)

What are the concepts of **convergent** and **divergent sequences**?

Convergent and divergent sequences are based on the limit of a sequence. A convergent sequence, the one most commonly worked on in calculus, means that one mathematical sequence gets close to another and eventually approaches a limit (convergence can also apply to curves, functions, or series). This is seen visually when a curve approaches the x or y axes but does not quite reach it. For example, take the sequence of numbers used above, or $\{x_n\}_{n \geq 1}$. Often the numbers come closer and closer to a

number we'll call L; written in calculus, $x_n \approx L$. If the numbers do come closer, the sequence is said to be *convergent* and has a limit equal to L. Conversely, if the sequence is not convergent, it is called *divergent*.

Most mathematicians and scientists are not only interested in how a sequence converges (or diverges), but also how fast it converges, which is called the *speed of convergence*. There are several basic properties of the limits of a sequence, including that the limit of a convergent sequence is unique, every convergent sequence is bounded, and any bounded increasing or decreasing sequence is convergent.

How are **limits** written in terms of **convergent and divergent sequences**?

Limits of convergent and divergent sequences are written as follows (the figure eight on its side is the symbol for infinity):

$$\lim_{n \to \infty} x_n = L$$

or $x_n \longrightarrow L$ when $n \longrightarrow \infty$.

What is a **series**?

A series is closely related to the sum of numbers. It is actually used to help add numbers; therefore, in a sequence it can be the indicated sum of that sequence. In general, the idea is to start with a number, then do something to that number to get the next number, then do the same to that number to get the next number, and so on. For example, a *finite series* with six terms is $2 + 4 + 6 + 8 + 10 + 12$, in which 2 is added to each number to get the next number. An example of an *infinite series* is one with the notation $1/2^n$, with $n \geq 1$, or $1/2 + 1/4 + 1/8 + \ldots$ (and so on).

To see a series written in notation, if set $\{x_n\}$ is a sequence of numbers being added, and set $s_1 = x_1$, then $s_2 = x_1 + x_2$; $s_3 = x_1 + x_2 + x_3$; and so on. And for $n \geq 1$, a new sequence is made, $\{s_n\}$, called the *sequence of partial sums*, or $s_n = x_1 + x_2 + \ldots + x_n$.

What is the **notation** commonly used for **series and sequences**?

When looking at a sequence and series, we need to distinguish between the ones we want to add, and the ones we do not want to add. The common symbol used for addition with a sequence or a series is the summation symbol, seen as the symbol Σ. When discussing a series, this symbol is used to mean the sum of numbers. For example, for the series called x_n:

$$\sum_{n \geq 1} x_n$$

Also note that for any given series Σx_n, we have to also remember to associate it with the sequence of partial sums ($s_n = x_1 + x_2 + \ldots + x_n$).

What are arithmetic series and sequences?

An arithmetic series—also called arithmetic progression—is one of the simpler types of series in mathematics. In such a series, each new term is the previous number plus a given number; it is usually seen in the form of $a + (a + d) + (a + 2d) + (a + 3d) +, \ldots, a + (n - 1)d$. An example of an arithmetic series would be $2 + 6 + 10 + 14 + \ldots$, and so on, in which d is equal to 4. The *initial term* is the first one in the series; the difference between each term (d, or 4 in this case) is called the *common difference*.

An arithmetic sequence is usually in the form of $a, a + d, a + 2d, a + 3d, \ldots$, and so on, in which a is the first term and d is the constant difference between the two successive terms throughout. An example of an arithmetic sequence is (1, 4, 7, 10, 13 ...), in which the difference is always a constant of 3. The notation for arithmetic sequences is:

$$a_{n+1} = a_n + d.$$

What does it mean if a **series is convergent**?

Convergence of a series is related to the convergence of a sequence, but don't confuse them. The convergence of the sequence of partial sums (usually written as $\{s_n\}$) differs greatly from the convergence of a sequence of numbers (usually written as $\{x_n\}$). For example, the series $\Sigma\, x_n$ (and its associated sequence of partial sums, or $\{s_n\}$), is convergent if and only if the sequence $\{s_n\}$ is convergent. Thus, the total sum of the series is the limit of the sequence $\{s_n\}$, seen as:

$$\lim_{n \to \infty} S_n = \sum_{n=1}^{n=\infty} x_n$$

What is a **geometric sequence**?

A geometric sequence—or geometric progression—is a finite or infinite sequence of real numbers with the ratio between two consecutive terms being constant (called the *common ratio* or r); in this case, each term is the previous term multiplied by a given number. In a geometric sequence, the formula for the nth term, in terms of the first number and the common ratio, becomes:

$$a_n = a_1 r^{n-1}$$

Here the first term a_1, r is the common ratio, and the number of terms is n. For example, for the numbers 2, 6, 18, 54, 162, a_1 is 2, r is 3, and n is 5. The equation then becomes:

$$2 \times 3^{5-1} = 162.$$

What is a **geometric series**?

If we add the numbers in a geometric sequence, we end up with a geometric series. A geometric series is obtained when each term is determined from the preceeding one by multiplying by a common ratio; there is a constant ratio between terms. For example, $1 + 1/2 + 1/4 + 1/8 + \ldots$, and so on, is a geometric series because each term is determined by multiplying the preceding term by 1/2. To find the sum of a geometric series, the formula is: Sum = $a(r^n - 1) / (r - 1)$ or $a(1 - r^n) / (1 - r)$, in which a is the first term, r is the common ratio, and n is the number of terms.

For example, to find the sum of the first six terms of a series represented by $2 + 6 + 18 + 54 + 162 + 486$, define $a = 2$; $r = 3$; and $n = 6$. Substitute the numbers: Sum = $2(3^6 - 1) / 3 - 1 = 729 - 1 = 728$. We could also have determined this number based on the first few numbers, such as $2 + 6 + 18 + 54$, as long as we knew the common ratio, the first number, and how many numbers in the series we wanted to add. This is something that can easily be determined based on just these four numbers.

CALCULUS BASICS

What is **"the" calculus**?

"The" calculus is a branch of mathematics that deals with functions; another name for calculus is infinitesimal analysis. It evaluates constantly changing quantities, such as velocity and acceleration; values interpreted as slopes of curves; and the area, volume, and length objects bounded by curves (remember, "curves" can also mean straight lines). It involves infinite processes that lead to *passage to a limit,* or the approaching of an ultimate, usually desired value. The tools of the calculus include differentiation (differential calculus, or finding a derivative) and integration (integral calculus, or finding the indefinite integral), both of which are foundations for mathematical analysis.

What is a **limit** in **calculus**?

A limit is a fundamental concept in calculus. Unlike a limit mentioned above (as in a series or sequence), a limit of a function in calculus takes on a somewhat different meaning. In particular, a limit of a function can be described as the following: If $f(x)$ is a function defined around a point c (but may not be at c itself), the formal limit equation becomes:

$$\lim_{x \to c} f(x) = L$$

Thus, the number L is called the limit of $f(x)$ when x goes to c.

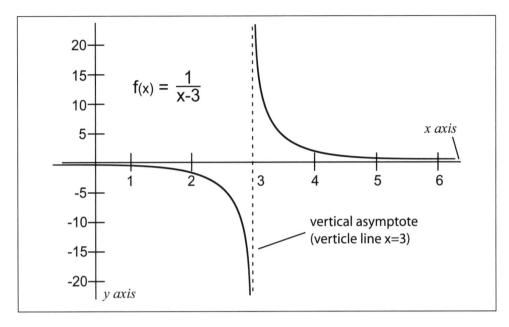

$$f(x) = \frac{1}{x-3}$$

x axis

vertical asymptote
(verticle line x=3)

y axis

Graph of an equation with infinity limited by x = 3.

What are **left** and **right limits**?

When a function is not defined around the point c (see the notation in the equation above), but only to the left or right of point c, then the limits are called the left limit and right limit at c. The formal left- and right-limit equations are the same as the usual limit equation, except below the "lim" sign, the $x \longrightarrow c$ is written as $x \longrightarrow c-$ for the left limit and $x \longrightarrow c+$ for the right limit.

How is **infinity treated** when **discussing limits**?

Defining infinity is a definite part of calculus, especially when discussing limits and "negative" and "positive" infinity. Whenever the inverse of a small number is taken, a large number is generated, and vice versa. In calculus, this is written as:

$1/0 = \pm\infty$

But $\pm\infty$ are no ordinary numbers, because they do not obey the usual rules of arithmetic, such as $\infty + 1 = \infty$; $\infty - 1 = \infty$; $2 \times \infty = \infty$; and so on. Therefore, in calculus functions, and thus limits, infinity is treated much differently. For example, for the function $f(x) = 1 / x - 3$, when $x \longrightarrow 3$, then $x - 3 \longrightarrow 0$. The limit function then becomes:

$$\lim_{x \to 3} f(x) = \tfrac{1}{0-} = -\infty$$

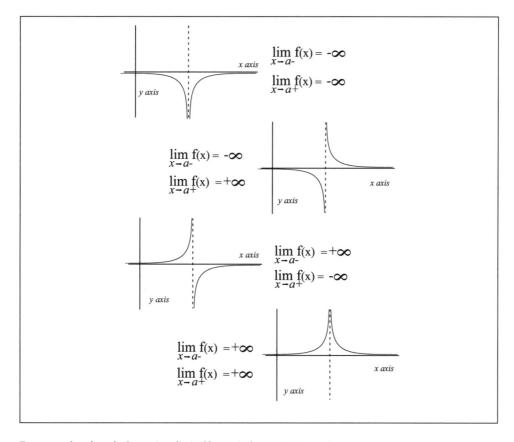

Four examples of graphed equations limited by vertical asymptotes.

$$\lim_{n \to 3+} f(x) = \tfrac{1}{0+} = +\infty$$

This example can be seen in the accompanying graph.

What are **vertical asymptotes** in association with limits?

Using the example above, as x gets closer to 3, the points on the graph get closer to the vertical line $x = 3$ (seen as a dashed line on the graph on p. 218). This line is called a *vertical asymptote*. The above chart shows four of the vertical asymptotes for the given function $f(x)$ and their notations.

What are **horizontal asymptotes** in association with limits?

A *horizontal asymptote* is similar to a vertical asymptote, but it is associated with the y axis. For example, for the function

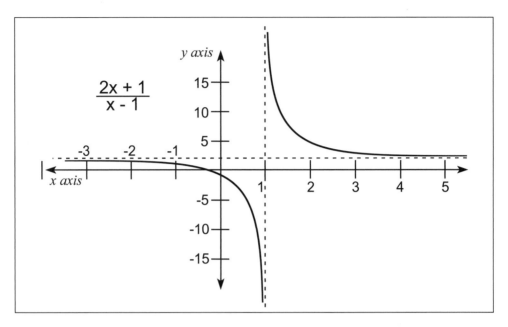

A graphed equation limited by horizontal asymptote y = 2.

$$f(x) = \frac{2x+1}{x-1}$$

gives a solution of:

$$\frac{2 + \frac{1}{x}}{1 - \frac{1}{x}}$$

and, thus, the implied limit is:

$$\lim_{x \to \pm\infty} f(x) = \frac{2+0}{1-1} = 2$$

This is because we know that:

$$\lim_{x \to -\infty} \frac{1}{x} = 0 \; and \; \lim_{x \to \pm\infty} \frac{1}{x} = 0$$

And as x gets larger (or closer to $\pm\infty$) the points on the graph come closer to the horizontal line $y = 2$ (seen as a dashed line on the graph above), which is called the horizontal asymptote.

What is **infinitesimal calculus**?

Infinitesimal may mean infinitely small in most people's dictionaries, or bring up thoughts of subatomic particles. To those who study arithmetic, however, it may mean numbers greater in absolute value than zero, yet smaller than any positive real number.

> ## What does the strange symbol "iff" mean in calculus?
>
> The symbol (or "word") *iff* actually is shorthand for "if and only if." It is not only mathematics that depends on the iff, but also philosophy, logic, and many technical fields. It is usually italicized; in addition, the phrase "P is necessary and sufficient for Q" is also seen as "Q iff P." The corresponding logical symbols for "if and only if" are ↔ and ≡.

But for those who study calculus, it is an area of mathematics pioneered by Gottfried Leibniz. His idea was based purely on the concept of infinitesimals; this was in opposition to the calculus of Isaac Newton, who based his calculus on the concept of the limit. Although historically the emphasis was placed on the minute, modern infinitesimal calculus actually has little to do with infinitely small quantities.

What does **continuous** and **discontinuous** mean in calculus?

When talking about polynomial functions, we know that a polynomial function $P(x)$ satisfies the limit function:

$$\lim_{x \to a} P(x) = P(a)$$

in which a represents all real numbers. This is called continuity.

But if $f(x)$ is a function on an interval around a, then $f(x)$ is continuous at a iff (see above for the definition of "iff"):

$$\lim_{x \to a} f(x) = f(a)$$

if it is not, then $f(x)$ is called discontinuous at a.

What is the **concept of bound**?

Similar to sequences (see elsewhere in this chapter), in calculus bounds are divided into the upper (either greater than or equal to every other number in a set of numbers; or greater than or equal to all the partial sums of a sequence) or lower (less than or equal to every other number). The symbol for infinity (∞) is used to denote a set of numbers *without bound,* or that increase or decrease "to infinity." In calculus, the bounds can be divided even more into greatest or least. For example, the greatest lower bounds and least upper bounds are of special interest to calculus, as those numbers may or may not be found within a set.

DIFFERENTIAL CALCULUS

What is **differential calculus**?

Differential calculus is the part of "the" calculus that deals with derivatives. It deals with the study of the limit of a quotient, usually written as $\Delta y / \Delta x$, as the denominator (Δx) approaches zero, with x and y as variables.

What is the **derivative of a function**?

One of the most important, core concepts in modern mathematics and calculus is the derivative of a function—or a function derived from another function. A derivative is also expressed as the limit of $\Delta y / \Delta x$, also said as "the derivative of y with respect to x." It is actually the rate of change (or slope on a graph) of the original function; the derivative represents an infinitesimal change in the function with respect to the parameters contained within the function.

In particular, the process of finding the derivative of the function $y = f(x)$ is called differentiation. The derivative is most frequently written as dy / dx; it is also expressed in various other ways, including $f'(x)$ (said as the derivative of a function f with respect to x), y', $Df(x)$, $df(x)$, or $D_x y$. It is important to note that the differentials, written as dy and dx, represent singular symbols and not the products of the two symbols. Not all derivatives exist for all values of a function; the sharp corner of a graph, in which there is no definite slope—and thus no derivative—is an example.

What is the **standard notation** for the **derivative**?

The following represents the definition of the derivative of $f(x)$ (note: in order for the limit to exist, both $\lim_{h \to 0+}$ and $\lim_{h \to 0-}$ must exist and be equal; thus, the function must be continuous):

For $f(x)$'s derivative at point x_0:

$$\frac{df}{dx}(x_0) = f'(x_0) = \lim_{x \to x_0} \frac{f(x) - f(x_0)}{x - x_0} = \lim_{h \to 0} \frac{f(x_0 + h) - f(x_0)}{h}$$

For $f(x)$'s derivative at $x = a$:

$$f'(a) = \lim_{x \to a} \frac{f(x) - f(a)}{x - a} = \lim_{h \to 0} \frac{f(a + h) - f(a)}{h}$$

Is there a **formula** for the **inverse of a derivative**?

Yes. In this case, the derivative of the inverse function represents the inverse of $y(x)$—or $x(y)$:

$$\frac{dy}{dx} = \frac{1}{\frac{dx}{dy}}$$

What are the **two ways** of **looking at the derivative**?

There are two major ways of looking at the derivative—the geometrical (or the slope of a curve) and the physical (the rate of change). The derivative was historically developed from finding the tangent line to a curve at a point (geometrically); it eventually became the study of the limit of a quotient usually seen as the change in x and y ($\Delta y/\Delta x$). Even today, mathematicians still debate which is the most useful and best way to describe a derivative.

Geometrically, after determining the slope of a straight line through two points on a graph of a function, and the limit where the change in x approaches zero, the ratio becomes the derivative dy/dx. This represents the slope of a line that touches the curve at a single point—or the tangent line.

Physically, the derivative of y with respect to x describes the rate of change in y for a change in x. The independent variable, in this case x, is often expressed as time. For example, velocity is often expressed in terms of s, the distance traveled, and t, the elapsed time. In terms of average velocity, it can be expressed as $\Delta s/\Delta t$. But for instantaneous velocity, or as Δt gets smaller and smaller, we need to use limits—or the instantaneous velocity at a point B is equal to:

$$\lim_{\Delta t \to 0} \frac{\Delta s}{\Delta t}$$

What are some **examples** of **derivatives of "simple" functions**?

The following lists some derivatives of "simple" functions:

$$\frac{d}{dx} x^n = nx^{n-1}$$

$$\frac{d}{dx} \ln|x| = \frac{1}{x}$$

What are some **simple derivatives** as functions of the **variable x**?

There are simple derivatives as functions of the variable x. In this case, u and v are functions of the variable x, and n is a constant:

$$\frac{d}{dx}(c) = 0$$

$$\frac{d}{dx}(x) = 1$$

$$\frac{d}{dx}(x^n) = nx^{n-1}$$

$$\frac{d}{dx}(u \pm v) = \frac{du}{dx} \pm \frac{dv}{dx}$$

$$\frac{d}{dx}(cu) = c\frac{du}{dx}$$

$$\frac{d}{dx}(xu) = u\frac{dv}{dx} + v\frac{du}{dx}$$

$$\frac{d}{dx}\left(\frac{u}{v}\right) = \frac{v\frac{du}{dx} - u\frac{dv}{dx}}{v^2}$$

What is an **example** of **computing the derivative**?

The following solves the derivative at $x = a$ for the function $f(x) = x^2$, using the definition for a derivative (see above):

$$\frac{f(a + h) - f(a)}{h} = \frac{(a + h)^2 - a^2}{h} = \frac{2ah + h^2}{h} = 2a + h$$

$$\lim_{h \to 0} \frac{f(a + h) - f(a)}{h} = 2a$$

Are there **higher derivatives** of a **function**?

Yes, there are higher derivatives of a function, often referred to as higher order derivatives. The "initial" derivative is often written as $f'(x)$, but the " ' " is assumed in most cases. The next derivative is the second derivative (second order derivative), most often written as $f'(x)$; the next is the third derivative, or third order derivative, most often written as $f'''(x)$; fourth derivative, or fourth order derivative, most often written as $f^{(4)}(x)$; and so on. The notation for the higher derivatives, or the nth derivative, is as follows:

$d/dx \, (d^{n-1}y/dx^{n-1})$
$= d^n y/dx^n$
$= f^{(n)}(x)$
$= y^{(n)}$

What is an example of the second derivative?

The second derivative is actually a function's derivative's derivative. In other words, the function's derivative may also have its own derivative, called the second derivative or second order derivative. If we let $y = f(x)$, the second derivative becomes $d/dx(dy/dx)$. This is equal to d^2y/dx^2, further represented by the symbols $f'(x)$ or y'. One good example of a second order derivative is acceleration—it is actually the second derivative of a change in distance. In other words, the first derivative gives instantaneous velocity (see above) while the second derivative gives acceleration.

This equation is also seen written as $D^n(y) = d^ny/dx^n$.

What is a **partial derivative**?

Partial derivatives (seen written as the symbol ∂) are derivatives of a function containing multiple variables that have all but the variable of interest held fixed during the differentiation. Thus, when a function $f(x, y, \ldots)$ depends on more than one variable, the partial derivative can be used to specify the derivative with respect to one or more variables. There are other terms, too: Partial derivatives that involve more than one variable are called *mixed partial derivatives*. And a differential equation expressing one or more quantities in terms of partial derivatives is called, logically, a *partial differential equation*. These equations are well known in physics and engineering, and most are notoriously difficult to solve.

What are the **product, quotient, power,** and **chain rules** for derivatives?

There are numerous rules for derivatives of certain combinations of functions, including the product, quotient, power, and chain rules. The following lists their common notation:

For a product:

$$\frac{d}{dx}[f(x)g(x)] = f(x)g'(x) + f'(x)g(x)$$

where f' is the derivative of f with respect to x.

For a quotient:

$$\frac{d}{dx}\left[\frac{f(x)}{g(x)}\right] = \frac{g(x)f'(x) - f(x)g'(x)}{[g(x)]^2}$$

where, again, f' is the derivative of f with respect to x.

For a power:

$$\frac{d}{dx}(x^n) = nx^{n-1}$$

For a chain:

$$\frac{dy}{dx} = \frac{dy}{du} \cdot \frac{du}{dx}$$

or

$$\frac{dz}{dt} = \frac{\partial z}{\partial x}\frac{dx}{dt} + \frac{\partial z}{\partial y}\frac{dy}{dt}$$

where $\partial z / \partial x$ is a partial derivative.

What are the **derivatives** of **trigonometric functions**?

There are also derivatives of the six major trigonometric functions—sine, cosine, cotangent, cosecant, tangent, and secant (for more information on these trigonometric functions, see "Geometry and Trigonometry"). The following lists the formulas for those derivatives (in this case, the functions are written with respect to the variable u):

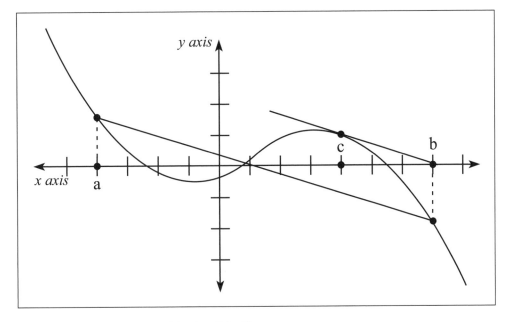

This graph illustrates the concept of the Mean-Value Theorem.

$$\frac{d}{dx}(\sin(u)) = \cos(u)\frac{du}{dx}$$

$$\frac{d}{dx}(\cos(u)) = -\sin(u)\frac{du}{dx}$$

$$\frac{d}{dx}(\tan(u)) = -\sec^2(u)\frac{du}{dx}$$

$$\frac{d}{dx}(\cot(u)) = -\csc^2(u)\frac{du}{dx}$$

$$\frac{d}{dx}(\sec(u)) = \sec(u)\tan(u)\frac{du}{dx}$$

$$\frac{d}{dx}(\csc(u)) = -\csc(u)\cot(u)\frac{du}{dx}$$

$$\frac{d}{dx}(\sin^{-1}(u)) = \frac{1}{\sqrt{1-u^2}}\frac{du}{dx}$$

$$\frac{d}{dx}(\cos^{-1}(u)) = \frac{1}{\sqrt{1-u^2}}\frac{du}{dx}$$

$$\frac{d}{dx}(\tan^{-1}(u)) = \frac{1}{1-u^2}\frac{du}{dx}$$

$$\frac{d}{dx}(\cot^{-1}(u)) = \frac{1}{1-u^2}\frac{du}{dx}$$

$$\frac{d}{dx}(\sec^{-1}(u)) = \frac{1}{|u|\sqrt{u^2-1}}\frac{du}{dx}$$

$$\frac{d}{dx}(\csc^{-1}(u)) = -\frac{1}{|u|\sqrt{u^2-1}}\frac{du}{dx}$$

INTEGRAL CALCULUS

What is the **integral calculus**?

The integral calculus is the part of "the" calculus that deals with integrals—both the integral as the limit of a sum and the integral as the antiderivative of a function (see below for more information). In general, the integral calculus is the limit of a sum of elements in which the number of the elements increase without bound, while the size of the elements diminishes. It is also considered the second most important kind of limit in the calculus (the first being limits in association with derivatives). It was originally developed by using polygons to approximate areas of geometrically shaped objects such as circles.

What are some **common integrals** in the calculus?

There are many standard integrals used in the calculus. The following lists only a few of those integrals in their common form:

$$\int a\,dx = ax$$

$$\int af(x)\,dx = a\int f(x)\,dx$$

$$\int (u \pm v \pm w \pm \cdots)\,dx = \int u\,dx \pm \int v\,dx \pm \int w\,dx \pm \cdots$$

$$\int u\,dv = uv - \int v\,du$$

$$\int f(ax)\,dx = \frac{1}{a}\int f(u)\,du$$

$$\int F\{f(x)\}\,dx = \int F(u)\frac{dx}{du}\,du = \int \frac{F(u)}{f'(x)}\,du$$

What is the **graphic representation** of the approximation of an **area under a curve** using integration?

It's easier to see the approximate area under a curve using integration by means of graphs—it all has to do with rectangles. The idea is to extend lines from the ends of the curve (here, $f(x)$) to the x-axis (or y-axis, depending on the curve); we'll call the total area under the curve Ω, then divide the entire area under the curve into equal-width sections (x_1, x_2, x_3, and so on) that are equal to parts of Ω (the subregions Ω_1, Ω_2, and so on). The next step is to figure out the area of a rectangle if each section was "cut off" below the curve and then above the curve. This creates rectangles defined by left- and right-end points. From the left- and right-sums, and a few more calculations, we can approximate the area of Ω. This can be seen graphically on the accompanying charts:

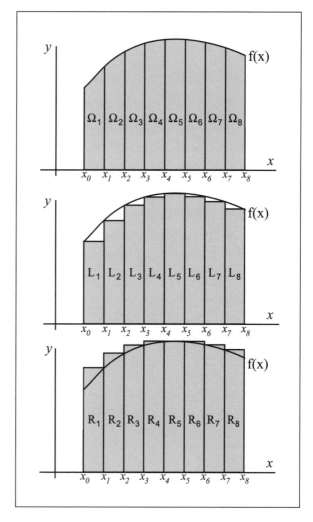

To calculate the area beneath a curve (top), you can first divide the area into equal parts with rectangles both beneath (middle) and above (bottom) the curve. Adjusting the width of the curves will result in an estimate that closely approximates the actual area.

What is the **definite integral**?

In actuality, the area is actually determined using limits. In the function $f(x)$, as n gets larger, the numbers determined by left (n) and right (n) will get closer and closer to the area Ω. This is seen as the following notation:

$$Area(\Omega) = \lim_{n \to \infty} LEFT(n) = \lim_{n \to \infty} RIGHT(n)$$

Thus, in the calculus terms, the area of the above graphic region is called the definite integral (also said as "the integral") of $f(x)$ from a to b, and is denoted by the following notation:

$$\int_a^b f(x)\,dx$$

The variable x can be replaced with any other variable. In other words, if the limits of integration (a and b) are specified, it is called a definite integral, and it can be interpreted as an area or a generalization of an area.

What are some **properties** of the **definite integral**?

There are several useful properties of the definite integral. *Theorem one* is based on the idea that if $f(x)$ and $g(x)$ are defined and continuous on $[a, b]$, except perhaps at a finite number of points, then the following apply:

$$\int_a^b (f(x) + g(x))\,dx = \int_a^b f(x)\,dx + \int_a^b g(x)\,dx$$

$$\int_a^b \alpha f(x)\,dx = \alpha \int_a^b f(x)\,dx$$

Theorem two is based on the idea that if $f(x)$ is defined and continuous on $[a, b]$, except at a finite number of points, then the following applies for any arbitrary numbers a and b, and any $c \in [a, b]$:

$$\int_c^c f(x)\,dx = 0$$

$$\int_a^b f(x)\,dx = \int_a^c f(x)\,dx + \int_c^b f(x)\,dx$$

$$\int_b^a f(x)\,dx = -\int_a^b f(x)\,dx$$

What is an **indefinite integral**?

From the above, we learned that when the limits of integration (in the case of a and b above) are specified, it is called a definite integral. Contrarily, if no limits are specified, it is called an *indefinite integral*. Thus, the indefinite integral is most often defined as a function that describes an area under the function's curve from some undefined point to another arbitrary point. This lack of a specified first point leads to an arbitrary constant (usually denoted as C) that is always part of the indefinite integral.

What is the Fundamental Theorem of the Calculus?

The Fundamental Theorem of the Calculus is the connection (or, more accurately, the bridge) between the integral and the derivative; in other words, it is another way of finding the area under a curve (see above) by evaluating the integral. In particular, if $F(x)$ is a function whose derivative is $f(x)$, then the area under the graph of $y = f(x)$ between the points a and b is equal to $F(b) - F(a)$. This is seen as the following notations:

$$\int_a^b f(x)\,dx = F(b) - F(a)$$

The number $F(b) - F(a)$ is also denoted by

$$[F(x)]_a^b \ \left(\text{or } F(x)\big|_a^b\right)$$

Thus,

$$\int_a^b f(x)\,dx = [F(x)]_a^b$$

What are **antiderivatives** and **antidifferentiation**?

An antiderivative is often interpreted as the same as an indefinite integral, but they actually do differ in definition. Using the notations above describing the Fundamental Theorem of the Calculus (see boxed text for definition of the Fundamental Theorem of the Calculus), the function $F(x)$ is an *antiderivative* of $f(x)$, described as also equal to the integral of $f(x)$. The actual process of finding $F(x)$ from $f(x)$ is called *integration* or *antidifferentiation*.

What is an **improper integral**?

An integral as seen above means that the function $f(x)$ needs to be bounded on the interval $[a, b]$ (both real numbers), and the interval also must be bounded. But an improper integral is one in which the function $f(x)$ becomes unbounded (called a type I improper integral) or the interval $[a, b]$ becomes unbounded ($a = -\infty$ or $b = +\infty$, which is called a type II improper integral).

Are there **double** and **triple integrals**?

Yes, there are double and triple integrals—and even multiple integrals in equations (\iint, \iiint, and so on). For example, the integration of a function of three variables, $w =$

$f(x, y, z)$, over a three-dimensional region R in xyz-space (three-dimensional space) is called a triple integral. The notation is as follows:

$$\iiint_R f(x,y,z)dV$$

To compute the iterated integral, we need to integrate with respect to z first, then y, then x. And when we integrate with respect to one variable, all the other variables are assumed to be constant.

DIFFERENTIAL EQUATIONS

What are **differential** and **ordinary differential equations**?

Logically, a *differential equation* is one that contains differentials of a function, with these equations defining the relationship between a function and one or more derivatives of that function. More specifically, differential equations involve dependent variables and their derivatives with respect to the independent variables. To solve such equations means to find a continuous function of the independent variable that, along with its derivatives, satisfies the equation. An *ordinary differential equation* is one that involves only one independent variable.

What do the **order** and **degree** of a **differential equation** mean?

The *order* of a differential equation is simply the highest derivative that appears in the equation. The *degree* of a differential equation is the power of the highest derivative term. (For more information about power, see "Math Basics.")

What are **implicit** and **explicit differential equations**?

As seen above, an ordinary differential equation is one involving x, y, y', y', and so on. Now add the idea that the order of the highest derivative is n. Thus, if a differential equation of order n has the form $F(x, y', y'', \dots y^{(n)}) = 0$, then it is called an *implicit differential equation*. If it is of the form $F(x, y', y'', \dots y^{(n-1)}) = y^{(n)}$, it is called an *explicit differential equation*.

What are some **first-order differential equations**?

A first-order differential equation is one involving the unknown function y, its derivative y', and the variable x. As seen above, these types of equations are usually referred to as explicit differential equations.

There are several types of first-order differential equations, including separable, Bernoulli, linear, and homogeneous (for explanations of the last two, see below). A first-order differential equation takes the form:

> ## What is an example of a differential equation?
>
> **A**n example of a differential equation involves letting y be some function of the independent variable t. Then a differential equation relating y to one or more of its derivatives is as follows:
>
> $$\frac{\partial}{\partial t} y(t) = t^2 y(t)$$
>
> In this equation, the first derivative of the function y is equal to the product of t^2 and the function y itself. This also implies that the stated relationship holds only for all t for which both the function and its first derivative are defined.

$$\frac{dy}{dx} = f(x, y)$$

What is a **linear differential equation**?

A linear differential equation is a first-order differential equation that has no multiplications among the dependent variables and their derivatives, which means the coefficients are functions of independent variables. Other terms associated with such differential equations are *nonlinear differential equations,* which have multiplications among the dependent variables and their derivatives, and *quasi-linear differential equations,* in which a nonlinear differential equation has no multiplications among all the dependent variables and their derivatives in the highest derivative term.

What are the **"solutions"** relative to **differential equations**?

There are three types of "solutions" when discussing differential equations:

General solution—This includes the solutions obtained from integration of the differential equations. In particular, the general solution of an nth order ordinary differential equation has n arbitrary constants from integrating n times.

Singular solution—Solutions that can't be expressed by the general solutions.

Particular solution—Solutions obtained from giving specific values to the arbitrary constants in the general solution.

What are the **"conditions"** in reference to **differential equations**?

In general, there are two "conditions" when discussing differential equations. The first are called *initial conditions,* in which constraints are specified at the initial point

What are first-order homogeneous and non-homogeneous linear differential equations?

These differential equations may be long-winded phrases, but they are actually types of *first-order differential equations*. A first-order homogeneous linear differential equation can be written in the notation as follows:

$$\left(\frac{\partial}{\partial t}y(t)\right) + a(t)y(t) = 0$$

The first-order homogeneous linear differential equations are those that place all terms that include the unknown equation and its derivative on the left-hand side of the equation; on the right-hand side, it is set equal to zero for all t.

The non-homogeneous linear differential equations are those that, after isolating the linear terms containing $y(t)$ and the partial differentials inside the above large parentheses on the left side of the equation, do not set the right-hand side identically to zero. It is often represented by one function, such as the $b(t)$ (see below). The standard notation is as follows:

$$\left(\frac{\partial}{\partial t}y(t)\right) + a(t)y(t) = b(t)$$

(usually in reference to time); such problems are called *initial value problems*. The other condition is the *boundary condition,* in which constraints are specified this time at the boundary points (usually in reference to space), with such problems called *boundary value problems*.

What techniques are used to **solve first-order differential equations**?

There are usually three major ways to solve first-order differential equations: analytically, qualitatively, and numerically. The analytical way includes the examples mentioned above, such as the linear and separable equations. Qualitative methods include such methods as defining the slope of a field of a differential equation. Finally, numerical techniques can be thought of as something close to Euler's method, a way of finding the largest divisor of two numbers.

What is **Euler's method**?

Swiss mathematician Leonhard Euler (1707–1783) was one of the most prolific mathematicians who ever lived. He developed Euler's method, which is a way of determining the largest divisor of two numbers. For example, if we want to find the largest divi-

sor of the numbers 6975 and 525, we consider one to be the large number and the other the small number. We already know both numbers are divisible by 0 and 5 (as they both end in 5), but how do we determine if they have a larger divisor? And if so, what is that number?

The key is to take the remainders of long division until we arrive at a remainder of zero. In this instance, we would first divide 6975 by 525; the answer comes out to 13 and a remainder:

$$
\begin{array}{r}
13 \\
525\overline{)6975} \\
-\ 525 \\
\hline
1725 \\
-1575 \\
\hline
150
\end{array}
$$

Take the remainder—in this case, 150—and then divide it into the 525. Take that remainder (which turns out to be 75) and divide it into the 150 remainder. The next iteration leads to no remainder (zero).

Large Number	Small Number	Remainder
6975	525	150
525	150	75
150	75	0

Thus, taking the number before we reached the zero—or 75—gives us the largest common divisor of both 6975 and 525.

What are **systems of differential equations**?

In real-life situations, quantities and their rate of change depend on more than one variable. For example, the rabbit population, though it may be represented by a single number, depends on the size of predator populations and the availability of food. In order to represent and study such complicated problems we need to use more than one dependent variable and more than one equation. Systems of differential equations are the tools to use. As with the first-order differential equations, the techniques for studying systems fall into the following three categories: analytic, qualitative, and numeric.

What is a **nonlinear differential equation**?

From above, we know that linear equations have specific rules; for example, the unknowns y, y', and so on, will never be raised to a power more than 1; they will not be in the denominator of a fraction; y times y' is never allowed (because two

Rabbit populations are affected not only by birth rates, but also by factors such as predation, disease, and available food supplies. Systems of differential equations may be used to take all these elements into consideration and estimate actual population numbers. *The Image Bank/Getty Images.*

unknowns multiplied together is, in a sense, a power 2 of unknowns); and they won't be inside another function (such as a sine).

But a nonlinear equation is much different, allowing for powers of $2(y' = y^2)$, multiplying differences ($y \times y' = x$); and even being inside another function ($y' = x \sin y$). Thus, the nonlinear equations are not as easy to solve as linear differential equations. But that does not mean they lack importance. In fact, because nonlinear equations are more realistic in describing real-life problems, they are much more interesting (and challenging) to mathematical and scientific researchers in many fields.

VECTOR AND OTHER ANALYSES

What is a **vector**?

A vector is considered to be an element of a linear or vector space. A vector is different than a point, as it represents the displacement between two points, not the physical location of a point in space. Vectors also define a direction; points do not. Vectors are usually represented by a line segment in a specific direction on a graph, with an arrow at one end of the segment. They can also be represented in several ways, including bold letters in an equation, for example, vectors \mathbf{A} and \mathbf{B}, and with arrows above the vector, such as \vec{x}.

What is the **component** of a **vector**?

A component of a vector is one with n numbers in a certain order. It is usually listed as $(x_1, x_2, ..., x_n)$, in which the numbers within the parentheses are called the components of the vector \vec{x}.

Logically, an infinite number of vectors can have the same components. For example, if the components are [3, 4], we know there are an infinite number of pairs of

points in the plane with x and y coordinates whose respective differences are 3 and 4. All these vectors are parallel to each other, equal, and have the same magnitude and direction. Therefore, any vector with components of a and b can be said to be equal to the vector $[a, b]$.

How is the **magnitude of a vector** determined?

The magnitude of a vector is equivalent to the length of a vector. Placing a pair of vertical lines (similar to the absolute value symbol) around a vector implies the magnitude of the vector. For example, if the variable V is used to represent a vector, then the expression |V| indicates the magnitude of the vector.

Vectors are used to indicate displacement between two points in space, such as, for example, the distance and direction travelled by a rocket from its launch point to its present position. *Reportage/Getty Images.*

What do **columns and rows** mean when discussing **vectors**?

Vectors can be described by columns and rows. For example, two-dimensional and three-dimensional vectors are usually represented as a single vertical column of numeric values. The following lists such columns in two and three dimensions:

A two-dimensional vector illustrated with columns and rows would look like this:

$$V = \begin{vmatrix} x \\ y \end{vmatrix}$$

And a three-dimensional vector would look like this using columns and rows:

$$V = \begin{vmatrix} x \\ y \\ z \end{vmatrix}$$

A row vector is usually used for problem solving in which a vector is described as $V = (x, y, z)$ during the specification of a problem. But note: Row vectors should not really be used with any mathematical descriptions.

How is the **length** of a **vector** calculated?

The length of a vector is calculated by taking the square root of the sum of the squares of each coordinate. For example, if the vector is defined by (x, y, z), then the length (L) of the vector is calculated this way:

$$L = \sqrt{x^2 + y^2 + z^2}$$

What is a **normalized vector**?

A normalized (or unit) vector is one in which the sum of the squares of all coordinates is equal to one. For example, the vector $(2, 2, 0)$ is not normalized; the vectors $(0.707, 0.707, 0.0)$ and $(1.0, 0.0, 0.0)$ are normalized. (An *outward normal* is another name for a normalized vector; it represents the direction that a polygon surface or vertex [endpoint] is facing.) Normalized (or unit) vectors are often seen written as \vec{x}, but more usually as \hat{x} (the $\hat{\ }$ is often referred to as a "hat").

A vector can be normalized by calculating the magnitude or length of the vector and dividing each coordinate by this value. For example, consider the following vector:

$$V = \begin{vmatrix} 3.0 \\ 4.0 \\ 0.0 \end{vmatrix}$$

The length of this vector is 5.0 (see above to determine how to solve for length of the vector); or $|V| = 5.0$. Thus, the value of the normalized vector is given by:

$$\frac{V}{|V|} = \begin{vmatrix} 3.0 \\ 4.0 \\ 0.0 \end{vmatrix} \cdot \frac{1}{5} = \begin{vmatrix} 0.6 \\ 0.8 \\ 0.0 \end{vmatrix}$$

In this case, 0.6 squared equals 0.36; 0.8 squared equals 0.64. Both added together with the zero equals 1. (If the vector is already normalized, then the value of |V| will be equal to one, and after division the vector will remain as it was before.)

How can **vectors** be **represented** in various **dimensional space**?

Vectors can be found in two-, three-, or multi-dimensional space. Two-dimensional vectors are seen visually on a graph as a line with an arrow connecting two points. A two-dimensional vector is defined by length and direction measured by the angles that the arrow makes with the x and y coordinate system axes; a vector in such a coordinate system is written as two components, (x, y).

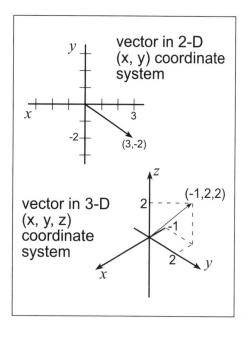

Vectors in a three-dimensional space are represented with three numbers, one along each coordinate axis. These are the coordinates of the arrow point, usually as (x, y, z) if the arrow starts at the origin. A more complex vector is one with multiple components, in which several different numbers in ordered n-tuples represent a vector. For example, $(4, 1, -2, 0)$ is an ordered 4-tuple representing a vector in four dimensions.

What is a **polar coordinate system**?

In effect, a polar coordinate system "wraps" a two-dimensional (Euclidean) coordinate system onto the surface of a sphere (for more information about coordinate systems, see "Geometry and Trigonometry"). A polar coordinate system examines a point in space defined in terms of its position and distance on a sphere with a unit radius. The center of the sphere is considered the origin; the first two coordinates are the longitude and latitude on the sphere; the third coordinate defines the distance of the point from the center of the sphere—the values of latitude, longitude, and height. In polar coordinates—it's easy to see on a globe of our own planet—latitude ranges from $+90$ to -90, longitude ranges from -180 to $+180$, and height ranges from zero to infinity. Height can also be negative: The North Pole is at coordinates of $(+90, \text{—}, +r)$ (the "—" means there is no longitude), the South Pole is at coordinates of $(-90, \text{—}, -r)$, and a point on the equator is at coordinates $(0, 0, r)$.

How are **vectors added**?

One way to combine vectors is through addition (or composition). This can be done algebraically or graphically. For example, to add the two vectors U [−3, 1] and V [5, 2], one can add their corresponding components to find the resultant vector R [2, 3]. One also can graph U and V on a set of coordinate axes, completing the parallelogram formed with U and V as adjacent sides, obtaining R as the diagonal from the common vertex of U and V.

How is the **product** of **two vectors** determined?

There are two distinct types of products of two vectors: scalar and vector products, sometimes called the inner and outer products (mostly in reference to tensor products; see below). The scalar (or dot) product of two vectors is not a vector because the product has a magnitude but not a direction. For example, if A and B are vectors (of magnitude A and B, respectively), their scalar product is: $A \bullet B = AB \cos \theta$, in which θ is the angle between the two vectors. This scalar quantity is also called the *dot product* of the vectors. These equations obey the commutative and distributive laws of algebra (for more information, see "Algebra"). Thus, $A \bullet B = B \bullet A$; $A \bullet (B + C) = A \bullet B + A \bullet C$. If A is perpendicular to B, then $A \bullet B = 0$.

The vector (or cross or skew) product of A and B is the length $C = AB \sin \theta$; its direction is perpendicular to the plane determined by A and B. In this case, this kind of multiplication does not follow the commutative law, as $A \bullet B = -B \bullet A$.

What is **vector analysis**?

Vector analysis is the calculus of functions with variables as vectors—a part of calculus also known for its derivative and integral equations. The components of a vector do not always have to be constants. They can also be variables and functions of variables, such as the position of a body moving through space represented by a vector whose x, y, and z components are all functions of time. In this case, the calculus can be used to solve such vector functions, which are also called vector analysis.

What is a **tensor**?

A tensor is a quantity that depends linearly on many vector variables. They are considered to be a set of n' components that are functions of the coordinates at any point in n dimensional space. Tensors are used in several fields of mathematics, such as the theory of elasticity (stress and strain) and mathematical physics, especially with regard to the theory of relativity.

What are some **other types** of **analysis**?

There are numerous other analyses in mathematics and the sciences besides vector analysis. At one time, the study of tensors was known as the absolute differential cal-

What is a linear combination?

A linear combination is a combination or the sum of two or more entities with each multiplied by some number (with not all the numbers being zero). Linear combinations of vectors, equations, and functions are common. For example, if x and y are vectors and a and b are numbers, then $ax + by$ is a linear combination. (For more information about equations and functions, see "Math Basics" and "Algebra.")

culus, but today it is simply called *tensor analysis.* Tensors were originally invented as the extensions of vectors. Tensor analysis is concerned with relations or laws that remain valid regardless of the coordinate system used to specify the quantities.

Complex analysis (or complex variable analysis) is the study of complex numbers and their derivatives, mathematical manipulations, and other properties. It is mostly used to find the solution to *holomorphic functions,* or those that are found in a complex plane, use complex values, and are differentiable as complex functions. Complex variables deal with the calculus of functions of a complex variable, incorporating differential equations and complex numbers; for example, one such variable of the form $z = x + iy$, in which x and y are real and the imaginary number $i = -1$. Complex-variable techniques have a great many uses in applied areas, such as electromagnetics.

Functional analysis is concerned with infinite-dimensional vector spaces and the mapping between them. It is also considered the study of spaces of functions. Finally, *differential geometry* is really a branch of geometry, but it includes the concepts of the calculus as applied to curves, surfaces, and other geometric entities. Originally it included the use of coordinate geometry; more recently it has been applied to other areas of geometry, such as projective geometry. In particular, differential geometry uses techniques of differential calculus to determine the geometric properties of manifolds (a topological space that resembles Euclidean space, but is not).

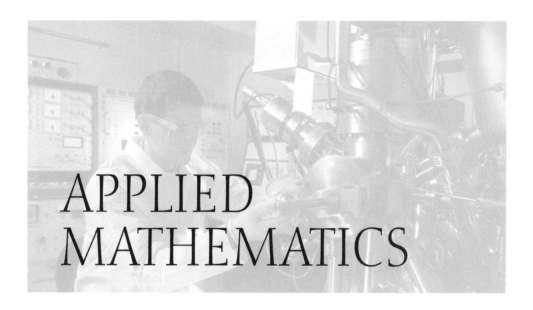

APPLIED
MATHEMATICS

APPLIED MATHEMATICS BASICS

What is **applied mathematics**?

Applied mathematics is not only concerned with using rigorous mathematical methods, but also applications of those methods. It entails a wide range of research in the worlds of biology, computer science, sociology, engineering, physical science, and many other fields, especially in the experimental sciences. In each case, applied mathematics is used by a researcher to more thoroughly understand a particular application or physical phenomena. The many uses of applied mathematics include numerical analysis, linear programming, mathematical modeling and simulation, the mathematics of engineering, mathematical biology, game theory, probability theory, mathematical statistics, financial mathematics, and even cryptography.

How did **applied mathematics grow** over time?

Historically speaking, applied mathematics was always concerned with using mathematics to solve problems in physics, chemistry, medicine, engineering, the physical sciences, technology, and biology. In fact, applied mathematics is older than pure mathematics, as it was used in areas that formed the core of early physics research, such as mechanics, fluid dynamics, and optics. As mathematical tools became more powerful, these areas of physics became more mathematically based. This mathematical analysis tie to science and engineering has always had a great place in history and has led to some of its greatest discoveries.

How has **applied mathematics changed** over the past few decades?

In the past few decades, applied mathematics has made tremendous strides in explaining our world. In particular, with the advent of more powerful computers and their

Using mathematics and computers, engineers have been able to create such technological advances as factory robots. *Taxi/Getty Images.*

mathematically driven software—from linked computers over a network to supercomputers—many disciplines that use applied mathematics have greatly advanced. For example, the use of giant wind tunnels to examine wind flow over aircraft only a few decades ago has been replaced by computer simulation. Now the design and testing of aircraft is accomplished by these simulations, making the expense of building physical prototypes a thing of the past because it is now merely a matter of mathematically "drawing" the aircraft on the computer in order to test new designs.

How do various **disciplines use applied mathematics**?

Depending on the discipline, researchers use applied mathematics in various ways. For example, some disciplines rely heavily on pure mathematics. Numerical analysis—a form of applied mathematics—is a field that uses pure mathematics to decipher partial differential equations and variational methods (for more information about numerical analysis, see below; for partial differential equations, see "Mathematical Analysis"). There are also areas of applied mathematics that overlap other fields of study. For example, there are mathematicians who use applied mathematics to study the structure of matter—especially the behavior of subatomic particles—a field that also overlaps the same types of studies done by subatomic physicists.

Why has the **combination** of **mathematical analysis** methods and **computers** been so important?

The combination of mathematical analysis and computers has been a strong alliance, especially with regard to engineering, technology, and the sciences. In the past few decades, researchers have used this combination to help predict the weather, describe in great detail nuclear fusion in the sun, understand the movement of space bodies around the solar system (orbital mechanics), and the flow of water in underground aquifers (fluid dynamics). There are also the studies of chaos—the unpredictable behavior of nonlinear systems—and quantum mechanics, or the physics of very small particles, both of which entail the use of applied mathematics and computers.

In addition, in engineering (and almost all technology) the mathematical analysis-computer combination has helped create structures that surround us every day.

One of many practical applications for mathematics includes processing images, which enhances the ability of electron microscopes like this one to produce more detailed visual information. *The Image Bank/Getty Images.*

This includes familiar modes of transportation and communication—from plans for airplanes and bridge construction to designing fiber optic cables and cellphone towers. It also includes how engineers design control systems, which are used in such diverse areas as robotics, aerospace engineering, and biomedical research.

Why is there such a demand for **applied mathematics** in the **processing of images**?

The field of image processing has grown immensely in the past few decades. There is a great demand for an increase in efficient processing of images, especially for multimedia, biology, and medicine, such as enhancing the quality of electron microscope and MRI (Magnetic Resonance Imaging) images. In addition, there is a great need to develop ways to store more and more information, especially regarding methods in transmitting and processing the information used in computers and networks. All of these things require the use of several mathematical techniques—and, thus, the use of applied mathematics.

PROBABILITY THEORY

What are **events** and **probability**?

Probability is a branch of mathematics that assigns a number measuring the "chance" that some "event"—or any collection of outcomes of an experiment—will occur. It is a quantitative description of the likely occurrence of the event, conventionally expressed on a scale from 0 to 1. For example, a very common occurrence has a probability of close to 1; an event that is more rare will have a probability close to 0. In more common usage, the word "probability" means the chance that a particular event (or set of events) will occur—all on a linear scale expressed as a percentage between 0 and 100 percent (%). An even more detailed way of looking at probability includes the possible outcomes of a given event along with the outcomes' relative likelihoods and distributions.

What is a **sample space**?

In any experiment, there are certain possible outcomes; the set of all possible outcomes is called the *sample space* of the experiment. Each possible result is represented by one and only one point in the sample space, which is usually denoted by the letter S. To each element of the sample space (or to each possible outcome) a probability measure between 0 and 1 is assigned, with the sum of all the probability measures in the sample space equal to 1.

What are **ratios** and **proportions**?

A ratio is the comparison of two numbers; it is most often written as a fraction or with a ":", as in 3/4 or 3:4 to separate the two numbers. For example, if we want to know the ratio of dogs in a shelter that houses 24 animals to the total count, we first determine the number of dogs (say, 10); then the ratio of dogs to animals in the shelter becomes 10/24, or 10:24, which is also said as "10 to 24." But there are rules to ratios. For example, order matters when talking about ratios; therefore, the ratio 7:1 is not the same as 1:7.

A proportion is an equation with a ratio on each side, and is a statement that two ratios are equal. For example, 1/2 = 4/8, or 1/2 is proportional to 4/8. In order to "solve the proportion"—or when one of the four numbers in a proportion is unknown—we need to use cross products to find the unknown number. For example, to solve for x in the following: 1/4 = x/8; using cross product, $4x = 1 \times 8$; thus, $x = 2$.

What are some **simple probability events**?

The probability measure of an event is sometimes defined as the ratios between the number of outcomes. There are many simple illustrations of probability events, many of which we are all familiar with. One of the simplest examples of probability is tossing a coin, with a sample space of two outcomes: heads or tails. If a coin were completely

What is subjective probability?

As the word implies, a subjective probability is thought of as a personal degree of belief that a particular event will occur. An individual's personal judgment is not based on any precise computation, but is most often a reasonable assessment of what will happen by a knowledgeable person. This is usually expressed on a scale of 1 to 0, or on the percentage scale. For example, if a person's baseball team has a winning streak, they might believe that their team has a probability of 0.9 of winning the division championship for the year. More likely, they will say their team has a 90 percent chance of winning, not because of any mathematical formula, but only because the team has had a winning record during the year.

symmetrical, the outcome would more likely be 0.5 (ratio of 1/2) for heads and 0.5 for tails. As we all know, it never comes out that way, which may or may not mean our coins are not in perfect balance.

Another example is weather records. Many of us keep track of weather over the years. But if one were to gather all the records for the day of May 10 over 30 years from the weather service, one could do some simple probability event measurements. For example, take a (fictitious) sampling of the cloud-covered days in a certain area for the last 30 years on May 10. Say there were 10 cloud-covered May 10s in 30 years; thus, the probability measure would be a ratio of 10/30 to the event that the day will be cloudy on May 10.

Insurance tables are also figured out in a similar way. For example, if, out of a certain group of 1,000 persons who were 25 years old in 1900, 150 of them lived to be 65, then the ratio 150/1,000 is assigned as the probability that a 25-year-old person will live to be 65. On the other hand, the probability of such a person not living to be 65 is 850/1,000 (because the sum of the two measures must be equal to 1). It is true that such a probability statement is valid only for a set group of people, but insurance companies get around this by using a much larger population sample and constantly revising the figures as new data are obtained. Thus, even though many people question the validity of such "broadbrush" results, the insurance companies believe that, probability-wise, the values they use are valid for most large groups of people and under most conditions of life.

How are **probabilities of compound events** determined?

Besides the probability of simple events, probabilities of compound events can also be computed. For example, if x and y represent two independent events, the probability **247**

that both x and y will occur is given by the product of their separate probabilities; and the probability that either of the two events will occur is given by the sum of their separate probabilities minus the probability that both will occur. For example, if the probability that a certain man will live to be 70 is 0.5, and the probability that his wife will live to be 70 is 0.6, the probability that they will both live to be 70 is $0.5 \times 0.6 = 0.3$; the probability that either the husband or wife will reach 70 is $0.5 + 0.6 - 0.3 = 0.8$.

What are the **definitions** of **chance**?

Chance is defined in many ways. For example, chance means opportunity to some, or the chance to do something. Chance can also mean luck or fortune, such as running into someone one has not seen in years by "pure chance." Chance also involves taking a risk, which may include some type of danger.

Mathematically speaking, chance is a measure of how likely it is that an event will occur—in other words, a probability. For example, if a meteorologist says that a hurricane on a certain path has struck the coast of Florida, say, about 4 times out of 10, then the ratio becomes 4 to 10, with the chance of striking under the same conditions being 40 percent.

What do the terms **random** and **stochastic** mean in probability?

When speaking about probability, the term *random* means the outcomes of an experiment have the same probability of occurring. Because of this, the outcome of the experiment produces a random sample.

Random also is commonly thought of as being synonymous with *stochastic,* which is from the Greek word meaning "pertaining to chance." It is usually used to indicate a particular subject seen from the point of view of randomness. Stochastic is often thought of as the "opposite" of *deterministic,* a term that means random phe-

nomena are not involved. In the case of modeling, stochastic models are based on random trials, while deterministic models always produce the same output for a given starting condition.

Are "random" numbers generated by a computer truly random?

For most general purposes, random numbers generated by a computer can be considered "random." But in reality, because a computer follows a set of rules in any program, the numbers generated are not truly random. In order for a sequence or specific numbers to be truly random, they must not follow any sort of rules. That's something to think about the next time you purchase a computer-generated, "random-numbers" lottery ticket.

Gamblers who play the lottery are well acquainted with the concepts of random numbers and probability. *The Image Bank/Getty Images.*

What is the concept of relative frequency?

Relative frequency is actually another term for proportion. It can be found by dividing the number of times an event occurs by the total number of times the experiment is done. In probability, this is often written in the notation $rfn(E) = r/n$, in which E is the event, n is the number of times the experiment is repeated, and r is the number of times E occurs. For example, a symmetrical coin can be tossed 50 times (n) in order to find out how many times tails will occur (E). If the result is 20 tails (r) and 30 heads, then the equation becomes 20/50, or 2/5 = 0.4; or, the relative frequency is 0.4 for tails. If this experiment is repeated over and over, the relative frequency will eventually get closer and closer to 0.5, which is the true value that should result when tossing a two-sided symmetrical coin.

How are the terms outcome, sample space, and event related?

These terms are definitely related: The *outcome* is the result of an experiment or other type of situation involving uncertainty; and the set of all possible outcomes is a sample space. Just as important are *events,* which are collections of the outcomes of an experiment, or any subset of the sample space. If there is only one single outcome in the sample space, it is called an elementary or simple event; events with more than one outcome are called compound events.

What is conditional probability?

Conditional probability is often phrased as "event A occurs given that event B has occurred." The common notation is a vertical line, or $A \mid B$ (said as "A given B"). Thus, P $(A \mid B)$ denotes the probability that event A will occur given that event B has occurred already.

Since there is always room for improvement—even in probability—conditional probability incorporates the idea that once more information becomes available, the probability of further outcomes can be revised. For example, if a person brings a car in for an oil change every 3,000 miles, it can be calculated that there is a probability of 0.9 that the service on the car will be completed within two hours. But if the car is brought in during a seatbelt recall, the probability of getting the car back in two hours might be reduced to 0.6. This is the conditional probability of getting the car back in two hours if there is a seatbelt recall taking place.

What are **independent events** in probability?

In probability theory, events are independent if the probability that they occur is equal to the product (multiply together) of the probabilities of either two or more individual events. (This is also often called statistical independence.) In addition, the occurrence of one of the events can give no information about whether or not the other event(s) will occur; that is, the events have no influence on each other.

For example, two events, A and B, are independent if the probability of both occurring equals the product of their probabilities, or P (A) | P(B) (the symbol "|" is often used to depict the product of the events in probability theory). One good example involves playing cards. If we wanted to know the probability of two people each drawing a king of diamonds (two independent events), it would be defined as $A = 1/52$ (the probability that one person will draw a king of diamonds) and $B = 1/52$ (the probability that the other person will draw a king of diamonds, assuming the first person puts the first drawn card back into the deck). Substituting the numbers into the equation, the result is: $1/52 \mid 1/52 = 0.00037$, or a slight chance that both people will draw the king of diamonds from the deck.

What is **Bayes's theorem**?

Bayes's theorem is a result that lets new information be used to update the conditional probability of an event. The theorem was first derived by English mathematician Thomas Bayes (1702–1761), who developed the concept to use in situations in which probability can't be directly calculated. The theorem gives the probability that a cer-

> ## How is set theory used to represent the relationships among events?
>
> **S**tatisticians often use the set theory to represent relationships among events (collections of outcomes of an experiment). This is usually written in the following notation, in which A and B are two events in the sample space S:
>
> - $A \cup B$, or "either A or B occurs, or both"; this is said as "A union B" in set theory;
> - $A \cap B$, or "both A and B occur"; this is said as "A intersection B" in set theory;
> - $A \subset B$, or "if A occurs, so does B occur"; this is said as "A is a subset of B" in set theory;
> - A', or "A does not occur."

tain event has caused an observed outcome by using estimates for all possible outcomes. Its simplest notation is as follows:

$$P(A \mid B) = \frac{P(A \cap B)}{P(B)} = \frac{P(B \mid A) \cdot P(A)}{P(B)}$$

What are **mutually exclusive events**?

Mutually exclusive events are those that are impossible to occur together. For example, a subject in a study of male and female humans can't be both male and female. Males are males; females are females. Thus, depending on what the study is about (such as a study of how many males versus females go to college from a certain high school), both "events" will be mutually exclusive.

What are the **addition rules** of probability?

In probability theory, the addition rule is used to determine the probability of events A or B occurring. The notation is most commonly seen in terms of sets: P $(A \cup B)$ = P (A) + P (B) − P (AB), in which P (A) represents the probability that event A will occur, P (B) represents the probability that event B will occur, and P $(A \cup B)$ is translated as the probability that event A or event B will occur. For example, if we wanted to find the probability of drawing a queen (A) or a diamond (B) from a card deck in a single draw, and since we know there are 4 queens and 13 diamond cards in the deck of 52, the equation and resulting probability becomes: 4/52 + 13/52 − 1/52 = 16/52 (the 1/52 is derived by multiplying 4/52 × 13/52).

Skilled card players have developed mathematical skills that help them estimate the chances of which cards might be held by their opponents based on what has been played and dealt already. *Stone/Getty Images.*

But there are also rules of addition for mutually exclusive and independent events. For mutually exclusive events, or events that can't occur together, the addition rule reduces to P $(A \cup B)$ = P (A) + P (B). For independent events, or those that have no influence on each other, the addition rule reduces to P $(A \cup B)$ = P (A) + P (B) − P (A) | P (B).

What are the **multiplication rules** of probability?

In probability theory, the multiplication rule is used to determine the probability that two events, A and B, both occur. As with the addition rules, the notation for multiplication rules of probability are most commonly seen in terms of sets: $P(A \cap B) = P(A|B) \cdot P(B)$ or $P(A \cap B) = P(B|A) \cdot P(A)$, in which $P(A)$ represents the probability that event A will occur, $P(B)$ represents the probability that event B will occur, and P $(A \cap B)$ is translated as the probability that event A and event B will both occur. In addition, $P(A|B)$ is the conditional probability that event A occurs given that event B has already occurred, and $P(B|A)$ is the conditional probability that event B occurs given that event A has already occurred. Similar to the addition rules, if there are independent events (or those that have no influence on one another), the equation reduces to $P(A \cap B) = P(A) \cdot P(B)$.

What is the **law of total probability**?

The law of total probability can be written as follows: The probability that an event A will occur, $P(A)$, is equal to the probability that event A and event B both occur, plus the probability that event A and event B' occur (or A occurs and B does not). Using the multiplication rule, this is written as:

$$P(A) = P(A|B) \cdot P(B) + P(A|B') \cdot P(B')$$

What was the **"gambler's ruin"**?

The gambler's ruin is an application of the law of total probability that was first proposed by Dutch mathematician and astronomer Christiaan Huygens (1629–1695), although many people before him, including astronomer Galileo Galilei (1564–1642), brought up the same probability problem, but phrased it differently. By 1656, Huygens

Dutch mathematician and astronomer Christiaan Huygens devised the notion of the "gambler's ruin." *Library of Congress.*

Galileo Galilei, best known for his work as an astronomer, had already discovered the ideas of probability later restated as the "gambler' ruin" by Christiaan Huygens. *Library of Congress.*

wrote a draft version of *Van Rekeningh in Spelen van Geluck,* a treatise about 15 pages long based on what he heard about the correspondence of French scientist and religious philosopher Blaise Pascal (1623–1662) and French mathematician Pierre de Fermat (1601–1665) the previous year. Of the fourteen problems he presents, the last five became known as the "gambler's ruin."

In particular, Huygens (and others) wanted to find the probability of a gambler's ruin. A common way of expressing the idea is by a game that has two players, with the game giving a probability q of winning one dollar and a probability $(1 - q)$ of losing one dollar. In the problem, if a player begins with 10 dollars and intends to play the game repeatedly until he either goes broke or increases his holdings to 20 dollars, the question asked is: "What is his probability of going broke?" The answer involves quite a bit of probability computation. (For more information about the gambler's ruin, see "Recreational Math.")

What are **permutations, combinations,** and **repeatables**?

In order to perform certain probability problems, specific counting techniques need to be used, including determining the number of permutations, combinations, or repeatables. The following explains each term; the examples are based on a set of five cats on a shelf—a, b, c, d, and e, for convenience. (Note: The cats can be arranged in 120 ways, expressed as $5 \times 4 \times 3 \times 2 \times 1 = 5!$ ["5 factorial"] $= 120$).

The number of *permutations* is the number of different ways specific entities within the cat group can be arranged, with the positions being important. For example, given five cats, how many unique ways can they be placed in three positions on the shelf if position is important? The answers include ade, aed, dea, dae, ead, eda, abc, acb, bca, bac, etc.—a total of 60 ways. The notation for this is 5 P3 = 5!/(5 − 3)! = 5 × 4 × 3 × 2 × 1 / (2 × 1) = 5 × 4 × 3 = 60 (P in this case stands for permutations).

Combinations mean the number of different ways specific entities can be grouped; but in this case, position does not matter. For example, in the problem of the cats, how many can be grouped into threes if position does not matter? The answers include abc, abd, abe, acd, ace, ade; but groupings such as cba are not allowed since it is equal to another combination: abc. The notation for this is 5 C3 = 5!/((5−3)! × 3!) = 5 × 4 × 3 × 2 × 1/(2 × 1 × 3 × 2 × 1) = 5 × 2 = 10 (C in this case stands for combinations).

With *repeatables,* position is important, too. But in this case, if one has five different cats, and many clones of each, how many unique ways can they be placed in three positions? This answer includes aaa, bbb, ccc, ddd, eee, eec, cee, etc.—a total of 125 ways. The notation for this is 5 R3 = 5^3 = 125 (R in this case stands for repeatables).

What are some **examples** in which **probability** is used?

There are thousands of examples in which probability is used, some are familiar, and some originate from the seamier side of life. For example, everyone has played at *coin tossing* at one time or another. Although there is no such thing as an idealized coin— a circular one of zero thickness—most coin tosses use the coins available, with either side face up ("heads" or "tails"; also phrased "heads up/down" or "tails up/down"). Thus, one can think of a coin as a two-sided die in lieu of the six-sided cubes we are all used to in a game of dice. If a coin is tossed with a good amount of spin, we can denote the two possible results as H for heads and T for tails. If we repeat the tosses N number of times, we obtain N (H) heads and N (T) tails. Thus, the fraction of N(H)/N and N(T)/N can be thought of as the chance (probability) to get a head or tail, respectively; P(H) and P(T) are the most common notations that represent the probability to get

heads and tails, respectively. If we toss the coin many, many times, the result should be close to 0.5. Of course, this means that if we bet on the chances of heads and tails, we will not be much of a winner if we play too many games—and we will have to have really good luck to win if we play fewer games.

Chuck-a-luck is a gambling game that has been played at many carnivals over the years. It involves a player who may bet on any one of the numbers 1 through 6, based on three dice placed in a cage. If a selected number appears, the gambler gets paid even money; if one of the numbers comes up twice, the gambler gets paid 2-1; and if the desired number comes up three times, the gambler gets paid 3-1. But if the player worked it out, he or she would find that the game's expected odds are worse than for almost any other table game of chance. The game is sometimes used as a fundraiser for charity, but the odds of coming away with more money than originally put in are quite small. And they are even worse if someone has loaded the dice.

Games such as coin tossing and dice rolling are common examples of the rules of probability in action. *Stone/Getty Images.*

Probability can also be associated with the violent "game" of chance called *Russian roulette.* In this bizarre sport, one or more of the six chambers of a revolver are filled with bullets, the magazine is rotated at random, and the gun is fired at the player's head. The risk taker "bets" on whether or not the chamber that rotates into place will be loaded. If it is, he loses not only his bet but also his life. Not to be outdone, there are people who have come up with a modified version of Russian roulette. In this case, the gun is loaded with a single bullet and two duelists alternately spin the chamber. Each duelist fires at the other until one is killed; the probability of the first duelist being killed is 6/11.

STATISTICS

What is **statistics**?

The analysis of events governed by probability is called statistics. In statistics, a group of facts is collected and classified in a methodical manner, which is why such a study is important to the fields of science, finance, social research, insurance, engineering, and sundry other areas. In general, the data are grouped according to their relative 255

A matched sample is a type of sampling method used in statistics. For example, in studying population IQs, identical twins could be paired up to measure and compare intelligence. *Photographer's Choice/ Getty Images.*

number, then certain other values are determined based on the characteristics of the group. The most important part of statistical theory is sampling. This is because in most applications, the statistician is not only interested in the characteristic of the sample, but also the characteristics of some much larger population. (For information about samples and populations, see below.)

Why are **populations** important to statistics?

A population is the entire collection of items—from people, animals, and plants to street numbers and various other things of any size—from which the statistician collects data. These data are of particular concern because in most cases the statistician is interested in describing or drawing conclusions about the population (also called the *target population*). For example, take a population of 10 cats. None of them are identical, but certain common features between the cats can be measured, such as color, fur length, and weight. The data collected about one of the common features, such as the fur length of the 10 cats, would be defined as the population.

What is a **sample** in **statistics**?

A sample is a generalization about a population and is represented by a group of units selected from the population (also called a subset of the population). The sample is meant to be representative of the population; thus, in many studies, there are many possible samples. There are also types of samples, such as a matched sample, in which two of the members are paired; an example would be the IQ of twins. There is often a good reason for taking samples of a population: Most of the time, a population is too large to study as a whole.

For example, take the "smaller" example of the above population of 10 cats. Again, none of them are identical, but certain common features between the cats can be measured, including color, fur length, and weight. If data is collected about the fur length of the 10 cats (the population), then if we chose only to take the cats with long fur, that would be a sampling. Another example is the population for a study of physical condition of all children born in the United States in the 1970's; the sample could be all children born on July 5 in any of those years.

> ## What is the major difference between a population and a sample?
>
> A population is examined to identify its certain characteristics; a sample is taken in order to make inferences about the characteristics of the population from which the sample was drawn.

How does one **sample a population**?

Sampling is the term used when one obtains a sample of a population; the number of members in the sample is called the *sample size*. As with most mathematical concepts, there are several types of sampling.

Random sampling is a technique involving a group of subjects (the sample) from a larger group (population); it is a method that reduces the likelihood of bias. In random sampling, each individual is chosen by chance, with each member of the population having a known (but often unequal) chance of being included in the sample.

Simple random sampling also involves a group of subjects (the sample) from a larger group (population), but in this case, each individual is chosen entirely by chance, with each member of the population having an equal chance of being included in the sample. In fact, each member of the population has an equal chance of being chosen at any stage of the sampling process.

Independent sampling comprises samples collected from the same (or different) populations that have no effect on one another. In other words, there is no correlation between the samples.

Stratified sampling includes random samples from various subgroups (also called subpopulation or stratum of the population) chosen to be representative of the whole population. It is often thought of as a better technique than simple random sampling. For example, if a sheep farmer wanted to determine the average weight (amount) of wool gathered from three types of sheep on his farm, he could divide his flock into the three subgroups and take samples from those groups.

In *cluster sampling* the entire population is divided into clusters (groups); then a random sampling is taken of the clusters. This technique is used when a complete list of the population's members can't be studied, but a list of population clusters can be gathered.

What is a **statistic** and a **sample statistic**?

A statistic is the measure of the items in a random sample. A sample statistic is meant to give information about a specific population feature (or parameter). For example, if a sample mean is gathered for a set of data, that would provide information about the overall population mean.

What type of sampling technique is often used in opinion polling?

We all know about opinion polls, especially during a major election. The statistical sampling method of most polling places is called *quota sampling,* a technique in which interviewers are given a quota of specific types of subjects to poll. For example, the sampling may involve asking 10 adult men, 10 adult women (both groups over 20 years of age) and 10 teenage (18 to 19 years old) voters for their opinion on the presidential election. But as we all know, these types of polls are not as accurate as they should be, mainly because the sample is not random.

What is the difference between **descriptive** and **inferential statistics**?

Descriptive statistics is a way to describe the characteristics of a given population by measuring each of its items, then taking a summary of the measurements in various ways. Inferential statistics, as the term implies, makes educated inferences (guesses) about the characteristics of a population by taking and analyzing data from a random sample.

What are **quantitative** and **qualitative variables** as used in statistics?

Variables are values used to come to conclusions in a statistical study. There are two main categories: *quantitative* and *qualitative* variables. Quantitative variables can be divided into three types. Ordinal variables are measured with an ordinal scale, in which higher numbers represent higher values, even though the intervals between numbers are not necessarily equal. For example, on a five-point rating scale measuring attitudes toward cutting back on air pollution, the difference between a rating of 2 and 3 may not be the same as the difference between a rating of 4 and 5. Interval variables are measured with an interval scale, in which one unit on the scale represents the same magnitude of the characteristic being measured across the whole range of the scale. For example, the Fahrenheit scale for temperature is an interval scale, in which equal differences on this scale represent equal differences in temperature, but a temperature of 30 degrees is not twice as warm as one of 15 degrees. The third type is the ratio scale variable. This is a scale similar to the interval scale, but with true zero points. For example, the Kelvin temperature scale is a ratio scale because it has an absolute zero. Thus, a temperature of 300 Kelvin is twice as high as a temperature of 150 Kelvin.

Qualitative variables are measured on a nominal scale, or a measurement that has assigned items to groups or categories. With these variables, there is no quantitative information and no ordering of the items is conveyed—it is qualitative rather than quantitative. Religious preference, race, and gender are all examples of nominal scales.

What **other variables** are often used in **statistics**?

When an experiment is conducted, variables manipulated by the experimenter are called *independent variables* (also independent factors), while others measured from the subjects are called dependent variables (also dependent measures). For example, consider a hypothetical experiment on the effect of lack of sleep on reaction time: Subjects either stayed awake, slept for 2 hours for every 24, 5 hours for every 24, or 8 hours for every 24; they then had their reaction times tested. The independent variables would be the hours slept by each person and the dependent variables would be the reaction time.

Some variables can be measured on a continuous scale—a *continuous variable* being one that, within the limits the variable ranges, can take on any value possible. For example, we can make the time to eat a lunch at a certain restaurant be the continuous variable because it can take any number of minutes or hours to finish the meal. But other variables can only take on a limited number of values—or dependent variables. For example, if the variables were a test score from 1 to 10, then only those 10 possible values would be allowed; these are called *discrete variables*.

What are the **measures of central tendency**?

The measures of central tendency are statistics that describe the grouping of values in a data set around a common value that in some way represents a typical member. They are broken down into four types: median, mode, average (also called the arithmetic mean), and the geometric mean.

What are the **arithmetic mean** and **geometric mean**?

The average, in general, means the average value of a group of numerical observations—or the statistical measure of the arithmetic mean (or just the mean). On a scale of measurement, it is usually the place in which the population is centered. Numerically, it equals the sum of the scores divided by the number of scores. For example, the sum of 3, 7, 10, 15, and 25 is 60; thus, the mean is 60 divided by 5, or 12. The geometric mean is another type of mean: For two quantities, it is the square root of the quantities' product; the notation is for n quantities, and the geometric mean is the nth root of their product.

What are **median** and **mode**?

The median is considered half the sum of the two numbers nearest the middle. In other words, in an ordered data set, the median is the value at the halfway point, above and below which lie an equal number of data values. (But note: There is an actual middle number for a set with an odd number of members, but no middle number for an even number of members.)

259

The mode is considered a single value that occurs more often than any other in a set of data. It is not the frequency of the most numerous number, but the value of the number itself. Often there is more than one mode if two or more values are commonly found in the set—often called a multi-modal population.

What is the **range** of a set of numbers?

The range of a sample (or data set) is used to characterize the spread or dispersion among observations in a given population, as it is the distance between the highest and lowest numbers. In statistics, it is often (logically) referred to as the statistical range. Numerically, it is represented by the highest score minus the lowest score. For example, for the range of the numbers 34, 84, 48, 65, 92, and 22, the range is $92 - 22 = 70$.

What is the **average deviation**?

The average deviation is a way of characterizing the spread (dispersion) among the measures in a given population. To determine the average deviation, compute the mean, then specify the distance between each score and that mean without regard to whether the score is above or below the mean. The following is the notation for this calculation (the symbol Σ stands for "sum of"; the symbol | | stands for absolute value):

$$\frac{\Sigma |x - \mu|}{N}$$

in which x is the various values of the samples, μ is the mean (or average) for the entire population, and N is the number of samples (the two vertical lines representing the absolute value means there are no negative numbers on top).

For example, if we have six people who weighed 166, 134, 189, 141, 178, and 150, the equation would read (μ is the average, or the total weight divided by the number of people, or $958/6 = 159.67$; $n = 6$, or the number of people; and x are the individual weights of the people):

$$\frac{\Sigma |166 - 159.67| + |134 - 159.67| + \cdots + |178 - 159.67| + |150 - 159.67|}{6}$$

The average deviation for this example is 18.

What is the **variance**?

The variance is the average of the squares of a set's deviations. It is used to characterize the spread among the measures of a given population. First, calculate the mean of the scores; then measure the amount that each score deviates from the mean. Finally, square that deviation (in other words, multiply it by itself, add all of them together, then divide by the total number of scores). An even easier way is to

square the numbers first. (Note: Taking the square root of the variance gives the standard deviation.)

For example, take the numbers 3, 5, 8, and 9, with a mean of 6.25 (the sum of the numbers divided by the total number of numbers). To calculate the variance, determine the deviation of each number from 6.25 (3.25, 1.25, 1.75, 2.75), square each deviation (10.5625, 1.5625, 3.0625, 7.5625), then take the average $22.75/4 = 5.6875$, which is the variance. An easier way to calculate the variance is to square all the numbers first (9, 25, 64, 81) and determine the mean ($9 + 25 + 64 + 81$ divided by $4 = 44.75$). Then subtract the square of the first mean ($6.25^2 = 39.0625$)—or $44.75 - 39.0625 = 5.6875$.

What is the **standard deviation**?

The standard deviation is considered by some to be the second most important statistic (or statistical measure) in the field; it is the measure of how much the individual observations are scattered about the mean. In general, the more widely values are spread out, the larger the standard deviation. For example, if the test results for two different exams taken by 50 people in a geology class range from 30 to 98 percent for the first exam and 78 to 95 percent for the second exam, the standard deviation is larger for the first list of exams. This spread (dispersion) of a data set is calculated by taking the square root of the variance; the notation for standard deviation is most commonly seen as follows: $\sqrt{V(x)} = s.d.$ (or simply s), in which $V(x)$ is the variance.

What is a **normal distribution**?

A normal distribution is an idealized view of the world, producing the familiar, symmetrically shaped "bell-shaped curve." It is usually based on a large set of measurements of one quantity—such as weights, test scores, or height—which are arranged by size. In a normal distribution, more than two-thirds of the measurements fall in the central region of the graph; about one-sixth of them are found on either side.

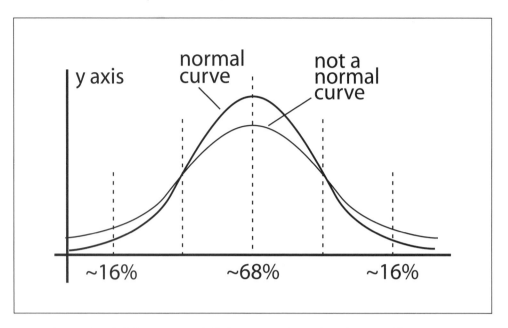

A "normal" curve has a bell shape to it into which the majority of a sampling of statistics falls within a certain average range.

Many of us are familiar with the normal distribution from standardized school test scores, as most result in a bell-shaped curve, with students hoping to at least fall in the middle of the curve.

What is a **cumulative distribution**?

A cumulative distribution is a plot of the number of observations that fall in or below an interval. For example, they are often used to determine where scores fall in a standardized test. For example, the x-axis (see bar graph illustration on next page) shows the intervals of scores (such as the interval labeled 35 shows any score from 32.5 to 37.5, and so on) and the y-axis shows the number of students scoring in or below each interval. This graphically illustrates for the students (and teachers) how well they did on the test compared to other students.

What are **skewness** and **symmetry**?

Skewness is when there is asymmetry in the distribution of the sample data values. In this case, the values on one side of the distribution seem to be further from the "middle" range than values on the other side. Symmetry essentially means balance: when the data values are distributed in the same way above and below (or on both sides of) the middle of the sample.

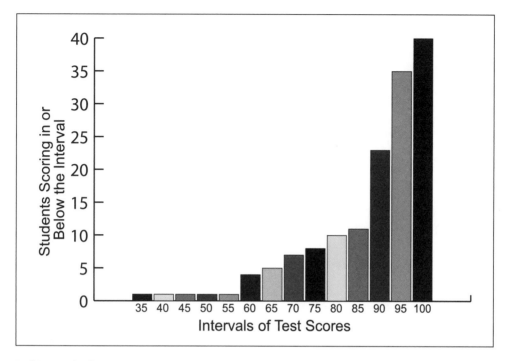

In this example of a cumulative distribution chart, all students (100 percent) fall within the maximum score of 100, while smaller numbers of students fall within each progressively lower score on a test.

What is a **statistical test**?

A statistical test is a procedure used to decide if an assertion (often called a hypothesis) about a population's chosen quantitative feature is true or false. These tests usually entail drawing on a random sample (or simple random sample) from the chosen population, then calculating a statistic about the chosen feature. From the result, a statistician can usually determine if the hypothesis is true, false, rare, common, or something in between. Overall, for a statistical test to be valid, it is necessary to choose what statistic to use, what sample size, and what criteria to use for rejection or acceptance of the tested hypothesis.

How is **statistical data presented**?

There are many ways to present statistical data, all of which involve graphical means to translate the results of statistical tests. A *histogram* is a graphical representation of a distribution function using rectangles. It is also most often constructed from a frequency table (see below). The widths of the rectangles usually indicate the intervals into which the range of observed values are divided; the heights of the rectangles indicate the number of observations that occur in each interval. The shapes of histograms vary depending on the chosen size of the intervals.

263

What does the term "statistically significant" mean?

For most of us, "significant" means important; in statistics it means probably true (not due to chance) but not necessarily important. In particular, significance levels in statistics show how likely a result is due to chance. The most common level—one thought to make it good enough to believe—is 0.95, which means the finding has a 95 percent chance of being true.

This can also be misleading, however. No statistical report will indicate a 95 percent, or 0.95, in its answer, but it will show the number 0.05. Thus, statisticians say the results in a "backward" manner: they say that a finding has a 5 percent chance of not being true. To find the significance level, all one has to do is subtract the number shown from 1. For example, a value of 0.05 means that there is a 95 percent $(1 - .05 = 0.95)$ chance of it being true; for a value of 0.01, there is a 99 percent chance $(1 - 0.01 = 0.99)$ of it being true; and so on.

Bar graphs are similar to histograms, but with the columns separated by each other by small distances. They are commonly used for qualitative variables. A *pie chart* is another way to represent data graphically. In this case, it is a circle divided into segments, or "pie" wedges. Each segment represents not only a certain category, but its proportion to the total set of data. Another type of graph is the *line graph,* which is similar to those seen in geometry: a representation of the data from connected point to connected point. They are one of the most common graphs seen for simple statistical data collection.

What is a **frequency table**?

A frequency table is a way of summarizing data. In particular, it is a way of displaying how entities (such as scores, number of people, etc.) are divided into intervals, and of counting the number of entities in each interval. The result shows the actual number of entities, and even the percentage of entities in each interval. For example, if we survey the number of people working in the 10 offices in a building, the data set might look like the following:

Number of people working in each office:

3	1	4	1	3	2	4	1	1	2

Another way to present this data is to note how many offices had 1, 2, 3, or 4 people working in them. This is known as finding the frequencies of each of the data values. An example of such a frequency table would be as follows:

number of people	0	1	2	3	4
frequency	0	4	2	2	2

This is a compact way of showing how the data values are distributed between the various number of people. It allows us to see, at a glance, the most "popular" number (also called the "modal class") of people working in an office for this building's 10 offices—in this case, a one-person office. Such data can further be visualized with the use of bar charts, histograms, or pie charts.

There are even more ways to see the data from this frequency table. We can also illustrate this chart in terms of percent. In particular, we can say that 40 percent of the offices contain 1 person, 20 percent have 2 people, 20 percent have 3 people, and 20 percent have 4 people.

What is a **percent**?

A percent (using the symbol %) is the ratio of one number to another. Percents are quantitative terms in which n percent of a number is n one-hundredths of the number;

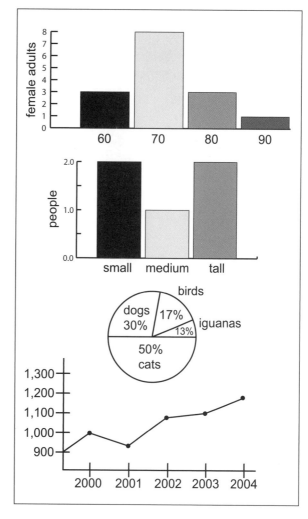

Here are four different types of charts used in statistics (from top to bottom): a histogram, a bar graph, a pie chart, and a line graph (statistics are not based on actual data here).

they are usually expressed as the equivalent ratio of some number to the number 100. For example, the ratio of 25 to 50 means the number 25 is 50 percent of 50. They are not true numbers; thus, percents can't be used in calculations, such as addition or multiplication. But operations can be conducted with percents when they are translated into ratios and fractions, such as 25 percent is equal to 0.25 or 1/4.

MODELING AND SIMULATION

What is a **mathematical model**?

Mathematical models are a way of analyzing systems by using equations (mathematical language) to determine how the system changes from one state to the next (usually using differential equations) and/or how one variable depends on the value or state of other variables. Simply put (although working through the equations is not always simple), the process of mathematical modeling includes the input of data and information, a way of processing the information, and an output of results.

A mathematical model can describe the behavior of many systems, including systems in the fields of biology, economics, social science, electrical and mechanical engineering, and thermodynamics. For example, modeling is usually used in the sciences to better understand physical phenomena; each phenomenon is translated into a set of equations that describe it. But don't think that all the results of models are indicative of the real world. Because it is virtually impossible to describe a phenomenon totally, models are considered to be merely a human construct to help us better understand our surrounding real-world systems.

What are some **basic steps** to **building** a **mathematical model**?

Just like building a physical structure, there are basic steps to building a mathematical model. The first steps are to simplify the assumptions, or to clearly state those assumptions on which the model will be based, including an understandable account of the relationships among the quantities to be analyzed. Second is to describe all the variables and parameters to be used in the model, and identify the initial conditions of the model. Finally, use step one's assumptions and step two's parameters and variables

to derive mathematical equations.

Can a mathematical model be too complex?

Yes, mathematical models can be too complex for a number of reasons. For example, if we wanted to model the development of a hurricane, we would take data from a forming storm—water vapor, pressure, temperature, and so on—and incorporate it into our model. In this way, we would try to develop as close to a white-box model of the hurricane system as possible.

But in reality a collection of such a huge amount of data—not to mention the computational cost—would effectively inhibit the use of such a weather model. There is also uncertainty because the development of a hurricane is an overly complex system, mainly because each separate part of a hurricane and its development causes some amount of variance in the model. For example, not only would we have to know about the details of the hurricane's development, but other factors would come into play, such as the ocean-interaction variables that contribute to the hurricane, variability of solar radiation, and even how periodic events such as El Niño—a periodic warming of the waters off the South America coast—affect the hurricane. Thus, meteorologists usually use some approximations to make the mathematical model more manageable, which is also why we still can't predict where and how much rain, wind, and tornadoes will occur during a hurricane.

What is a common **example** of **mathematical modeling**?

One of the most common examples of a mathematical model has to do with the growth of populations—human, animal, or otherwise. In the sciences, this type of mathematical modeling often leads to intelligent hypotheses about the future and past of a population—from a possible population explosion to even an extinction. (For more information about the use of mathematical modeling in the sciences, see "Math in the Natural Sciences.")

What does *a priori* refer to in **mathematical modeling**?

The term *a priori* refers to the amount of information available in a system. Based on the amount of a priori information, mathematical modeling problems are often classified into white-box (a system in which all necessary information is available) or black-box (a system in which there is no a priori information available) models. Practically all systems are somewhere between the white-box and black-box models; thus, this concept only works as an intuitive guide on how to approach a mathematical problem.

How can the **accuracy** of a **mathematical model** be determined?

There is usually one good way to determine the accuracy of a mathematical model: Once a set of equations has been built and solved, if the data generated by the equa-

Many different types of data, including air pressure, temperature, humidity, and so on, must be taken into account when constructing computer models of hurricanes. *The Image Bank/Getty Images.*

tions agree (or come close to) the real data collected from the system, then we can determine its accuracy. In fact, the set of equations and models are only "valid" as long as the two sets of data are close. If a model result leads to conclusions that are not close to the real-world scenario, then the equations are further modified to correct for the discrepancies as much as possible.

For example, in weather prognostication, meteorologists use various numerical models to make long-term predictions of weather systems (for more information about models and weather prediction, see "Math in the Natural Sciences"). It is interesting to see how meteorologists use a combination of several of the weather models to forecast the weather in certain spots around the United States and the world—mainly because no weather forecasting model has all the right answers. Every day, researchers are tweaking their respective weather models (based on more collected data) in hopes of eventually understanding our weather a bit better.

What is a **simulation**?

A simulation is an imitation of some real event or device. It is often used interchangeably with the word modeling (as in modeling of natural systems). A simulation tries to represent certain features (or behaviors) of a complex physical system based on the underlying computational models of the phenomenon, environment, or experience.

Simulations are used to understand the operation of real-world, practical systems; for instance, the modeling of natural systems such as the human body. They can be used to simulate a certain type of technology, such as a new type of airplane, or model an engineering concern, such as the stability of a building during an earthquake. *Simulators,* interactive, physical devices that simulate reality, are also part of simulation. For example, space shuttle simulators are used to help astronauts understand the intricacies and various scenarios associated with flying the space shuttle.

OTHER AREAS OF APPLIED MATHEMATICS

What is **numerical analysis**?

Numerical analysis is a branch of applied mathematics that studies certain specialized techniques for solving mathematical problems. It employs mathematical axioms, theorems, and proofs, and often uses empirical results to further analyze or examine new methods.

What are some **characteristics** of **numerical analysis methods**?

Characteristics of numerical analysis methods include accuracy (the numerical approximation should be as accurate as possible), robustness (the algorithm should be able to solve many problems and should relate to the user when the results are inaccurate), and speed (the faster the computation, the better the method).

What are some **areas** that **numerical analysis addresses**?

Areas addressed by numerical analysis include computing the values of functions, solving equations, optimizing functions, evaluating integrals, and solving for differential equations. (For more about these areas, see "Mathematical Analysis.")

What is **operations research**?

Operations research, more commonly called optimization theory, is a form of applied mathematics designed to determine the most efficient way of doing something by using mathematical models, statistics, and algorithms in the decision-making process. It is a branch of mathematics that entails the many diverse areas of optimization and minimization, including the calculus of variations, control theory, decision theory, game theory, linear programming, and many others. Operations research is most often used to analyze complex, real-world systems, stressing the improvement in or optimization of performance.

What is the Monte Carlo method?

The Monte Carlo method gives approximate numerical solutions to a number of problems that are too difficult to solve analytically by performing specific statistical sampling experiments. Although forms of the method have been known for a while, it was initially developed for numerical integrations in statistical physics problems during the early days of electronic computing. It was named after the city in the Monaco principality, some say, because of the simple random number generator of roulette played in the Monaco casinos; others say the method's creator was honoring a relative who had a propensity toward gambling.

But there is more to the Monte Carlo method than meets the computation. For one thing, there is more than one Monte Carlo method. For example, one method, called the Markov chain Monte Carlo method, has played a critical role in such diverse fields as physics, statistics, computer science, and structural biology. And the list of applications continues: As recently as the late 1990s, researchers and statisticians began to realize the usefulness and power of Monte Carlo methods for prediction.

What is **optimization**?

Optimization problems deal with the point at which a given function is maximized (or minimized). It is divided into several subfields, depending on the form of the objective function and the constraints that the maximized point often has to satisfy.

What are some **applications** of **operations research**?

Applications of operations research span a wide variety of fields, especially with regard to reducing costs or increasing efficiency. The following lists just a few examples: constructing a telecommunications network at low cost and top efficiency, especially when under high demand or after being damaged; determining routes of school buses so fewer vehicles are needed; and designing a computer chip that will reduce manufacturing time.

What is **information theory**?

Information theory is a branch of the mathematical theory of probability and statistics, allowing it to quantify concepts of information. It was formulated primarily by American scientist Claude E. Shannon (1916–2001; who was also called "the father of information theory") to explain the aspects and problems inherent in information and communication. In particular, it involves efficient and accurate storage, transmission, and representation of information, such as the engineering requirements—and limi-

What is some of the history behind operations research?

In the United Kingdom, operations research is known as operational research; other English-speaking countries most often use the term "operations research" (OR). The idea of operations research started during World War II, when military planners in the United States and United Kingdom were searching for ways to make better decisions in the fields of logistics and training schedules. Unlike its origins, today's applications in operations research have little to do with its traditional sense. In other words, it's not applied to the battlefield, but to logistical, scheduling, and other such decisions in industry.

tations—of communication systems. (Note: Information theory has nothing to do with library and information science or with information technology.)

In information theory, the term "information" is not used in the traditional sense. Here it is used to mean a measure of the freedom of choice with which a message is selected from the set of all possible messages. Because it is possible for a string of non-sense words and a meaningful sentence to be equivalent with respect to information content, "information" in this sense takes on a different meaning.

What is **game theory**?

As the words imply, game theory has to do with the mathematics and logic used in the analysis of games—or the situations that involve groups with conflicting interests. Another way to look at a game is as a conflict that involves gains and losses between two or more opponents who all follow a set of formal rules.

In particular, game theorists study the predicted and actual behavior of individuals in specific games, as well as the optimal strategies used in the games. Thus, the principles of game theory can be applied to games such as cards, checkers, and chess; or, they can be applied to real-world problems in economics, politics, psychology, and even warfare. For example, game theory is used to determine optimal policy choices of presidential candidates, or even to analyze major league baseball salary negotiations.

MATH IN SCIENCE AND ENGINEERING

MATH IN THE PHYSICAL SCIENCES

PHYSICS AND MATHEMATICS

What **science fields use** an abundance of **mathematics**?

The sciences that are "heavy users" of mathematics are the so-called physical sciences, including physics, chemistry, geology, and astronomy. These scientific fields are often contrasted with the natural or biological sciences. The physical sciences analyze the nature and properties of energy and non-living matter, and they often need the help of mathematics to determine the complex relationships between their interactions.

What is **physics**?

Physics is often described as the science of the interactions between matter and energy. It includes the subfields of atomic structure, heat, electricity, magnetism, optics, and many other phenomena.

Traditionally speaking, physics is divided into classical and modern—although many subdivisions of the two overlap—and both are ruled by mathematics. Classical physics includes Newtonian mechanics, thermodynamics, acoustics, optics, electricity, and magnetism. Modern physics includes such fields as quantum field theory and relativistic mechanics.

Other common divisions of physics are experimental and theoretical physics. Theoretical physicists use mathematics to describe the physical world and predict how it will behave; they depend on experimental results to check, understand, change, or eliminate theories. Experimental physicists test their predictions with practical experiments, often using mathematics to conduct the experiments.

What is **mathematical physics**?

Mathematical physics uses the concepts of statistical mechanics and quantum field theory. But it is not the same as theoretical physics. Mathematical physics studies physics on a more abstract and meticulous level. Theoretical physics entails less mathematics than mathematical physics and has more to do with experimental physics. But like many fields of science, defining mathematical physics is not easy. For example, still another definition of modern mathematical physics states that it entails all areas of mathematics other than classical mathematical physics.

CLASSICAL PHYSICS AND MATHEMATICS

How is **mathematics** used in physics **to describe motion**?

Everything in the universe is in motion, from the rotating Earth to subatomic particles. Motion in physics is described mainly through mathematics, including speed, velocity, acceleration, momentum, force (something that changes the state of rest or motion of an object), torque (when a force causes rotation or twisting around a pivot point), and inertia (a body at rest remains at rest, and a body in motion remains in motion, until acted upon by an outside force).

Unlike what most people think, speed and velocity are not the same. Speed is the rate at which something moves; velocity is speed in a certain direction. Speed is also called a scalar quantity, described by the following formula: speed = distance/time. For example, if you drive 200 miles in 2 hours, and your speed is constant, your average speed is 200/2, or 100 miles per hour. On the other hand, velocity is known as a vector quantity (for more about vectors, see "Mathematical Analysis"). That gives velocity both speed and direction—and that leads directly to acceleration.

When an object's velocity changes, we say that it accelerates. Acceleration—also a vector like velocity—is represented as the change in velocity divided by the time it takes for the change to occur. We define the formula for acceleration—or the change in

velocity per unit time—as $a = \Delta v/\Delta t$. In this equation, a is acceleration, Δv is change in velocity of an object (the delta symbol stands for change), and Δt is the change in time needed to reach the velocity. For example, if acceleration is constant, and a person drove from a standing point to 60 miles per hour in 5 seconds, the equation becomes: 60 miles per hour / 5 seconds (or final speed minus the initial speed, all divided by the elapsed time). This means that the acceleration is equal to 17.6 feet per second squared (you have to change the miles per hours to feet and seconds, respectively).

Momentum relates to the amount of energy maintained by a moving object; it is also defined as the force necessary to stop an object from moving. It depends on the mass and velocity of an object, and is represented as: $M = mv$, in which M is momentum, m is the mass of the object, and v is the object's velocity.

Whether you are standing on the Moon or on Earth, your *mass* will always be the same; however, you will *weigh* less on the Moon because the gravitational pull on your body is less. *Taxi/Getty Images.*

What is the **mathematical distinction** between **weight** and **mass**?

There is a definite difference between weight and mass, which is easily represented in the formula $W = mg$, in which W is weight (or the gravitational pull on an object), m is mass (or the quantity of matter in an object), and g is the gravitational pull. For example, if you wanted to lose weight, you could move further from the Earth or live on the Moon (in both cases, the gravitational pull would be less). But remember, no matter where you travel in the universe, you will always have the same mass—unless you diet!

Can **Newton's three laws** be expressed **mathematically**?

Yes, Newton's three laws are all based on mathematical formulas, but the equations would be too complex for this text. They are defined (some with applicable mathematical notation) as follows: *Newton's First Law (Law of Inertia)* states that without any forces acting on an object, it will maintain a constant velocity. In other words, an object will stay still or keep moving in a straight line until something pushes it to change its speed or direction. The one force that stops almost everything from standing completely still or moving in a straight line is gravity. This is easily seen when you

277

Yes, mathematical equations can be used to describe work and energy. Energy comes in many forms, but its basic definition is in terms of work. Work is done when a force moves a body a certain distance. It is expressed in the simple equation: $W = Fd$, in which W is work, F is the force, and d is the distance. In this definition, only force in the direction of the object's motion counts.

throw a baseball in the air: The ball does not move in one, continuous, straight direction because the Earth's gravity pulls it downward toward the surface.

Newton's Second Law (Law of Constant Acceleration) states that if a force acts on an object, the object accelerates in the direction of the force; the force creates an acceleration proportional to the force (and inversely proportional to the mass). This is written in the following notation: $F = ma$, in which F is the force, m is the mass, and a is the acceleration. Newton actually expressed this in terms of the calculus—a form of mathematics he created to explain these physical laws (for more about the calculus and Newton, see "History of Mathematics" and "Mathematical Analysis"). He wrote the equation as follows:

$$F = m\frac{\Delta v}{\Delta t}$$

in which m is mass, Δv is the change in velocity, and Δt is the change in time. This is because instantaneous acceleration is equal to the instantaneous change in velocity in an instance in time (or change in time).

Newton's Third Law (or Law of Conservation of Momentum) states that forces on an object are always mutual. To put it another way, if a force is exerted on an object, the object reacts with an equal and opposite force on the phenomenon that initially exerted the force. Simply stated, objects exert equal but opposite forces on each other. This is often phrased, "For every action, there is an equal and opposite reaction." The mathematical equations for these forces are complex and beyond the scope of this text.

What is **Newton's Law of Universal Gravitation**?

As hard as it is to comprehend, (almost) everything in the universe is attracted to everything else. This physical law is not only one of the most well-known but also one of the most important. Newton's law states that the gravitational force between two masses, m and M, is proportional to the product of the masses and inversely proportional to the square of the distance (r) between them. In formula form, this is written as follows (note: in some texts, the masses of the two objects are written as m_1 *and* m_2):

$$F = G \times \frac{M \times m}{r^2}$$

G is a constant in nature (also called a universal constant), indicating how strong a gravitational force exists. In other words, the farther away the objects, the less the attraction between the objects.

What is **statistical mechanics**?

Statistical mechanics applies statistics to the field of mechanics—the motion of particles or objects when subjected to a force. It is used to understand the properties of single atoms and molecules of liquids, solids, gases—even the individual quanta of light that make up electromagnetic radiation—to the bulk properties of everyday materials. Because statistical mechanics mathematically helps to understand the interactions between a large number of microscopic elements, it is used in a wide range of fields. In a way, it is also the "opposite" of thermodynamics, which approaches the same types of systems from a macroscopic, or large-scale, point of view.

What is **Ohm's Law**?

Ohm's Law is important to the field of electrical studies. It states that direct current flowing in a conductor is directly proportional to the potential difference between its ends. First summarized by German physicist Georg Simon Ohm (1789–1854), it is usually seen in formula form as: $V = IR$ (or $I = V/R$), in which V is the potential difference (voltage), I is the current (also written in some texts as i), and R is the resistance

of the conductor. This can also be written in terms of electric quantities (voltage = current × resistance) and with units of measure (volts = amps × ohms).

MODERN PHYSICS AND MATHEMATICS

What is **modern physics**?

In contrast to classical physics—although there are overlapping topics—modern physics includes relativistic mechanics, atomic, nuclear and particle physics, and quantum physics.

How did **Albert Einstein** use **mathematics**?

German-born American theoretical physicist Albert Einstein (1879–1955) is recognized as one of the greatest physicists of all time, but he was also a notable mathematician. In 1905 he developed the special theory of relativity, mathematically demonstrating that two observers moving at great speeds with respect to one another will experience different time intervals and measure lengths differently, that the speed of light is the "speed limit" for all objects having mass, and that mass and energy are equivalent.

By around 1915, Einstein completed a mathematical formulation of a general theory of relativity, this time adding gravitational effects to determine curvature of a time-space continuum. He further tried to discover a unified field theory, which would combine gravity, electromagnetism, and subatomic phenomena under one set of rules, but he, and no one since, has ever found such a theory. (For more about Albert Einstein, see "History of Math.")

What is **relativity**?

Relativity refers to the idea that the velocity of an object can be determined only relative to the observer. For example, if a fly moves around the inside of a car at about 1 mile per hour, inside the car's frame of reference, the fly is moving at 1 mile per hour. But if the car goes past you at 65 miles per hour, it will appear as if the fly is traveling at 66 miles per hour, not 1 mile per hour. In other words, it's all a matter of reference, and it's "relative" to your viewpoint.

What new ideas came out of Einstein's (and others') study of relativity? He showed that space and time could no longer be viewed as separate, independent entities, forming a four-dimensional continuum called *space-time* (also written as spacetime). It is not easy to verbally explain the intricacies of Einstein's theory. The best way to interpret his works is with the use of formulations from certain mathematical branches, such as tensor calculus (for more about tensors, see "Mathematical Analysis"). But such complex equations are beyond the scope of this book.

What does Einstein's famous equation $E = mc^2$ signify?

Among his other accomplishments, Albert Einstein showed mathematically that there was a connection between mass and energy: energy has mass, and mass represents energy. This equation, also called the energy-mass relation, is expressed as $E = mc^2$, in which E is energy, m is mass, and c is the speed of light. Because c is a very large number, even a very small amount of mass represents an enormous amount of energy.

Did **Einstein's theories** change our concept of **dimensions**?

Yes, our ideas about dimensions changed dramatically thanks to Einstein's theories, not to mention the mathematics involved in producing those theories. In particular, one can't distinguish space and time as elements in the description of events. Instead, they are joined in what is called the *fourth dimension*—also called a four-dimensional manifold known as space-time (see above).

Although it sounds like something out of the television show *Star Trek,* in space-time events in the universe are described in terms of the four-dimensional continuum. Simply put, each observer locates an event by three spacelike coordinates (positions) and one timelike coordinate. The choice of the timelike coordinate in space-time is not unique; hence, time is not absolute but is relative to the observer. The strange effects go on: In general, events at different locations that are simultaneous for one observer will not be simultaneous for another observer. (For more information about dimensions, see "Geometry and Trigonometry.")

Why was **Max Planck** important to **quantum theory**?

Quantum theory (or physics) entails the emission and absorption of energy by matter and the motion of material particles. It is a special situation in which very small quantities are involved. When added to the theory of relativity—in which great speeds are involved—both form the theoretical basis of modern physics.

One of the most important aspects of quantum theory is the quanta. In 1900 German physicist Max Karl Ernst Ludwig Planck (1858–1947) proposed that all forms of radiation—such as light and heat—come in bundles called quanta. These bundles are further emitted and absorbed in small, discrete amounts, thus behaving in some situations like particles of matter. For example, a bundle of light energy is known as light quanta or *photons*. Planck devised the equation: $E = hv$, in which E is the amount of energy in a single particle, v is the frequency of the wave, and h is the constant now known as *Planck's constant.*

It is interesting to note that some people divide physics using Planck's discovery: The term classical physics is often referred to as "before Planck"; while the term for modern physics is often referred to as "after Planck."

What is **quantum mechanics**?

Quantum mechanics is a branch of quantum theory that simply determines the probability of an event happening, although the mathematical calculations to prove such things are very rigorous and complex. In fact, quantum mechanics is often called the "final mathematical formulation of the quantum theory." Developed during the 1920s, it accounts for matter at the atomic level and is considered an extension of statistical mechanics, but is based on quantum theory. One important part of quantum mechanics is

Physicist Max Planck was the first to propose the idea that energy existed in bundles called "quanta." His theories later led to the development of quantum mechanics and modern physics. *Library of Congress.*

wave mechanics, which is an extension of quantum mechanics based on Schrödinger's equation. This idea states that atomic events can be explained as interactions between particle waves. (For more on Erwin Schrödinger, see "History of Mathematics.")

What is the **Pauli exclusion principle**?

Quantum theory also relies on the Pauli exclusion principle. This principle, developed by Austrian-born Swiss physicist Wolfgang Pauli (1900–1958) in 1925, states that two particles of a certain class—called femions, and otherwise known as electrons, neutrons, and protons—can never be in the same energy state. For example, two electrons with the same quantum number can't occupy the same atom.

What is the **Heisenberg uncertainty principle**?

German physicist Werner Karl Heisenberg (1901–1976) not only helped with the quantum theory of light waves, he also developed the Heisenberg uncertainty principle. This states that it is impossible to determine, at the same time, both the energy and velocity of a particle.

How else is **mathematics** used in **physics**?

There are hundreds of other applications of mathematics in physics. The following
lists only a few:

How do quantum physicists regard light waves?

Added to the mix of the mathematically rich quantum theory was an idea developed by French physicist Prince Louis Victor Pierre Raymond de Broglie (1892–1987), who discovered the wave nature of electrons and of particles in general (and also devised a mathematical explanation of the kinetic theory of heat). He determined that not only do light waves often exhibit particle-like properties, but particles also often exhibit wave-like properties.

This opened a can of quantum worms. From there, two different formulations of quantum mechanics developed. First was the wave mechanics of Austrian physicist Erwin Schrödinger (1887–1961), who used a mathematical entity (the wave function) related to the probability of finding a particle in space at a given point. Schrödinger also developed a model of the atom that differed from the traditional Niels Bohr model. Second and mathematically equal to Schrödinger's theory was the matrix mechanics of German physicist Werner Karl Heisenberg (1901–1976).

Fluid mechanics—This is the study of air, water, and other fluids in motion. It includes the mathematics of turbulence, wave propagation, and so on.

Geophysics—This is a geological study with a physics basis. Much of the field is dominated by the mathematics of large scale movement of materials, such as earthquakes, volcanic activity, and fluid mechanics (for example, underground molten volcanic material).

Optics—This is the mostly mathematical study of the propagation and evolution of electromagnetic waves, such as diffraction and the path of light rays. Optics requires a great knowledge of geometry and trigonometry, not to mention complex equations.

CHEMISTRY AND MATH

What is **chemistry**?

Chemistry is the science of matter. It studies the composition, structure, and properties of substances (matter) and its reactions and changes. Because chemistry includes all materials in the universe, it is useful for studying many things—from the chemical composition of gases in galaxies to the chemical reactions within living cells. It also includes mathematics in many forms, such as when determining chemical compositions and understanding relationships between certain chemicals.

283

What are the **atomic number and mass** of an **element**?

The atomic number is the number of protons in an atomic nucleus. The atomic mass of an atom—usually measured in atomic mass units—is the total mass of the atom, or the combined mass of its protons and neutrons (the mass of the electrons is negligible). The importance of atomic numbers and mass is simple: The atoms of each element has a specific atomic number and mass—each determined by "adding" or "subtracting" protons and neutrons within the atom.

What is an **ion**?

Scientists know that atoms can gain or lose electrons, thus acquiring a negative or positive electrical charge (determined by the number of protons minus the number of electrons). For example, if there are 4 protons and 6 electrons, the net charge is -2 (often called the valence). An ion is an atom—or group of atoms—that takes on the net electric charge, and can be positive (cation) or negative (anion). Based on the "mathematics" of losing or gaining electrons, ions can be formed or destroyed. This is one example of where math comes in handy in chemistry.

What is an **angstrom**?

An angstrom (Å) is a unit of measurement often used in chemistry most often in reference to molecules; for example, the average molecule diameter is between 0.5 to 2.5 angstroms. An angstrom is equal to about 3.937×10^{-9} inch or one hundredth of a millionth of a centimeter (10^{-8} centimeter). (For more information about measurement, see "Mathematics in History.")

What is **density**?

Density (usually abbreviated as d or r) is a mathematical concept used to describe the ratio between the mass of an object and its volume. The actual formula is the density times the volume is equal to an object's mass, or $d \times v = m$. In the standard (American) measurement system, density is measured in pounds per cubic foot. But in the sciences, the metric system is usually used and density is measured in grams per cubic centimeter (or grams per milliliter). For example, the density of water is 1 gram per cubic centimeter, lead is 11.3 grams per cubic centimeter, and gold is 19.32 grams per cubic centimeter. (Note: In the majority of cases, the higher the density, the "heavier" it feels to us on Earth.)

What are **formulas** and **equations** in **chemistry**?

Formulas and equations in chemistry don't always mean the same as in mathematics. Formulas in chemistry are representations of a chemical compound using symbols for the elements and subscripts for the number of atoms present. For example, the chem-

What is Avogadro's number?

Avogadro's number (also called Avogadro's constant or Avogadro's figure) was determined by Italian physicist Lorenzo Romano Amedeo Carlo Avogadro, Count of Quarengna and Cerreto (1776–1856), who was also the first one to use the term "molecule" in chemistry. It represents the number of elementary entities, such as atoms, molecules, or formula units, in a mole of any chemical substance (a mole is approximately $6.02214199 \times 10^{23}$ atoms, according to the most recent number from the National Institute of Standards and Technology). To translate even further, a mole is the molecular weight of a substance in grams; one mole is the amount of a substance that contains Avogadro's number. For example, the number of carbon atoms in 12 grams of the substance carbon-12 is equal to one mole.

ical formula for water is H_2O, in which there are two atoms of hydrogen (H) bonded to an atom of oxygen (O). The subscript 2 indicates that there are two atoms of hydrogen in the molecule; if there is no subscript number, as with the oxygen (O), a subscript of 1 is implied. (Remember, not all compounds are molecular; for example, NaCl, or sodium chloride [regular table salt] is called an ionic compound. In these cases, the formula shows the proportion of the atoms of each element making up the compound.) There are other types of formulas in chemistry, but this is the most familiar.

Equations in chemistry also differ from those in mathematics. Chemical equations represent the reaction relationship between two or more chemical compounds—along with the products of the chemical reaction. For example, the chemical equation $2H_2 + O_2 \longrightarrow 2H_2O$ is the reaction of hydrogen with oxygen to form water. The arrow indicates the direction of the reaction toward the product; the reactants (or the substances that react) are hydrogen and oxygen. There is also a methodology in writing chemical equations. Simply put, first determine the reactants and outcome; next, determine the formula for each substance; and finally, balance the equation.

What is the **pH scale**?

The "pH" scale stands for p(otential of) H(ydrogen) scale, or the logarithm of the reciprocal of hydrogen-ion concentration in gram atoms per liter. In simpler terms, the pH is merely the measure of the hydrogen ion concentration of a solution. The pH numbers are based on a scale from 0 to 14, in which numbers less than 7 represent acidic solutions and numbers greater than 7 represent alkaline (base) solutions. A reading of 7 is considered neutral.

Mathematically speaking, once the concentration of hydrogen ions is determined chemically (based on moles per liter), the pH value is established by taking the expo-

What is radioactive decay?

Mathematics can also be applied to radioactive substances found within certain rocks. *Radioactive decay* is the disintegration of a radioactive substance and the emission of certain ionizing radiation (such as alpha or beta particles—or even gamma rays). Simply put, when rocks form, the minerals within the rock often contain certain radioactive atoms that decay at a specific rate.

Radioactive decay is especially important in radioactive dating, in which the original and decayed radioactive elements are used to determine the age of the rock. This is because certain radioactive elements will decay to a mixture of half the original element and half another element (or isotope) in a specific timeframe; this is also called the half-life of the original element. For example, "half" of the Uranium-238 in a rock will decay into Lead-207 in 704 million years (thus, the half-life of Uranium-238 is said to be 704 million years). Statistically, this change follows a specific decay function for each isotope of an element. And in each of these exponential functions, the time for the function's value to decrease to half is constant, making radioactive dating perfect in determining the age of certain rocks.

nent used in expressing this concentration and reversing its sign. It is most often expressed as the notation $pH = -\log_{10}[H^+]$. For example, if the hydrogen ion concentration of a solution is determined to be 10^{-4} (or 0.0001) moles per liter, the pH is 4.

Many people are familiar with the pH scale from high school, especially the practice of using special whitish paper called litmus paper to check for pH. The paper contains a powder extracted from certain plants, allowing the user to determine acidity (the paper turns red), neutrality (the paper stays white), or alkalinity (the paper turns blue) of various solutions. The stronger the acid or base, the more intense the red or blue, respectively. And pH isn't just for use in chemistry class. For example, it is also important to people who work the soil. All plants need a certain soil pH to grow and flourish, which is why most gardeners and farmers determine the acidity or alkalinity of their soil in order to grow better crops.

What is the **Universal Gas Law**?

The Universal Gas Law (also called the Universal Gas Constant or the Perfect Gas Law), is a chemical law that can also be looked at mathematically. It is represented by the equation $PV = nRT$, in which P is pressure, V is volume, n is the number of moles of gas, R is the gas constant, and T is the temperature in Kelvin. (For more about the Kelvin temperature scale, see "Mathematics throughout History.")

What are **calories**?

To most people, calories are usually associated with a very large piece of chocolate cake. But in that case, they are called nutritionist's calories, or the unit of energy-producing potential equal to the amount of heat that is contained in food and released upon oxidation by the body. The body needs the calories in the foods we eat to use as energy. This is why nutrition and weight-control texts often contain such entries as "a 140 pound person walking for one hour at a moderate pace burns off 222 calories."

In chemistry, a calorie also refers to a unit of energy. But in terms of chemical experiments, a calorie is the amount of heat required to raise the temperature of 1 gram of water by 1 degree Celsius from a standard initial temperature at a pressure of 1 atmosphere (sea level). The unit measurement for energy is a joule, in which 1 calorie equals 4.184 joules; 1 joule is translated (most often in metric) as the energy needed to lift 2,000 grams a distance of 10 centimeters.

ASTRONOMY AND MATH

What are **astronomy** and **astrophysics**?

Astronomy is the study of matter in outer space. It is usually considered a branch of physics. But because it encompasses (literally) an astronomical number of subjects—everything from the study of a star's surface to the end of the universe—it is often considered a field of its own. Astrophysics is a branch of astronomy dealing with the physics of celestial bodies and the universe as a whole. It deals with problems that range from the structure, distribution, evolution, and interaction of stars and galaxies to the orbital mechanics of a near-Earth asteroid.

Who was **Hipparchus**?

Hipparchus of Rhodes (also seen as Hipparchus of Nicaea, as he was born there; c. 190–c. 120 BCE) was one of the greatest Greek astronomers. A partial list of his discoveries includes: being the first to discover the precession of the equinoxes,

The famous astronomer Nicholaus Copernicus (pictured here) is commonly thought of as the first person to propose a heliocentric (Sun-centered) model of the solar system, but actually Aristarchus surmised the truth centuries before him. *Library of Congress.*

287

compiling an extensive star catalogue, assigning "magnitudes" as a measure of stellar brightness, and calculating the length of the year to within 6.5 minutes of the correct value. His planetary models were mathematical, not mechanical. And although Hipparchus did not invent it, he was probably the first person to systematically use trigonometry, which was a necessity for most of his discoveries.

What was *De revolutionibus orbium coelestium*?

In the year of his death, astronomer Nicolaus Copernicus (1473–1543; in Polish, Mikołaj Kopérnik) published *De revolutionibus orbium coelestium* (*On the Revolutions of the Heavenly Spheres*). This manuscript gave a full account of his theory that the Sun, and not the Earth, was at the center of the solar system (or universe). Although this theory was not new, Copernicus offered the idea in all its mathematical detail. This heliocentric (versus geocentric) view of the heavens, now known as the Copernican system, is the foundation of modern astronomy.

What are **Kepler's Laws of Planetary Motion**?

A great deal of mathematics went into the formulation of Kepler's Laws of Planetary Motion. These laws were devised by German astronomer and mathematician (and Danish astronomer Tycho Brahe's [1546–1601] assistant) Johannes Kepler (1571–1630). He presented the first and second laws in his work *Astronomia nova* (*New Astronomy*) in 1609; the third law was published in 1619 in *Harmonice mundi*. The three laws are as follows:

What astronomical event with mathematical significance occurred on December 25, 1758?

On December 25, 1758, the appearance of a comet we now call "Halley's Comet" (or Comet Halley) proved a famous astronomer's predictions (unfortunately, it was 16 years after his death). From around 1695, Edmond Halley (1656–1742; also seen as Edmund Halley) carefully studied comets, especially those with parabolic orbits. But he also believed that some comets had elliptical orbits, and he thus theorized that the comet of 1682 (now Comet Halley) was the same comet that appeared in 1305, 1380, 1456, 1531, and 1607. In 1705 he predicted that the comet would appear again 76 years later—in 1758—a prediction that came true.

Such a calculation was a great feat in those days, with Halley even taking into account the comet's orbital perturbations produced by the planet Jupiter. Even today, the comet maintains its 76-year cycle. Its last appearance was in 1986; it will again appear in the year 2062.

Kepler's first law (or law of elliptic orbits)—Each planet moves about the Sun in an orbit that is an ellipse, with the Sun at one of the two foci of the ellipse.

Kepler's second law (or the law of areas)—An imaginary straight line joining a planet to the Sun will sweep out equal areas of the ellipse in equal periods of time.

Kepler's third law (or the harmonic law)—The square of the period of a planet's revolution is directly proportionate to the cube of the semi-major axis of its orbit.

How did **Pierre-Simon de Laplace** apply mathematics to **astronomy**?

French mathematician, astronomer, and physicist Marquis Pierre-Simon de Laplace (1749–1827) was one of the first to work out the gravitational mechanics of the solar system using mathematics. In his *Mécanique Céleste* (*Celestial Mechanics*), Laplace translated the geometrical study of mechanics used by Isaac Newton to one based on calculus (or physical mechanics). He also proved the stability of the solar system, but only on a short time scale. Laplace is also known for his theory about the formation of the planets. He believed they originated from the same primitive mass of material, a theory now known as Laplace's nebular hypothesis. His other studies included major contributions to differential equations and to the theory of probability.

What are **astronomical units** and **light years**?

An astronomical unit is one of the more common measurements used in astronomy. It is a distance equal to the average distance from the Earth to the Sun, or 92,960,116

289

Halley's Comet, named after Edmond Halley, last appeared near Earth in 1986 and will be seen again in the night sky in 2062. *Stone/Getty Images.*

miles (149,597,870 kilometers); it is often seen rounded off to 93,000,000 miles (149,598,770 kilometers) and used in reference to great astronomical distances. For example, the Earth is 1 AU from the Sun; the planet Venus is 0.7 AU; Mars 1.5 AU; Saturn 9.5 AU; and the farthest planet, Pluto, is 39.5 AU from the Sun.

A light year is an even larger unit. As the name implies, it is the distance light travels in one year, or about 5.88 trillion miles (9.46 trillion kilometers). In most cases, light year measurement is reserved for deep space objects. (For more about measurement, see "Mathematics throughout History.")

What is the **Hubble constant**?

Astronomers have always been interested in the age of our universe and the speed of various objects in space. The Hubble constant was devised by American astronomer Edwin Hubble (1889–1953). It is the ratio of the recessional speed of a galaxy—because the universe is expanding—to its distance from the observer. In other words, the velocity at which a typical galaxy is receding from Earth, divided by its distance from Earth.

The reciprocal of the Hubble constant is then thought to be the age of the universe, usually written in terms of kilometers per second per million light years. If the number is high, the universe would be very young; if the number is low, the universe would be much more ancient. Although there have been numerous theories, the true age of the universe is usually considered to be somewhere between 12 and 20 billion years old.

The most recent agreed-upon rate at which the universe is expanding is approximately 20 kilometers per second per 106 light years of distance. That makes the universe about 15 billion years old.

How long does it take for the Sun's light to reach the Earth?

Because the Sun is an average of 93,000,000 miles (149,598,770 kilometers) from the Earth, and the speed of light is approximately 186,000 miles per second, it is easy to determine the approximate time (t) it takes for the Sun's light to reach the Earth using mathematics:

$$t = 93,000,000 \text{ miles} / 186,000 \text{ miles per second}$$
$$= 500 \text{ seconds (miles cancel each other out)}$$
$$= 8.3 \text{ minutes}$$

What is the **Titius-Bode Law**?

The Titius-Bode Law was developed by German astronomer Johann Daniel Titius (1729–1796); Titius's idea was brought to the forefront by German astronomer Johann Elert Bode (1747–1826). The law actually represents a simple mathematical rule that allows one to determine the distances (also called the semi-major axis) of the planets in astronomical units. It is determined using the equation $a = 0.4 + (0.3)2^n$, in which n is an integer and a is the astronomical unit. Interestingly enough, most of the planets—and even the asteroids in the Asteroid Belt—adhere to the law. The only exception is Neptune, the second-to-last planet in our solar system.

Distances of the Planets from the Sun in Astronomical Units

Planet	n	Titius-Bode Law*	Actual Semi-Major Axis**
Mercury	$-\infty$	0.4	0.39
Venus	0	0.7	0.72
Earth	1	1	1
Mars	2	1.6	1.52
asteroid belt	3	2.8	2.8
Jupiter	4	5.2	5.2
Saturn	5	10	9.54
Uranus	6	19.6	19.2
Neptune	—	—	30.1
Pluto***	7	38.8	39.4

* The original formula was $a = (n + 4)/10$, in which $n = 0, 3, 6, 12, 24, 48 \ldots$; a is the mean distance of the planet to the sun.

** This is based on the formula $a = 0.4 + (0.3)2^n$, in which $n = -\infty, 0, 1, 2, 3, 4, 5, 6, 7$. The results can also be found using $a = 0.4 + 3 \times n$, in which $n = 0, 1, 2, 4, 8, 16, 32, 64, 128$. Both formulas are "modern versions" of the Titius-Bode Law.

*** Pluto is a modern addition; the planet was unknown during Bode and Titius's time.

Neptune is the only planet in our solar system that does not fit the model set by the Titius-Bode Law. *The Image Bank/Getty Images.*

What is the **Hertzsprung-Russell diagram**?

The Hertzsprung-Russell diagram is a "two dimensional" graph of the mathematical relationship between the absolute magnitude, luminosity, stellar classification, and surface temperature of stars—all resulting in a diagram of the stellar life cycle. It was plotted by Danish astronomer Ejnar Hertzsprung (1873–1967) in 1911 and independently by American astronomer Henry Norris Russell (1877–1957) in 1913.

How do **astronomers determine** the **distances** to other **planets**?

Astronomers need mathematics, of course, to determine the distances to the planets and satellites of our solar system. One of the first astronomers to work this out was Nicholaus Copernicus (see above) using simple observations of planetary positions.

One of the earliest methods to determine such distances was to use the orbital period of a planet. This varies as the square root of the cube of the distance from the Sun: $T = k \times r^{(3/2)}$, in which T is the time for one revolution, r is the distance between the centers of the Sun and the planet, and k is a constant.

How do **astronomers** determine the **distances to the stars**?

Relatively nearby objects beyond the solar system appear to shift position relative to more distant objects as the Earth moves from one side of the Sun to the other—a phenomenon called *parallax*. You can use parallax to determine the distance to stars, as long as these stellar objects are within a few dozen light years of Earth. (More distant objects in the sky do not change their position enough as the Earth orbits from one side of the Sun to the other.) First, measure the position of the star in the sky; then, measure it again in six months when the Earth is on the opposite side of its orbit. If the distance *a* and the angle *ac* are known (as seen in the diagram below), using trigonometry, *c* can be determined as *a* / cos (*ac*).

Is it easy to figure out the **size of a distant object**?

Yes, it is often possible to figure out the size of distant objects, as long as they aren't too small. The key is in knowing the distance to the object. For example, if someone holds a nickel at arm's length and then has someone hold the nickel 200 yards away,

the coin may appear to be smaller, but its size really hasn't changed. If a person knows how large an object appears to be, and how distant it is, they can work backward to determine the true size of the object. Simply put, this is also how astronomers work out the size of distant objects in outer space.

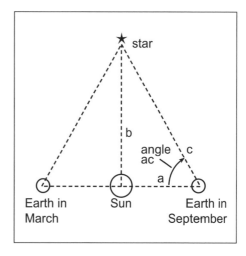

Using parallax to determine the distance to a star is done by calculating the apparent shift of a star as the Earth orbits the Sun.

What "mathematical measurement" error once occurred when a spacecraft reached Mars?

The Mars Climate Orbiter spacecraft, a joint effort between Lockheed Martin and NASA's Jet Propulsion Laboratory (JPL), was supposed to go into orbit around the red planet on September 23, 1999. Instead, the Martian craft lost all contact with Earth. After much deliberation, a review panel for the incident came to a disconcerting conclusion: A thruster error developed when project teams used different measuring systems for the navigation commands—NASA used metric units; Lockheed Martin used English standard units, and no one caught the discrepancy.

To this day, scientists can only speculate as to what happened to the orbiter, a craft sent to study the climate and weather patterns of the Martian atmosphere. Some say that because the orbiter dropped down to within 36 miles (60 kilometers) of the planet—about 62 miles (100 kilometers) closer than planned—atmospheric friction probably overheated the propulsion system and tore the vehicle apart. Others believe the craft was propelled through the atmosphere (or bounced off) and out into space again and is now perhaps circling the Sun like an artificial comet.

How was math used to discover extrasolar planets?

Astronomers have always dreamed about detecting other planets outside our solar system, or "extrasolar planets." In 1994, Polish astronomer Alekzander Wolszczan (1946–) announced the discovery of the first extrasolar planet—actually, it was two planets with masses 3.4 and 2.8 times that of Earth's mass—orbiting the pulsar PSR B1257+12. (A pulsar star sends out a periodic pulse of light detected from Earth.) Wolszczan found the planets by measuring the periodic variation in the pulse arrival time.

There are several major methods used to search for extrasolar planets, and all of them entail using mathematics. For example, the Doppler shift method measures the change in wavelength (color) of light coming from a star over the course of days, months, **293**

and years. The change in wavelength—or the Doppler shift of the light—is caused by the star orbiting a common center of mass with a companion planet. An example in our own solar system is the gas giant Jupiter: Its massive gravitational pull causes the Sun to wobble around a circle with a velocity of 39.4 feet (12 meters) per second.

Another detection method is called astrometry, which measures the periodic wobble that a planet causes in the position of its parent star. In this case, the minimum detectable planet mass gets smaller in inverse proportion to the planet's distance from the star. These methods work, and by 2005 more than 150 such planets have been discovered.

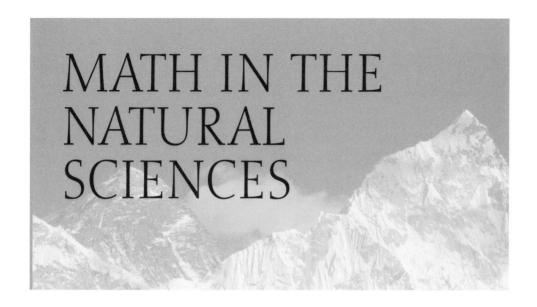

MATH IN THE NATURAL SCIENCES

MATH IN GEOLOGY

What is **geology**?

The word "geology" comes from the Greek *geo,* meaning "the Earth," and the *ology* suffix comes from *logos,* meaning "discussion." Overall, geology is considered to be the study of the Earth. In modern times, thanks to space probes reaching into the solar system, geology also now entails the surface features on other planets and satellites.

Who first made some of the **first accurate measurements** of the **Earth**?

Hellenic geographer, librarian, and astronomer Eratosthenes of Cyrene (276–194 BCE) made several accurate measurements of the Earth, which is why he is often known as the "father of geodesy" (the science of Earth measurement). Although he was not the very first to deduce the Earth's circumference, Eratosthenes is thought by most historians to be the first to accurately measure it.

Eratosthenes knew the Sun's light at noon reached the bottom of a well in Syene (now Aswan on the Nile River in Egypt) (meaning the Sun was directly overhead) on the summer solstice. He compared it to a well's shadow at the same time in Alexandria. Knowing that the zenith distance (the angle from the zenith [point directly overhead] to the point where the Sun was at noon) was 0 degrees at Syene, this meant that at Alexandria it was about 7 degrees. By measuring these angles and the distance between the two cities, Eratosthenes used geometry to deduce that the Earth's circumference was 250,000 stadia. The number was later revised to 252,000 stadia, or 25,054 miles (40,320 kilometers).

The actual circumference of the planet is 24,857 miles (40,009 kilometers) around the poles and 24,900 miles (40,079 kilometers) around the equator, because the Earth

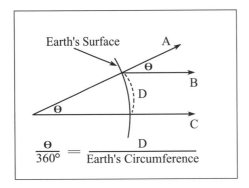

$$\frac{\Theta}{360°} = \frac{D}{\text{Earth's Circumference}}$$

Eratosthenes brilliantly used his knowledge of angles and mathematics to be the first to determine the Earth's circumference accurately.

is not completely round. From his data Eratosthenes also determined another accurate measurement: the Earth's diameter. He deduced the Earth was 7,850 miles (12,631 kilometers) in diameter, which is close to the modern mean value of 7,918 miles (12,740 kilometers).

How do scientists **measure** the **Earth's rotational speed**?

The Earth's rotational speed is based on the sidereal period of the Earth's rotation, but it differs depending on where the observer is located. By dividing the distance traveled once around the Earth by the time it takes to travel that distance, the speed can be determined.

For example, a person on the Earth's equator will travel once around the Earth's circumference—or 24,900 miles (40,079 kilometers)—in one day. To get the speed, divide the miles by the time it takes to get back to the same place (around 24 hours), or just over 1,000 miles (1,609 kilometers) per hour. A person at one of the poles is hardly moving at any speed. This is because there is so little distance traveled in a day (a stick stuck vertically in the ice exactly at the North or South Pole will only travel about 0.394 inch [1 centimeter] per day.)

What about other places on Earth? Traveling north or south from the equator toward the poles decreases one's tangential rotational speed. Thus, the rotational speed at any point on the Earth can be calculated by multiplying the speed at the equator by the cosine of the point's latitude.

But remember that the rotation of the Earth is not always consistent from year to year or even season to season. Scientists know there are other factors in the rotational equation, including the differences caused by widespread climatic conditions. For example, during El Niño years (the periodic upwelling of warmer waters around the equator in the Pacific Ocean off South America), the rotation can slow down. This happened between 1982 and 1983, when the Earth's rotation slowed by 1/5,000th of a second.

What is the **geologic time scale**?

The geologic time scale is a way of managing large amounts of time in a convenient chart. The scale is actually a measurement encompassing the entire history of the Earth—from its beginnings some 4.55 billion years ago to the present day. The largest divisions include eons, eras, and periods; the smaller time divisions include epochs, ages, and subages.

The difference between the Earth's sidereal and solar days has to do with angles and the Earth's rotation. The mean solar day is equal to 24 hours, or the average of all the solar days in an orbital year. The mean sidereal day is 23 hours, 56 minutes, and 04.09053 seconds. It is not exactly equal to a solar day because by the time the Earth has rotated once, it has moved a little in its orbit around the Sun. Thus, it rotates for about another four minutes before the Sun is considered to be back in exactly the same place in the sky as it was the day before.

The actual divisions of geologic time are not arbitrary, or uniform. The larger divisions are based on major events that occurred sporadically over the Earth's long history. For example, the end of the Permian Period, about 240 million years ago, was marked by a major catastrophe. Some scientists estimate that close to 90 percent of all species on the Earth died at that time, resulting in a major extinction event that may have been caused by huge volcanic eruptions or even a space object striking the Earth. The smaller divisions are usually based on specific local structures or fossils found within the rock. Most often they are named after local towns, people, and sundry other nearby associations.

What is the **longest span of time** measured on the **geologic time scale**?

The longest span of time measured on the geologic time scale is the Precambrian Era (also called the Precambrian Eon). It represents the time between 4.55 billion years to about 544 million years ago, or about seven-eighths of the Earth's history. This time period includes the beginning of the Earth's formation, its cool-down, its crust's formation, and, within the last billion years of the time period, the evolution of the first single-celled to multi-celled organisms. The demarcation of 544 million years ago represents a burst in the evolution of multi-celled organisms, including the first plant and animal species.

How do **geologists use angles** to understand **rock layers**?

Mathematics—especially geometry—is instrumental in understanding rock layers. In a branch of geology called stratigraphy, scientists measure angles and planes in rock in order to know the location of certain rock layers and the possible geologic events that affected the layers over time. In particular, geologists measure strike and dip. *Strike* is the angle between true north and a horizontal line contained in any planar feature, such as a fault (usually caused by an earthquake) or inclined bed (often caused by the uplift of hot molten rock around a volcano). *Dip* is the angle at which a bed or rock vein is inclined to the horizontal; it is measured perpendicular to the strike and in the vertical plane (as opposed to the strike's horizontal line).

297

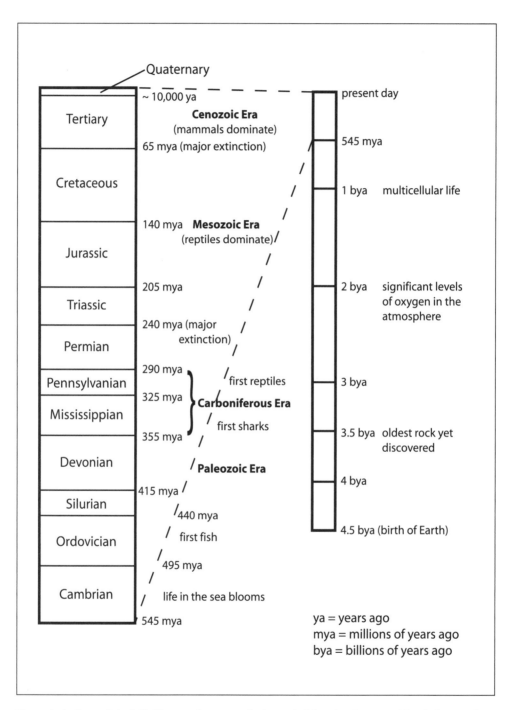

The geologic time scale is divided into epochs, ages, and other periods based on important historical events that radically changed life on Earth.

298

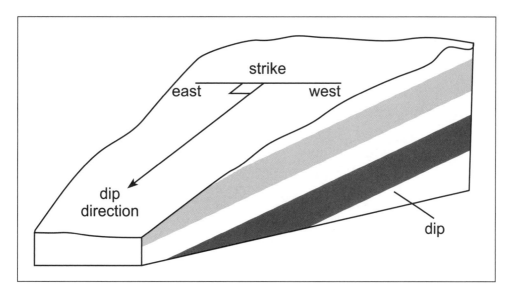

A "dip" is the angle at which a layer of rock or vein is inclined, while a "strike" is the angle made between the direction of true north and the direction of the planar feature, such as an incline or fault.

How are the **shapes of crystals** classified?

Geometry plays an important part in the study of minerals. This is because certain minerals exhibit specific shapes called crystals, with specific crystalline forms occurring when a mineral's atoms join in a particular pattern or internal structure. This arrangement is determined by several factors, including the chemistry and structure of the mineral's atoms, or even the environment in which the crystal grew.

Overall, there are specific angles between corresponding faces of all crystals. Mineralogists (scientists who study minerals) divide these crystalline forms into 32 geometric classes of symmetry; they use this information to identify and classify certain minerals. The crystals are also subdivided into seven systems on the basis of an imaginary straight line that passes through a crystal's center (or axis). The seven groups include cubic (or isometric), tetragonal, orthorhombic, monoclinic, triclinic, hexagonal, and trigonal (or rhombohedral). For example, a crystal in the cubic system has three axes that intersect at right angles; the axes are also of equal lengths. The best way to envision this crystal is to think of a box with equal sides—or a cube.

What is a **carat**?

A carat is a unit of measurement representing the weight of precious stones, pearls, and certain metals (such as gold). It was originally a unit of mass based on the carob seed or bean used by ancient merchants in the Middle East. In terms of weight measurement, a carat equals three and one-fifth grains troy, and it is also divided into four

The geometric arrangement of molecules within the mineral determines the shapes of its crystals. *The Image Bank/Getty Images.*

grains (sometimes referred to as carat grains). Diamonds and other precious stones are estimated by carats and fractions of carats; pearls are usually measured by carat grains (for more about grains and measurement, see "Mathematics throughout History").

Carats of gold are measured based on the number of twenty-fourths of pure gold. For example, 24-carat gold is pure gold (but for a goldsmith's standard, it is actually 22 parts gold, 1 part copper, and 1 part silver, as real gold is too malleable to hold its shape), 18-carat gold is 75 percent pure, 14-carat gold is 58.33 percent pure, and 10-carat gold is 41.67 percent pure gold.

How is **modeling and simulation** used in **geology**?

Like so many other fields of science, mathematical modeling and simulation is used in geology to understand the intricacies of physical events in the past, present, and future. For example, hydrologists (geologists who study water flow above and below the Earth's surface) often use models to simulate the effects of increased groundwater pumping of wells. They may also use a simulation to determine how much water can be presently pumped out of a well, or how much can be pumped out in the future without harm to the environment. Other hydrologists may use modeling to understand the flow of water in a river, bay, or estuary, for example, to determine how the water erodes a shoreline. Still other researchers may model how snow on a volcanic mountain melts, gathers debris, and potentially flows toward populated areas during an eruption event. (For more about modeling and simulation, see "Math in Computing.")

How do geologists **measure** the **intensity of earthquakes**?

Geologists measure the intensity of earthquakes in order to compare and judge potential damage. One of the first standard ways to measure intensity was developed in 1902 by Italian seismologist Giuseppe Mercalli (1850–1914) and is called the Mercalli Intensity Scale (it was later modified and renamed the Modified Mercalli Intensity Scale). The numbers, in Roman numerals from I to XII, represent the subjective measurement of an earthquake's strength based on its effects on local populations and structures. For example, Roman numeral V on the scale represents a quake felt by nearly

What is Mohs' Scale of hardness?

Mohs' Scale of hardness (also seen as Mohs Hardness Scale, Mohs Scale, or even erroneously as Moh's Scale) was invented by German mineralogist Friedrich Mohs (1773–1839). This arbitrary scale measures hardness or the scratch resistance of minerals and is often used as a quick way to help identify minerals in the field and laboratory. But the numbers assigned to the various minerals are not proportional to their actual scratch resistance. Thus, the main reason for using the scale is to know that a mineral with a lower number can be scratched by a mineral with a higher number.

Mineral	Hardness
talc	1
gypsum	2
calcite	3
fluorite	4
apatite	5
orthoclase	6
quartz	7
topaz	8
corundum	9
diamond	10

everyone, with some dishes and windows broken, unstable objects overturned, and disturbances of trees, poles, and other tall objects sometimes noticed.

But scientists wanted a more solid, less subjective scale. One of the first scales developed to measure the true magnitude was invented by American seismologist Charles Francis Richter (1900–1985) and German-born seismologist Beno Gutenberg (1889–1960). In 1935 these scientists borrowed the idea of magnitude from astronomers (stellar brightness is measured by magnitude), defining earthquake magnitude as how fast the ground moved as measured on a particular seismograph a specific distance from the quake's epicenter.

The Richter Scale is not a physical scale like a ruler, but rather a mathematical construct—it is not linear, but logarithmic. Thus, an increase in each whole number on the scale represents a ten-fold increase in power. Its numbers represent the maximum amplitude of seismic waves that occur 62 miles (100 kilometers) from the epicenter of an earthquake. Because seismographs are usually not located at this exact interval, the magnitudes are deduced using the arrival of specific waves of energy given off when an earthquake occurs.

Mt. Everest (right peak, with Mt. Nuptse at left) in Nepal, rising to a height of 29,022 feet, is the tallest mountain on Earth when measuring height compared to sea level. *National Geographic/Getty Images.*

Although the Richter Scale is mentioned most often in the media when a quake occurs, there is a more precise scale in use today that is based on the mathematics of motions caused by the earthquake. Called *moment magnitude,* this method uses a physical quantity related to the total energy released in the quake, which is called a moment. Seismologists can also deduce moment magnitude from a fault's geometry in the field or a seismogram reading. Scientists occasionally use moment magnitude when describing an earthquake event to the public, but because the concept is so difficult to explain the number is often translated into the Richter Scale.

What is **sea level**?

Sea level is the height of the ocean's surface at a certain spot and depends on changing conditions. It is also the basis for most Earth surface measurements, because sea levels are used as a reference point in determining land elevations and ocean depths.

Scientists have averaged out the highest and lowest altitudes and depths from sea level locations: The highest is Mount Everest (Nepal-Tibet), which measures 29,022 feet, 7 inches (8,846 meters) above sea level; the lowest on land is the Dead Sea (Israel-Jordan), which measures 1,299 feet (396 meters) below sea level. The greatest depth below sea level is the Mariana Trench in the Pacific Ocean, a deep chasm measuring 36,201 feet (11,033 meters) below sea level.

What is **mean sea level**?

Mean sea level (MSL) is the average water level (height of the sea) for all stages of a tide. Locally, MSL is measured by tidal gages at one or more points over a given period of time. The resulting numbers average out wind waves and other periodic changes in sea level. The overall values of a MSL are measured with respect to level marks on land called benchmarks. Thus, scientists know a true change in MSL is either from a change in sea level from, for example, possible global warming effects, or changes in the gage's height on land, such as in the case of local uplift.

There is also a more mathematically intensive way to determine the MSL. To a geodesist (a person who studies the shape of the Earth), MSL is determined by comparing measured heights of the global Mean Sea Surface (MSS) above a level reference surface called a geoid—a mathe-

A forester uses a global positioning system (GPS) to accurately pinpoint his location within an undeveloped landscape. Using satellite technology, a GPS can not only determine latitude and longitude, but also heights above sea level. *Stone/Getty Images.*

matical model of an ellipsoid shape that approximates Earth's mean sea level. This comparison is done because the Earth does not have a geometrically perfect shape (for example, the Atlantic Ocean north of the Gulf Stream is about 3.3 feet (1 meter) lower than it is farther south). The MSS is not a "level" surface, thanks to such factors as currents created by wind, as well as atmospheric cooling and heating that cause differences in sea levels around the world. But interestingly enough, it never differs from the global geoid by more than about 6.56 feet (2 meters).

How do scientists use **mean sea level** in connection with **global climate change**?

Many scientists are interested in the long-term mean sea level change, especially in connection with global climate change. By taking such long-term measurements, these scientists are hoping to confirm the predictions of several climate models, including the idea that global warming is a result of the "greenhouse" gases from either human or natural sources.

There are two major ways to determine such sea level variations. The first estimates sea level changes using tide gauge measurements, mathematically averaging the numbers. Graphs of the most recent estimates using this method show a 0.669 to 0.960 inch (1.7 to 2.44 millimeter) rise in sea level per year. The second method uses global positioning system (GPS) devices and satellite altimeter measurements, both of

which accurately pinpoint global ocean heights quickly and more efficiently. For example, from 1994 to 2004, scientists mathematically constructed graphs from satellite altimeter measurements, showing that the global mean sea levels have risen anywhere between 1.10 and 1.18 inches (2.8 and 3.0 millimeters).

No matter what the method, scientists do know the global mean sea levels are slowly rising. Many believe that about one quarter of the rise is caused by thermal expansion as the oceans warm, and another one quarter by small glaciers melting around the world. Some rise may also be caused by such human activities as burning trees, pumping ground water, and draining wetlands. Currently, scientists are not quite certain about the true rate of sea level rise, mainly because of the intensity of working on the data: Ocean-tide gauge records must be averaged, over many decades, and corrected for variable ocean dynamics and distortions of Earth's crust.

MATH IN METEOROLOGY

What is **meteorology**?

Meteorology is the study of atmospheric phenomena, their interactions, and processes. It is often considered part of the Earth sciences and is most commonly associated with weather and weather forecasting.

What is the **composition** of the **air**?

Meteorologists determine the composition of air by analyzing its various constituents; these are mainly displayed in terms of percent of the atmosphere. The first 40 to 50 miles (64 to 80 kilometers) above the surface contains 99 percent of the total mass of the Earth's atmosphere. It is generally uniform in composition, except for a high concentration of ozone, known as the ozone layer, at 12 to 30 miles (19 to 50 kilometers) in altitude.

In the lowest part of the atmosphere—the area in which humans, other animals, and plants live—the most common gases are nitrogen (78.09 percent), oxygen (20.95 percent), argon (0.93 percent), carbon dioxide (0.03 percent), and minute traces of such gases as neon, helium, methane, krypton, hydrogen, xenon, and ozone. Water vapor is also present in the lower atmosphere, although variable and at a very low percent. Higher in the atmosphere, the composition and percentages change as the atmosphere thins. (For more about percents, see "Math Basics.")

How is **air temperature measured**?

Simply put, there are two ways to look at air temperature: On the micro-scale, it is the small scale measure of gas molecules' average kinetic energy; on a larger scale, it is the action of the atmospheric gases as a whole. In physics, an entire branch is devoted

> ## Why do weather reports sometimes say the humidity is 100 percent when there is no rain or snow falling?
>
> A reading of 100 percent humidity usually means there is a high probability that rain will be or is occurring, but not always. It might be 100 percent RH because clouds are forming. If the RH near the ground is much less—for example, if a relatively dry air mass is in place—there will be no rain at the surface. This is why Doppler radar sometimes shows rain or snow in an area when none is actually reaching the ground.

to objects' temperatures and the transfer of heat between objects of differing temperatures. Called thermodynamics, it is a study that entails a great deal of mathematical knowledge.

No matter what the type of temperature discussed, the most common apparatus for measurement is the thermometer. The most familiar thermometers are thin, long, closed glass tubes containing some type of liquid—most often alcohol or mercury. When the temperature increases—or the air around the tube heats up—it causes the liquid to expand, moving it up the tube. Air temperature measurements are most commonly read in Centigrade or Fahrenheit (for more about Centigrade [Celsius] and Fahrenheit, see "Mathematics throughout History").

How are **absolute** and **relative humidity** determined?

Like many other facets of meteorology, mathematics comes in handy when determining absolute and relative humidities. The absolute humidity is the mass of water vapor divided by the mass of dry air in a specific volume of air at a specific temperature. In this instance, the warmer the air, the more water vapor it contains.

On the other hand, relative humidity (RH) is the ratio of the absolute humidity to the highest possible absolute humidity, which in turn depends on the current air temperature. Mathematically, RH is often defined as the ratio of the water vapor density (mass per unit volume) to the saturation water vapor density, usually expressed as a percent. The equation for relative humidity is: RH = actual water vapor density / vapor saturation density × 100 percent.

More commonly, RH is thought of as the amount of water vapor in the air at a given temperature in comparison to the amount that the air *could* contain at the same temperature. For example, if an area is experiencing 100 percent relative humidity, that usually means the air is saturated with (can't hold anymore) water vapor.

What is the **heat index**?

Our bodies dissipate heat by varying the rate of blood circulation, losing water through the skin and sweat glands—and, as a last resort, by panting—when the blood is heated above 98.6 degrees Fahrenheit (37 degrees Centigrade), the average body temperature. Sweating cools the body through evaporation. You can get the same feeling when you put alcohol on your skin, because as the alcohol evaporates, the skin is cooled.

The heat index (HI) is an index that combines air temperature and relative humidity to estimate how hot it actually feels. It is based on a mathematical concept called the heat index equation, a long equation that includes the dry air temperature, relative humidity (in percent form), and many biometeorological factors too long to list here. The resulting heat index table represents the apparent, or "feels like," temperature. For example, if the air temperature is 90 degrees Fahrenheit, with the relative humidity at 60 percent, it will feel like 100 degrees Fahrenheit.

Why do meteorologists want people to pay attention to the heat index? The major reason involves how the body responds to high heat-value numbers: If the relative humidity is high, it curtails evaporation on the skin, and the body is unable to effectively cool itself (and a person will perceive that the air is warmer). When heat index values grow higher, conditions exceed the level a body can remove heat, causing the body temperature to rise. This can cause heat-related illnesses, such as sunstroke or heat exhaustion. For example, according to the United States National Weather Service, exposure to direct sunlight can increase the HI by up to 15 degrees Fahrenheit (9.4 degrees Centigrade). And when a heat index between a mere 90 degrees Fahrenheit (32.2 degrees Centigrade) to 105 degrees Fahrenheit (40.6 degrees Centigrade) can cause possible sunstroke, heat exhaustion, and heat cramps, it is easy to see the meteorologists' concerns.

The following table shows how the heat we actually experience changes with temperature and humidity (humidity is expressed as a percentage; temperatures are in degrees Fahrenheit).

The Heat Index

Actual °F	90%	80%	70%	60%	50%	40%
80°F	85°F	84°F	82°F	81°F	80°F	79°F
85°F	101°F	96°F	92°F	90°F	86°F	84°F
90°F	121°F	113°F	105°F	99°F	94°F	90°F
95°F		133°F	122°F	113°F	105°F	98°F
100°F			142°F	129°F	118°F	109°F
105°F				148°F	133°F	121°F
110°F						135°F

According to the National Weather Service, sunstroke, heat cramps, and heat exhaustion are possible above 90 degrees Fahrenheit; temperatures above 105 degrees can also lead to heat stroke; and above 130 degrees heat stroke is likely if exposure to such temperatures is prolonged.

How is **barometric** (or air) **pressure measured**?

Barometric (or air) pressure, named after the instrument used to measure this pressure, is caused by the weight of the atmosphere pressing down on the land, ocean, and air below, with gravity creating the downward force. Because pressure is dependent on the amount of air above a certain point, pressures are greatest at the surface and less at higher altitudes. On the average, at sea level, the air has a pressure of 14.7 pounds per square inch (a one-inch square has 14.7 pounds of air pressure on each side).

The United States National Weather Service does not measure pressure in pounds per square inch, but in terms of inches of mercury—or how high the pressure pushes mercury in a sealed tube. Air pressure aloft is reported in millibars (or hectopascals [hPa], a term most often used by scientists to measure air pressure).

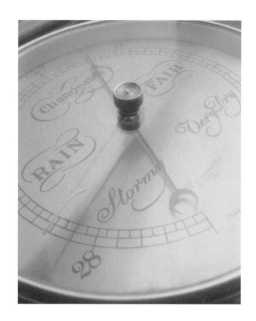

A barometer is a type of gauge that measures atmospheric pressure, which makes it useful in detecting high and low pressure systems that can predict changes in the weather. *Taxi/Getty Images.*

Most of us are very familiar with air pressure as it changes with the weather. For example, the use of the terms "high pressure system" and "low pressure systems" are often indicators of the types of weather fronts traveling through a region. In general, falling air pressure (seen on a barometer) means that clouds and precipitation are more likely; rising air pressure means that clear weather is more likely. In addition, many people experience "personal" changes in air pressure. For example, the discomfort or even pain felt in a person's ears as they ascend or descend in an airplane, a large hill, or even in an elevator is evidence of changing air pressure.

How are **millibars converted** to **inches of mercury**?

Mathematically, the conversion is simple. The air pressure at sea level is 29.92 inches of mercury, or 1,013.2 millibars. For example, if you see air pressure of 1,016 millibars on a weather map, convert to inches of mercury by multiplying by 29.92, and then divide by 1,013.2. The result is 30.00 inches of mercury.

How much does **air pressure decrease** with **altitude**?

It takes mathematics to figure out how much air pressure decreases with altitude. Close to the surface, and due to the pull of gravity, the air pressure exerted by air mol-

307

The expression "thin air" is actually a reference to the atmosphere's density—or how "thick" the air molecules are near the Earth's surface. In chemistry terms, density is merely the mass of anything (including air) divided by the volume the mass occupies. For example, the density of dry air at sea level is high, mainly due to the pull of gravity. In metric system terms, sea level density is about 1.2929 kilograms/meter3, or about 1/800th the density of water. But as altitude increases, the density drops dramatically. Mathematically speaking, the density of air is proportional to the air pressure and inversely proportional to temperature. Thus, the higher up one is in the atmosphere, the lower the air pressure and the lower the air density.

There is somewhat of an athletic advantage to higher elevations—at least for players of such sports as football and baseball. Because the air density is lower, a ball thrown in high-elevation places like Denver, Colorado, will travel even farther than a ball thrown in a close-to-sea-level city, such as Miami, Florida. In fact, the air at the Denver stadium allows balls to travel almost 10 percent farther.

ecules is greatest (around 1,000 millibars at sea level). From there, it declines quickly with altitude to 500 millibars at around 18,000 feet (5,500 meters). At 40 miles (64.37 kilometers), it will be 1/10,000th of the surface air pressure.

This can also be interpreted another way: For altitudes of less than about 3,000 feet (914.4 meters), the barometric air pressure decreases about 0.01 inches of mercury for each 10 feet (3 meters) of altitude (or a decrease of 1 inch of mercury for each 1,000 foot [304.8 meters] gain in altitude). If millibars are used, it is 1 millibar for every 26.25 foot (8 meter) altitude gain. That means if a person takes a ride in an elevator, hits the button for the 50th floor—and coincidentally has a barometer in his or her pocket—the pressure would fall by approximately 0.5 inch (1.27 centimeters) during the ascent. This also means that higher-altitude cities have major differences in barometric readings. For example, the air pressure in almost mile-high Denver, Colorado, is only 85 percent that of cities that reside at sea level.

How is **wind measured**?

Wind speed is the measurable motion of air with respect to the surface of the Earth. It is measured in terms of a unit distance over a unit time, such as miles per hour. The wind direction is also an indication of the wind's source. For example, a southerly wind means the wind is blowing toward north—it is coming from a southerly direction.

What is the **new formula** used for calculating **wind chill**?

Most people know about wind chill: the temperature your body feels when it is exposed to a certain air temperature combined with a particular wind speed. The higher the wind speed, the "colder" the wind chill temperature and the faster the exposed areas of a person's body will lose heat (a process known as transpo-evaporation; when moisture evaporates, the surface from which it evaporates loses some heat).

The newest wind chill chart—called the Wind Chill Temperature Index—took over from the old chart (developed in 1945) in the 2001–2002 winter season. The reason for the change was simple: The original wind chill index revolved around heat loss, with a standard set at the chill experienced while standing outside in air moving 4 miles (2 kilometers) per hour. Based purely on temperature and wind—and on how water freezes in plastic containers—the charts were developed in Antarctica by Paul Siple and his fellow explorer, P. F. Passel, back in 1939, partly with the intention of being used in World War II battlefield planning.

Not everyone was thrilled with this simplistic, two-factor interpretation, however. There were pieces missing from the wind chill puzzle, such as the fact that humans constantly generate heat to the lack of wind measurements above 40 and below 5 miles per hour (64 and 5 kilometers per hour). Passel and Siple's wind speeds were also taken about 33 feet (10 meters) above the ground, making the chart more valuable for a third-floor office than ground level. But the biggest problem overall was that the old wind chill chart could not accurately predict how humans perceive temperature.

Thus, the new wind chill index was created. This chart includes such changes as wind speeds calculated at the average height of a human head (about 5 feet [1.52 meters] above the ground); it is based on a human face model and sundry other more "modern" considerations. The actual general formula for the wind chill has now changed to the following: Wind chill in degrees Fahrenheit = $35.74 + 0.6215T - 35.75(V^{0.16}) + 0.4275(V^{0.16})$, in which T is the air temperature (in degrees Fahrenheit), and V is the wind speed (in miles per hour).

The biggest difference between the old and new indexes is that the new index usually registers warmer temperatures than the old index. Still, no matter what the equation or chart, when temperatures are icy cold and winds are high, everyone should be careful and bundle up.

Calculating the Wind Chill

Wind Speed (mph)

0	5	10	15	20	25	30	35	40	45	50	55	60
40°F	36°F	34°F	32°F	30°F	29°F	28°F	28°F	27°F	26°F	26°F	25°F	25°F
35°F	31°F	27°F	25°F	24°F	23°F	22°F	21°F	20°F	19°F	19°F	18°F	17°F
30°F	25°F	21°F	19°F	17°F	16°F	15°F	14°F	13°F	12°F	12°F	11°F	10°F

Wind Speed (mph)												
0	5	10	15	20	25	30	35	40	45	50	55	60
25°F	19°F	15°F	13°F	11°F	9°F	8°F	7°F	6°F	5°F	4°F	4°F	3°F
20°F	13°F	9°F	6°F	4°F	3°F	1°F	0°F	−1°F	−2°F	−3°F	−3°F	−4°F
15°F	7°F	3°F	0°F	−2°F	−4°F	−5°F	−7°F	−8°F	−9°F	−10°F	−11°F	−11°F
10°F	1°F	−4°F	−7°F	−9°F	−11°F	−12°F	−14°F	−15°F	−16°F	−17°F	−18°F	−19°F
5°F	−5°F	−10°F	−13°F	−15°F	−17°F	−19°F	−21°F	−22°F	−23°F	−24°F	−25°F	−26°F
0°F	−11°F	−16°F	−19°F	−22°F	−24°F	−26°F	−27°F	−29°F	−30°F	−31°F	−32°F	−33°F
−5°F	−16°F	−22°F	−26°F	−29°F	−31°F	−33°F	−34°F	−36°F	−37°F	−38°F	−39°F	−40°F
−10°F	−22°F	−28°F	−32°F	−35°F	−37°F	−39°F	−41°F	−43°F	−44°F	−45°F	−46°F	−48°F
−15°F	−28°F	−35°F	−39°F	−42°F	−44°F	−46°F	−48°F	−50°F	−51°F	−52°F	−54°F	−55°F
−20°F	−34°F	−41°F	−45°F	−48°F	−51°F	−53°F	−55	−57°F	−58°F	−60°F	−61°F	−62°F
−25°F	−40°F	−47°F	−51°F	−55°F	−58°F	−60°F	−62°F	−64°F	−65°F	−67°F	−68°F	−69°F
−30°F	−46°F	−53°F	−58°F	−61°F	−64°F	−67°F	−69°F	−71°F	−72°F	−74°F	−75°F	−76°F
−35°F	−52°F	−59°F	−64°F	−68°F	−71°F	−73°F	−76°F	−78°F	−79°F	−81°F	−82°F	−84°F
−40°F	−57°F	−66°F	−71°F	−74°F	−78°F	−80°F	−82°F	−84°F	−86°F	−88°F	−89°F	−91°F
−45°F	−63°F	−72°F	−77°F	−81°F	−84°F	−87°F	−89°F	−91°F	−93°F	−95°F	−97°F	−98°F

What are the **major scales** used in interpreting **hurricanes and tornadoes**?

There are two major scales used to interpret hurricane and tornado intensity—and thus potential damage. The *Saffir-Simpson Hurricane Damage-Potential Scale* is a hurricane force scale using the numbers 1 through 5 to rate a hurricane's intensity. The scale was developed by engineer Herbert Saffir (1917–) and pioneer hurricane expert Robert Simpson (1912–) in 1971. A number on the scale is assigned to a hurricane based on its peak wind speed; it is also used to give an estimate of the potential property damage and flooding expected along the coast from a hurricane landfall.

Saffir-Simpson Hurricane Scale

Category	Wind Strength	Pressure (mb)	Effects
1	74-95 mph (119-153 kph; 65-83 knots)	>980 mb	Oceans surge 4 to 5 feet above normal; minimal damage; some bushes and trees lightly damaged; unanchored mobile homes may shift; some coastal flooding and pier damage
2	96-110 mph (154-177 kph; 84-95 knots)	980-965 mb	Waters surge 6 to 8 feet above normal; considerable vegetation damage; some door, window, and roof damage; mobile homes and poorly built piers receive significant damage; coastal and low-lying areas flood from 2 to 4 hours before center hurricane reaches shore

3	111-130 mph (178-209 kph; 96-113 knots)	964-945 mb	Surges are 9 to 12 feet above normal; small homes and utility buildings receive structural damage; leaves blown off trees and some large trees blown over; mobile homes destroyed; coastal and low-lying areas flood 3 to 5 hours before center of hurricane reaches shore; evacuations of low-lying areas may be needed
4	131-155 mph (210-249 kph; 114-134 knots)	944-920 mb	Waters surge 13 to 18 feet above normal; small homes may lose roofs; trees and signs blown down; mobile homes destroyed; many doors and windows destroyed; low-lying and coastal areas flood 3 to 5 hours before center of hurricane reaches shore; all terrain lower than 10 feet above sea level should be evacuated before flooding; homes as far as 6 miles inland can be affected
5	>155 mph (>249 kph; >135 knots)	<920 mb	Waters surge more than 18 feet above normal levels; many house and commercial buildings lose their roofs; some small buildings and small utility structures completely destroyed; all vegetation leveled; major damage to lower levels of homes and other buildings that are less than 15 feet above sea level and within 500 yards of shoreline; massive evacuations required

The *Fujita-Pearson Tornado Intensity Scale* (or "F-Scale") is used to measure tornado wind speeds. It was developed in 1971 and named after Tetsuya Theodore Fujita (1920–1998) of the University of Chicago and Allan Pearson, who was then head of the National Severe Storms Forecast Center in Kansas City. It was Fujita who came up with a system to rank tornadoes according to how much damage they cause. He developed his categories by connecting the twelve forces of the Beaufort wind scale (knots based on what the sea surface looks like—from smooth to waves over 45 feet) with the speed of sound (Mach 1). Then, for each category he estimated how strong the wind must be to cause certain observed damages. Fujita's scale was later combined with Pearson's scale, which measures the length and width of a tornado's path, or its contact with the ground.

Fujita Tornado Scale

F-Scale	Intensity	Wind Strength	Frequency	Damage Description
F0	Gale tornado	40-72 mph (64-116 kph; 35-62 knots)	29%	Some branches break off trees; some damage to buildings, such as to roof shingles and windows; billboards and other signs show damage

F-Scale	Intensity	Wind Strength	Frequency	Damage Description
F1	Moderate tornado	73-112 mph (117-180 kph; 63-97 knots)	40%	Trees uprooted; cars overturned or pushed off roads; mobile homes shoved off foundations; roof surfaces torn off houses
F2	Significant tornado	113-157 mph (181-253 kph; 98-136 knots)	24%	Mobile homes destroyed; sheds and other small buildings destroyed; roofs on houses completely torn off; large trees broken and uprooted; train boxcars tilt over; some light objects become projectile missiles
F3	Severe tornado	158-206 mph (254-332 kph; 137-179 knots)	6%	Walls and roofs on houses blown down; metal buildings severely damaged or collapsed; forests and farmland destroyed; large cars lifted into the air and thrown; trains overturned
F4	Devastating tornado	207-260 mph (333-419 kph; 180-226 knots)	2%	Well-built homes nearly completely destroyed; cars thrown long distances; some structures with weaker foundations lifted and thrown; large pieces of concrete and steel become deadly projectiles
F5	Incredible tornado	261-318 mph (420-512 kph; 227-276 knots)	<1%	Homes destroyed, with even those with good foundations lifted and thrown long distances; large office and other buildings have considerable damage; trees debarked; cars and other car-sized objects become projectiles

| F6 | Inconceivable tornado | 319-379 mph (513-610 kph; 277-329 knots) | <1% | Such tornadoes are highly unlikely; the areas they damage would not be recognizable and evidence for them would probably only be detected through engineering studies; such areas would likely be surrounded by areas damaged by F5 forces, with cars, large appliances, and other large objects causing additional damage |

What is **numerical weather prediction**?

Numerical weather prediction is forecasting the weather using numerical models. Because of the complexity of the mathematics involved—not to mention the number of variables needed to predict the weather—all numerical model studies are run on high-speed computers. The computer solves a set of equations, resulting in a computer model of the atmosphere showing how weather conditions will change over time.

How do **computer models** attempt to **predict the weather**?

In general, computer models used to predict weather use around seven equations that govern how the basic parameters—temperature, pressure, and so on—change over time in the atmosphere. Scientists call the study of how they can physically and mathematically represent all the processes in the atmosphere *dynamics*.

In reality, everyone knows computer models can't perfectly predict the weather at this time. This is because of several factors, including errors in the initial conditions (or the observations the model gets to begin making its forecast) and errors inherent in the model (a computer model can't take into consideration all the factors controlling the weather). Long-term forecasts are even more inaccurate because these two errors are compounded mathematically over time.

What are some **examples** of **weather prediction models**?

Because there is more than one group carrying out weather predictions, there

The strength of a tornado is rated on the Fujita-Pearson Tornado Intensity Scale, which takes into account wind speeds and damage created by the twister. *The Image Bank/Getty Images.*

are many computer models used by meteorologists around the world. For example, the United States National Weather Service's weather predictions are carried out at the National Centers for Environmental Prediction (NCEP). The NCEP runs several different computer models each day to determine the best weather forecasts. Some are used for short-term forecasting, others for the longer term; and some are used for global or hemispherical predictions, while others are only regional. They include several mathematically intensive computer models (for more information about computer modeling, see "Math in Computing").

NGM—NGM, or the Nested Grid Model, is one in which observations are converted to values at various points that are evenly spaced, making it easy for the computer programs to plug them into equations. This model is now considered to be obsolete.

ETA—The ETA model was named after the ETA coordinate system, which is a mathematical coordinate system that takes into account topographical features, such as mountains. It is similar to the NGM model and forecasts the same atmospheric variables, but its smaller grid gives a more detailed forecast.

AVN, MRF, and GSM—The AVN model, MRF (Medium Range Forecast), and the GSM (Global Spectral Model) convert data into a large number of mathematical waves; they then return the waves in a manner that will produce a forecast map.

ECMWF—The ECMWF (European Center for Medium-Range Weather Forecasts) is considered to be one of the most advanced weather forecast models in the world; it is mostly used for the Northern Hemisphere.

UKMET—The UKMET (United Kingdom meteorology offices) model also gives forecasts for the entire Northern Hemisphere.

MM5—The MM5 (Mesoscale Model #5) and WRF (Weather Research Forecast model) are actually the same models. The MM5 has long been a research computer model for smaller geographic forecast regions such as Antarctica; the WRF is the name for MM5 as an operational model, not just for research.

MATH IN BIOLOGY

What is **biology**?

Biology is the science of life. It includes the study of the characteristics and behaviors of organisms; how a population, species, or individual comes into existence and evolves; and the interaction of organisms with the environment and each other.

What is **mathematical biology**?

Mathematical biology is another word for biomathematics, the interdisciplinary field that includes the modeling of natural biological processes using mathematical tech-

niques. Mathematical biology is carried out by mathematicians, physicists, and biologists from various disciplines within their fields. These scientists work on such problems as modeling blood vessel formation, with possible applications to drug therapies; modeling the electrophysiology of the heart; exploring enzyme reaction within the body; and even developing models that track the spread of disease.

A simple Mendel matrix.

What is **population dynamics**?

One major area of interest in mathematical biology is population dynamics. A population is the number of individuals of a particular species in a certain area; population dynamics deals with the study of short- and long-term changes in certain biological variables in one or several populations.

Population dynamic studies have actually been around for centuries. For example, weight or age comparisons of human or other animal populations—or even how such populations grow and shrink over time—have long been areas of study. With regard to human populations, the two simplest kinds of input in a population study are birth and immigration rates, and the two basic outputs are death and emigration rates. If the inputs are greater than the outputs, the population will grow; if the outputs are greater than the inputs, the population will shrink.

How does **population dynamics** use **mathematics**?

Population dynamics combines observations and mathematics, especially the use of differential equations (for more about differential equations, see "Mathematical Analysis"). For example, to determine what the population of a certain country will be in ten years, scientists use a mathematical model commonly called the exponential model, or the rate of change of a population as it is proportional to the existing population. (For more about population growth mathematics and the environment, see below.)

Who was **Gregor Mendel**?

Austrian monk Gregor Johann Mendel (1822–1884) performed experiments with pea plants from 1857 to 1865 that eventually led to his discovery of the laws of heredity. Gathering 34 different kinds of peas of the genus *Pisum* (all tested for their purity), he attempted to determine the possibility of producing new variants by cross-breeding. By self-pollinating the plants—and covering them over so there was no unplanned

cross-pollination—he determined the detailed characteristics of their offspring, such as height and color.

Prior to Mendel, scientists believed that heredity characteristics of a species were the result of a blending process, and that over time various parental characteristics were diluted. Mendel showed that characteristics actually followed a set of specific hereditary laws. He worked out what can be described as a mathematical matrix of the characteristics, thus determining what characteristics were dominant and recessive in the plants. (For more about matrices, see "Algebra.")

But Mendel had a hard time getting his results published. Even after publication by a local natural history society, his work was ignored. Mendel gave up both gardening and science when he was promoted to abbot. Coincidentally, and amazingly, by 1900 three different biologists working in three different countries—Hugo de Vries in the Netherlands, Erich Tschermak von Seysenegg in Austria, and Karl Correns in Germany—determined the hereditary laws independently. But they all knew about Mendel's work, graciously giving the credit for the findings to him. Mendel, rightfully, is now commonly considered the "father of genetics."

What were some contributions **John Haldane** made to **genetics**?

Scottish geneticist John Burdon Sanderson Haldane (1892–1964), along with Sir Ronald Aylmer Fisher (1890–1962) and Sewall Green Wright (1889–1988), developed population genetics. Among other contributions, Haldane's famous book *The Causes of Evolution* (1932) was the first major work of what came to be known as the *modern evolutionary synthesis*. It made use of Charles Darwin's theory of the evolution of species by natural selection, presented in terms of the mathematical consequences of Gregor Mendel's theory of genetics, to form the basis for biological inheritance.

What is **computational biology**?

Computational biology refers to biological studies that include computation, mainly with computers. Many biologists study computational biology to develop algorithms

and software to manipulate and analyze biological data; they also use computers to develop and apply certain mathematical methods to analyze and simulate molecular biological processes.

Another good reason for this marriage of biology and computers is obvious in today's world: the genome—human and otherwise. To take on the giant task of mapping genomes (the entire collection of genes in a species), scientists have turned to the computer, using it for such studies as genomic sequencing, computational genome analysis, and protein structure analysis.

Computational power is needed for a plethora of other tasks, too. For example, it is being used to develop methods to predict the structure and function of newly discovered proteins and structural RNA

Charles Darwin took into account Mendel's ideas about genetics in forming his theory of evolution. *Library of Congress.*

sequences in humans and other organisms, to group protein sequences into families of related sequences, and to generate phylogenetic trees (or lineage trees, such as the human relationship to apes) to examine evolutionary connections.

What is **bioinformatics**?

Bioinformatics is a field that evolved by joining biology and information science. In the past few decades, advances in molecular biology and the increase in computer power have allowed biologists to accomplish tasks such as mapping large portions of genomes of several species. For example, a baker's yeast called *Saccharomyces cerevisiae* has been sequenced in full.

Humans have not been exempt, either: The Human Genome Project was completed in 2003. It determined the complete sequence of the three billion DNA subunits (bases) for humans, identified all human genes, and made all the associated information accessible for further biological study. Since that time, other universities and agencies have taken on the task of analyzing the results, such as determining the gene number, exact locations, and functions. Such a deluge of information has also made it necessary to store, organize, and index all the sequence data, which is where information science, or the method to store and work on such large amounts of data, comes in the form of bioinformatics. The computer experts who deal with such information are known as bioinformatics specialists.

How many bases are in a human's genome sequence?

There's a good reason why computers are so important to biologists working on the human genome. The amount of data is staggering, and would take scientists generations to analyze without the benefit of computers. For example, it would take about 9.5 years to read out loud (without stopping) the three billion bases in a person's genome sequence. This is calculated on a reading rate of 10 bases per second, equaling 600 bases per minute, 36,000 bases per hour, 864,000 bases per day, and 315,360,000 bases per year.

One million bases (called a megabase and abbreviated Mb) of DNA sequence data is roughly equivalent to one megabyte of computer data storage space. Because the human genome is three billion base pairs long, three gigabytes of computer data storage space are needed to store the entire genome. This includes something called nucleotide sequence data only and does not include other information that can be associated with sequence data. Because of such numbers, scientists working on the human genome are grateful they have computers on their side!

MATH AND THE ENVIRONMENT

What is **ecology**?

Ecology (also known as bionomics) is a branch of biology that deals with the abundance and distribution of organisms in nature, as well as the relations between organisms and their environment. It is an inherently quantitative science, with ecologists using sophisticated mathematics and statistics to describe and predict patterns and processes in nature.

How is **mathematics** used to **describe population growth** of organisms in a **certain environment**?

In general, a certain population of organisms—from rabbits to humans—will grow exponentially if it is left unchecked. That means that in a "perfect world" a population's rate of increase is constant. This can be seen in the following equations:

- after 1 year $= P_0(1 + r)$
- after 2 years $= P_0(1 + r)^2$
- after 3 years $= P_0(1 + r)^3$

 ...

- after n years $= P_0(1 + r)^n$

in which P_0 is the population today and r is the rate of increase. These equations are further modified for population growth statistics, but such intricate calculations are not within the scope of this book. (For more about population growth and biology, see above.)

Who were **John Graunt** and **Sir William Petty**?

English statistician John Graunt (1620–1674) is generally considered to be the founder of the science of demography, which is the statistical study of human populations. In 1661, after analyzing some major statistics of the London populace, he wrote what is considered the first book on statistics, *Natural and Political Observations upon the Bills of Mortality*.

Sir William Petty was a practical mathematician who wanted to establish a national statistics office in England that, among other studies, could calculate economic losses and benefits due to the plague. *Library of Congress.*

The "Bills of Mortality" refers to the collections of mortality figures in London, a city that had suffered greatly from the outbreak of several plagues. Because the king wanted an early warning system of new outbreaks, weekly mortality records were kept, along with the causes of death. Based on this information, Graunt made an estimate of London's population that is thought to be the first time anyone interpreted such data; it is therefore considered by some to mark the beginnings of population statistics.

Graunt's work influenced his friend, Sir William Petty (1737–1805). (He also influenced Edmond Halley, the astronomer and discoverer of Comet Halley; for more about Halley, see "Mathematics in the Physical Sciences.") Petty's work was a bit more practical (and political): he wanted to set up a central statistical office for the English crown in order to make estimates about the sum of England's overall wealth. His unusual approach was to assume that the national income was the same as the total national consumption. He didn't forget about the plague, but added estimates of losses to the national economy due to the plague. From there, he suggested that modest investment by the state to prevent deaths from plague would produce abundant economic benefits.

What is a **logistic equation**?

A logistic equation (resulting in a curve on a graph) represents the exponential increase in numbers of a species until it reaches the carrying capacity in its specific 319

environment. This carrying capacity, usually referred to by the letter K, is the maximum population size that can be regularly sustained by an environment. Change the environment and K changes, for example, by such events as adding a predator, removing a competitor, or adding a parasite. The notation that follows (in the form of a differential equation) represents a rate of population increase that is limited by interspecific competition (see above):

$$\frac{dN}{dt} = rN\frac{(K-N)}{K}$$

in which N is the population size, t is time, K is the carrying capacity, and r is the intrinsic rate of increase.

What are **survivorship curves**?

Survivorship curves record and plot the fate of the young, and their chances of survival in key age categories. Significant factors affecting all populations are birth rates, death rates, and longevity. By recording the numbers of births and deaths over a period of time, researchers can determine the average longevity of organisms in each age class; these numbers tell a great deal about a population.

There are three basic survivorship curves. Type I curves represent species that have offspring with a high survival rate, with most living to a certain age and then dying; humans are an example. Type II curves represent organisms with a steady death rate from the time they are born or hatch until they die; their survivorship varies and includes such species as deer, large birds, and fish. Type III curves include those organisms that have a low survivorship shortly after being born, but with a high longevity for the individual organisms that survive; maple and oak trees can be included in this category.

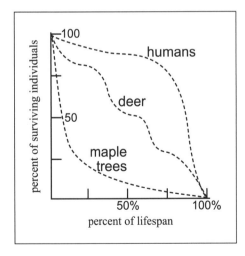

In this sample graph of a survivorship curve, it is easy to see how the survival rates of maple trees, deer, and people vary greatly over time.

What is the **air quality index**?

Mathematics plays an important part in the air quality index (AQI), a scale developed by the U.S. government to measure how much pollution is in the air. The AQI measures five specific pollutants: ozone, particulate matter, carbon monoxide, sulfur dioxide, and nitrogen dioxide. The levels range from 0 (good air quality) to 500 (hazardous air quality); the higher the index, the higher the level of pollutants and the greater the likelihood of detrimental health effects.

What are the Lotka-Volterra Interspecific Competition Logistic Equations?

The Lotka-Volterra Interspecific Competition Logistic Equations are concerned with the predator-prey relationships between species in the environment, and are based on differential equations (for more on differential equations, see "Mathematical Analysis"). Such predator-prey theories were developed independently by then-Austrian (now the Ukraine) chemist, demographer, ecologist, and mathematician Alfred James Lotka (1880–1949) and Italian mathematician Vito Volterra (1860–1940) in 1925. They refer to *interspecific competition*, or the competition between two or more species for some limiting resource, such as food, nutrients, space, mates, nesting sites, or anything in which the demand is greater than the supply.

Most people think about the AQI in terms of being outdoors—and most weather broadcasts include air quality listings, especially in larger cities. When the readings are high, people are warned not to participate in strenuous activities like sports or hard work outside; people with asthma or other lung problems are urged to stay inside.

What is **environmental modeling**?

As with most of the sciences, mathematical modeling and computer simulations also come in handy for environmental applications on a local, regional, and global scale. For example, scientists model environmental landscape changes, global climate change and the impacts on ecosystems, watershed and reservoir interactions, and forest management and sustainability.

What is **computational ecology**?

Computational ecology can be considered a subset of environmental modeling, because it addresses practical questions arising from environmental problems using mathematics. For example, in the field of ecotoxicology, mathematical models are used to predict the effects of environmental pollutants on populations. Natural resource management uses mathematics to set quotas for fish and game. And conservation ecologists use mathematical models to determine the effects of various recovery plans for threatened species, and even to design nature preserves.

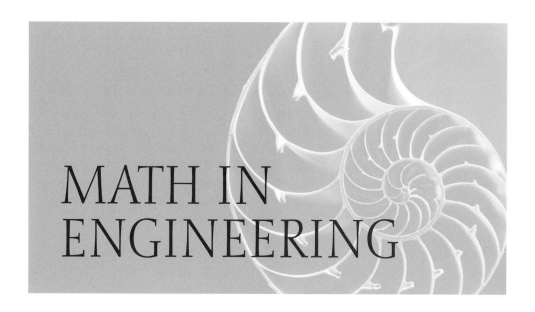

MATH IN ENGINEERING

BASICS OF ENGINEERING

What is **engineering**?

Engineering is a discipline that deals with the "art" or science of applying scientific knowledge to solve practical problems, usually in the areas of commerce and industry. Scientists ask the "why" of a question, then research the answer; in contrast, engineers want to know how to solve the problem and then how to implement the solution. But it's not always easy to separate the two. Often a scientist has to use engineering basics (such as building special equipment for research), and engineers often have to do scientific research.

The word engineer (as well as engine) developed from the Latin root *ingeniosus* ("skilled"); in some languages, such as Arabic, the word for engineering also means "geometry." The various branches of engineering include aerospace, agriculture, architectural, biomedical, computer, civil, chemical, electrical, environmental, mechanical, petroleum, and material science.

What **types of mathematics** are used in **engineering**?

Mathematics is definitely a necessity in engineering, especially the fields of algebra, geometry, calculus, and statistics. Certain divisions of engineering rely on variations of mathematics, including combinations of arithmetic, algebra, geometry, calculus, differential equations, probability and statistics, complex analysis, and others. For example, civil and structural engineers use a great deal of linear algebra and work with matrices. Mechanical engineers use logs and exponents, calculus, differential equations, and probability and statistics. And a chemical engineer uses such mathematics as algebra and geometry, logs and exponents, integral calculus, and differential equations.

The accomplishments of French mathematician and physicist Baron Jean Baptiste Joseph Fourier (1768–1830) prove that not all famous mathematicians did just math. Fourier was a teacher, became involved in the messy French Revolution, and was arrested for his views and imprisoned in 1794. For a time, he even feared the guillotine, but political changes resulted in Fourier being freed. By 1798 Fourier joined Napoleon's army in its invasion of Egypt as a scientific adviser. After Napoleon lost the Battle of the Nile to Nelson and was confined to Malta, Fourier continued his work in Egypt, establishing educational facilities there and carrying out archeological explorations. Back in France with Napoleon in 1801, he supervised the draining of the swamps of Bourgoin and the construction of a highway from Grenoble to Turin. He also spent time writing *Description of Egypt,* a book that Napoleon edited, and included some historical rewrites (by the second edition of the book, Napoleon himself would be completely edited out of the text).

As if he wasn't busy enough, during this time Fourier wrote his now-famous 1807 paper, *On the Propagation of Heat in Solid Bodies,* a mathematical work on the theory of heat that presented one of his major contributions: the Fourier series. But it was an uphill battle to get approval from his peers. In 1811 he submitted his 1807 ideas for a mathematics prize, along with additional work on the cooling of infinite solids and on terrestrial and radiant heat. Only one other entry was received, making Fourier's work the obvious winner. Finally, by 1822, he published his 1811 essay, making the techniques of Fourier analysis available to everyone. To this day, the functions that he worked out have a multitude of applications in engineering, science, and mathematics.

What are **interpolation** and **extrapolation**?

Interpolation in mathematics involves finding a value (or outcome) of a function between already known values; in other words, it is a method of estimating the values in between sampled data points. Extrapolation in mathematics is estimating the value of a problem beyond the range covered by the existing data. Both methods are used a great deal in engineering.

What is a **Fourier series**?

The idea for the Fourier series was developed by French mathematician and physicist Baron Jean Baptiste Joseph Fourier (1768–1830) as an alternate method of expressing a function by the expansion of the function. A Fourier series is actually a specific type

of infinite mathematical series that involves trigonomic functions. More simply put, it is essentially an infinite sum of sine waves. The Fourier series is used in applied mathematics. In engineering and physics, it is used to split up a periodic (or continuous) function into a group of simpler terms; in electronics, it is used to express the periodic functions seen in waveforms of communications signals.

What is a **finite element analysis**?

A finite element analysis (also known as FEA or finite element method) is a powerful tool to solve problems in engineering, especially for heat transfer, fluid mechanics, and mechanical system problems. The FEA consists of a computer model of a material or design that is stressed; the outcome is then analyzed for specific results. In reality, the computer is conducting a numerical analysis technique used for solving differential equations, and relating it to stress in the engineering problem.

When Napoleon Bonaparte led his army into Egypt, French mathematician and physicist Baron Jean Baptiste Joseph Fourier joined him and was a scientific adviser. Fourier also worked for Napoleon when he was involved in draining the Bourgoin swamps and building a new highway. *Library of Congress*.

This technique was first developed in 1943 by Richard Courant (1888–1972), who used a form of FEA to find approximate solutions to vibrational systems. Early in the 1970s, only companies that owned expensive mainframe computers were using FEA, including the aeronautic, automotive, defense, and nuclear industries. Since the mid-1990s, however, use of FEA has grown with the advent of faster and cheaper computers with more memory. The results are more accurate, too, allowing various industries to analyze new product designs and refine existing products.

What **types of analyses** interest engineers?

There are several types of analyses that interest engineers, all of which involve mathematical modeling. *Structural analysis* deals with linear and nonlinear models and stresses on a material. The linear models assume the material does not plastically deform (the remaining deformation after the load causing it is removed); nonlinear models stress the material past its elastic capabilities. The stresses in the material then vary with the amount of deformation. *Vibrational analysis* deals with possible resonance and subsequent failure. It is used to test a material that may experience

Why is finite element analysis important to many industries?

Finite element analysis (FEA) is important to various industries—especially those that need to predict failure of a structure, object, or material when under unknown stresses—because it allows designers to understand all of the theoretical stresses within the structure. This cuts manufacturing costs that would occur if a sample of the structure was actually built and tested.

FEA uses a complex system of points (nodes), making up a grid called a mesh. The mesh is programmed to contain all the material, properties, and other factors that constitute the structure and determine how it will react to certain load conditions, such as thermal, gravitational, pressure, or point loads. The nodes are then assigned a density throughout the material, all depending on the stress levels anticipated in a certain area. In general, points with more stress (such as corners of a building or contact points on a car frame) will usually have a higher node density than those with little or no stress. As researchers examine the results of the FEA, they learn how the structure responds to the various stresses. In this way, a prototype of the structure won't have to be built until the majority of the theoretical "kinks" are worked out of the system.

random vibrations, impacts, or shocks. *Fatigue analysis* is used to determine the life of a material or structure. It shows the effects of occasional (periodic) or cyclic loading on a structure or object, pointing out where cracks or fractures are most likely to occur. Engineers measure *heat transfer* to determine a material or structure's conductivity or thermal fluid dynamics. In this way, researchers understand how a material will respond to various hot and cold conditions—or even how it diffuses heat and cold—over time.

What is **dimensional analysis**?

Simply put, dimensional analysis is a way of manipulating unit measures using algebra to determine the proper units for a quantity that is being computed. For example, the units of length over time represent velocity in feet per second; acceleration is velocity over time. Thus, acceleration will then have units of feet per second per second, or feet per second squared.

What is the **least squares method**?

This mathematical procedure, called either the least squares method or the method of least squares, finds the best-fitting curve for a given set of points by minimizing the sum of the squares of all deviations from the curve. It is often used in engineering for

fluid flow, certain elasticity problems, and diffusion and convection in materials.

Why is **Laplace transform** important in engineering?

Laplace transform is a way to solve linear differential equations and translate them into simple algebraic problems that are easier to solve. It was developed by French mathematician and theoretician Marquis Pierre-Simon de Laplace (1749–1827). Although it carries his name, the Laplace transform seems to have been first used by Denis Poisson (1781–1840) in 1815. Today, it is used extensively in electrical engineering problems.

How are **modeling** and **simulation** used in engineering?

Modeling and simulation have become an essential part of engineering on both a small and large scale. Because building any size of structure takes time and money, engineers often develop a mathematical model, a set of equations that

The Petronas Towers in Malaysia, shown here standing at 1,483 feet, were the tallest structure in the world in 1998. But in 2004 Taipei 101, at 1,670 feet high, became the tallest. To construct such massive buildings, engineers need to understand a lot about stress and pressure so that the structures don't collapse under their own weight. *The Image Bank/Getty Images.*

describe what may happen to a structure if it is built the way it is represented by the model. Using a computer (or graphic) representation gives the engineers a three-dimensional view.

For example, before the space shuttle was built, engineers used mathematical modeling to simulate what the craft would look like in three dimensions. In this way, the engineers learned how the shuttle would fly, how strong the heat-resistant tiles had to be in order to reenter the Earth's atmosphere, and even how to maneuver the

327

English electrical engineer Oliver Heaviside (1850–1925) was a self-taught genius who made several contributions to the field of electricity and even atmospheric studies. In 1902 Heaviside predicted that there was a conducting layer in the atmosphere that allowed radio waves to follow the Earth's curvature—a layer now named after him.

In electrical engineering, Heaviside was best known for *operational calculus,* a tool for solving linear differential equations with constant coefficients. It was usually applied to brief or fleeting (called transient) phenomena and was very similar to Laplace transform in its calculations. Although Laplace had developed his ideas almost a century before, Heaviside knew nothing of them, because they were not well-known during his time.

But Heaviside's operational calculus did have its problems, as well as its critics. It was severely limited because of its lack of mathematical theory. This not only limited its applications, but also created many uncertainties and ambiguities in the equations and solutions. Today, operational calculus has been replaced by Laplace transform, especially in fields such as electrical engineering.

shuttle under various conditions as it landed. The computer was the only way to solve such problems without real-life testing. It quickly and easily solved a plethora of mathematical equations—especially calculus and differential equations—that represented how the shuttle would take off, fly, and land.

What is **fluid mechanics**?

Fluids are substances that flow, including gases and liquids. Fluid mechanics, or hydraulics, is the study of the physical behavior of these gases and liquids and their role in engineering systems. This includes the mathematics of the forces in and motion of substances, turbulence, wave propagation, and so on.

Most fluid mechanics problems in engineering are mathematically modeled using differential equations. These models can also be applied to other engineering areas, such as electromagnetism and the mechanics of solids (because solids still "move," albeit slowly). Other fluid mechanical studies include the compressibility of substances. In most cases, liquids are considered to be incompressible and gases are considered to be compressible. But there are exceptions in some everyday engineering applications, and they can easily be explored using mathematical modeling.

Engineers also use special mathematical equations to determine certain characteristics of fluid flow, such as whether the flow is slow and smooth (laminar) or turbu-

lent. For example, the ratio of inertia forces to viscous forces within a fluid can be expressed by what is called the Reynold's number; laminar fluid flow can be described by the Navier-Stokes equations. For no viscosity (or an ideal flow called inviscid flow), the Bernoulli equation can be used. Finally, when the flow is zero (or static), the fluid is governed by the laws and equations of fluid statics.

How **old** is the **study** of **fluid mechanics**?

According to many historians, fluid mechanics may be the oldest subfield in physics and engineering. In particular, ancient civilizations needed to control water flow for agricultural development, drinking-water supplies, and transportation. Thus, the development of fluid mechanics, which is the study of the

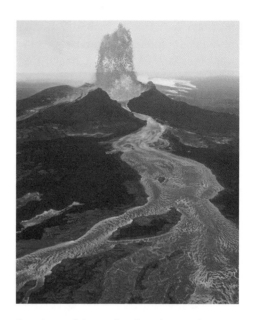

To understand the way lava flows from a volcano, mathematicians apply their knowledge of fluid mechanics. *Science Faction/Getty Images.*

motion and behavior of fluids, led to even more (and complex) improvements. For example, agricultural requirements led to irrigation waterways, dams, weirs, pumps, and even crude forms of "sprinkler systems"; the need for a potable (drinkable) water supply led to better wells, fountains, and water storage systems; and water transportation innovations included improved sails and rigs, as well as methods to build and waterproof sailing vessels.

But early fluid mechanical studies did not end there. Over time, they extended into almost every realm of science and engineering. For example, mechanical engineering uses fluid mechanics because of the need to know about fluids used in combustion (ships and automobiles), lubrication (from the smaller inner workings of a wheel to larger mechanisms such as locks along a canal), and energy systems (hydroelectric power). Civil engineering utilizes fluid mechanical studies to interpret how fluid systems traveled over structures (aqueducts and pipes carrying drinking or waste water). Electrical engineers use fluid flow to analyze how to cool electronic devices with either air or water. Even early (and current) aeronautical engineers needed to know how air flowed over an airplane wing, providing the much-needed lift that allows a plane to become airborne.

How are studies of **fluid mechanics** used **today**?

The list of engineering uses of fluid mechanics in the modern world seems endless—and no wonder, since it is one of the most widely applied areas of mathematics and engineer-

ing. Some of today's uses of fluid mechanics in various fields include: understanding the movement of molten (liquid) rock, or lava, in volcanic eruptions; studying the flow of air over objects to help design airplanes, the space shuttle, and even spacecraft that fly through the atmospheres of other planets; air flow studies in the automobile industry to design cars with more aerodynamic profiles; analyzing the ups and downs of the stock market; examining natural hazards, such as snow conditions that result in an avalanche; interpreting turbulent flow in sewer and water pipes, and in river channels; studying complicated flow of weather patterns in the atmosphere; researching the effects of gravity (and other) waves in space; and using fluid mechanics applications to study the deep oceans and coastal shorelines, including waves and currents.

CIVIL ENGINEERING AND MATHEMATICS

How do **civil engineers** use mathematics?

Although most civil engineers actually spend only a small portion of their time doing calculations, mathematics is essential to this field. For example, civil engineers may use math to carry out the technical calculations in order to plan a construction project. They may use math to model and simulate the possible behavior of a structure before it is actually built. They also use math to understand the necessary chemistry (the strength of a material) and/or physical components (how strong parts need to be) of a construction project.

How do **surveyors** use mathematics?

Surveyors use mathematics—especially geometry and trigonometry—because they need to measure angles and distances on the ground. They then interpret the data, accurately plotting such information as boundaries and locations of structures on a map. These maps are then used for personal or legal means, such as a survey of a person's lot showing ownership boundaries in order to obtain a mortgage. The traditional method of surveying is called plane surveying, which does not take into consideration the curvature of the Earth because, for most small projects, this curvature doesn't really matter. When it does, especially for projects measuring greater distances, the method used is called geodetic surveying.

How do **surveyors** make **measurements**?

Most of a surveyor's measurements are gathered with a theodolite, an instrument that acts as a telescope, ruler, and protractor. The theodolite is set up over a known spot, such as a previously surveyed corner of a lot; its telescope then sights a specific spot, such as another corner of a lot, and the distance is measured (most modern theodolites use lasers to measure distance), supplemented by angular measurements in both the horizontal and vertical planes. The surveyor then uses trigonometry to analyze the

data, converting it to a more usable form, usually in *x, y,* and *z* coordinates. For example, depending on the desired outcome, the vertical angle and slope distances can be converted from polar measurements to show differences in elevations and horizontal distances. The horizontal distances and angles can also be converted from polar measurements to rectangular coordinates. (For more about coordinates, see "Geometry and Trigonometry.")

How is **linear algebra** used to determine the **stability of structures**?

Structural engineers use linear algebra a great deal, mainly because there are numerous equations with many unknowns associated with the analysis of a structure in equilibrium. Most of the time, these equations are linear, even when bending

Geometry and trigonometry are essential mathematical disciplines that surveyors must understand to measure property boundaries accurately. *Stone/Getty Images.*

(material deformation) is involved. Linear algebra can also be used for other structural concerns, because it deals with the study of vectors, vector spaces, linear transformations, and systems of linear equations. Of course, linear algebra is not only used to understand structures; almost every subfield in engineering uses these types of mathematical calculations. (For more about linear algebra and linear equations, see "Algebra.")

How is math used to calculate the **pressure behind a dam**?

There are many engineering considerations and calculations needed when building a dam, the most important being the water pressure behind the structure. Engineers know that as the elevation of water behind a dam increases, the height and density of the water causes higher pressures at the bottom of the dam. Thinking in mathematical terms, the horizontal force acting on the dam is the integral of the water pressure over the area of the dam that is in contact with the water. The force exerted by the water pushes horizontally on the dam face, and this is resisted by the force of static friction between the dam and the bedrock foundation on which it rests. The water also tries to rotate the dam about a line running along the base of the dam; the torque resulting from the weight of the dam acts in the opposite sense.

For example, take the water pressure on a dam, such as the Hoover Dam on the Colorado River between Arizona and Nevada. Before building that dam, engineers

331

Hoover Dam, an impressive structure on the Colorado River, could only have been made possible by engineers designing it in such a way that the walls were thick enough to withstand massive water pressure. *Stone/Getty Images.*

needed to know the pressure not only along the entire structure, but also—and especially—at its base. In general, the pressure exerted by water equals the density times the depth, in which the density of water is 62.4 pounds per cubic foot. For Hoover Dam, this gives a pressure of 37,440 pounds per square foot, or 18.72 tons per square foot; the pressure calculated at half the height of the dam is 9.36 tons per square foot. This is why the base width of the dam is 1,660 feet (201.2 meters)—it is thicker to compensate for the increase in pressure at the bottom of the dam—while the width of the crest of the dam is only 45 feet (13.7 meters).

How is **mathematics** used to enable **buildings to withstand earthquakes**?

It is not usually the quake that kills people, but the collapse of structures. In particular, the horizontal shaking during a quake is mostly responsible for causing building or road damage and collapse. Most structures are designed to carry heavy loads, so they are strong in the vertical direction. Designing structures to withstand the horizontal earthquake shaking can save buildings and lives.

There may be other ways to mitigate the amount of structural collapse during quakes—all include a healthy dose of simple and complex mathematics. One expensive way would be to design all buildings to withstand the largest ground shaking an area can expect. This could be done using mathematics familiar to designers and engineers; the math involved analyzes how large quake frequency waves travel through an area. Yet another, more practical, solution might be to design buildings to withstand the specific types of shaking expected in a region. Again, mathematics could be used to determine the frequency at which each building vibrates (or the number of times a building sways per second) versus the potential type of quakes that roll through the area.

MATHEMATICS AND ARCHITECTURE

What is **architecture**?

Simply put, architecture is the design of structures, mainly buildings, by architects. But the definition does not end there. An architect not only builds the structures, he or she also takes into consideration the form, symmetry, spaces, and beauty of the

What is the world's longest suspension bridge?

Ten years after construction began, Japan's Akashi Kaikyo Bridge—also known as the Pearl Bridge—was finally opened on April 5, 1998. It is the longest suspension bridge in the world, stretching 12,828 feet (3,910 meters) across the Akashi Strait to link the city of Kobe with Awaji-shima Island. Its main span length (or center section, which is the way "world's longest" status is determined) reaches 6,532 feet (1,991 meters) between support columns. The span length is almost a quarter mile longer than the previous record holder, the StoreBaelt (Great Belt East Bridge) in Denmark, which also opened in 1998.

But this bridge may eventually lose its top-dog status. The Italian government has approved plans to begin building what will be the longest suspension bridge in the world between mainland Italy and Sicily. It will be quite an engineering feat, with the main span stretching just over 10,827 feet (3,300 meters). Interestingly enough, Japan and Italy are known to be tectonically active, with both receiving more than their fair share of volcanic eruptions and earthquakes, as well as, in Japan, tsunamis.

building. In order to do this, mathematics is needed to work out such building factors as angles, distances, shapes, and sizes.

How has **math** been used **historically in architecture**?

Historically, there has been a great connection between architecture and mathematics. Ancient mathematicians were architects and vice versa, using their skills to build pyramids, temples, aqueducts, cathedrals, and a range of other architectural structures we find beautiful and awesome today. For example, in ancient Greece and Rome, architects were *required* to also be mathematicians. During medieval times, most buildings and structures carried some symbolic reference to the church; the mathematical end of architecture was almost forgotten during this time. By the European Renaissance around 1400, a new kind of architecture developed that emphasized mass and interior space to produce aesthetically pleasing "pictures" similar to those found in paintings and sculptures. This led to an entirely new way of looking at architecture and altered its connection to mathematics.

What is the **golden ratio**?

The golden ratio (also known as extreme and mean ratio, golden section, golden mean, or divine proportion) is a number that has many interesting properties; it is associated with the balance between symmetry and asymmetry used in art and design.

333

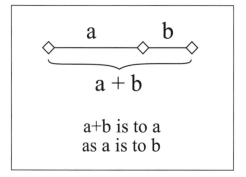

a+b is to a
as a is to b

The idea of the Golden Ratio is illustrated here by the relationships between a, b, and a + b.

Two quantities are said to be in the golden ratio if the "whole is to the larger as the larger is to the smaller." Euclid expressed it as, "A straight line is said to have been cut in extreme and mean ratio when, as the whole line is to the greater segment, so is the greater to the less." This is seen in the accompanying illustration, in which for two segments "a" and "b," the entire line is to the "a" segment as "a" is to the "b" segment.

The symbol for the golden ratio is φ (the Greek letter "phi," or a circle with a vertical slash through it); it is equal to about 1.6103398 and is considered an irrational number. The calculation to reach the golden ratio is as follows:

$$\phi/1 = (1 + \phi)/\phi$$
$$\phi^1 = 1^2 + 1\phi$$

This equals the quadratic equation:
$$\phi^2 - \phi - 1 = 0$$

which results in:
$$\phi = 1/2 + \sqrt{5}/2 \approx 1.6103398$$

What is the **historical significance** of the **golden ratio**?

It is thought that over the centuries many architects and painters used the golden ratio in their works. Some historians believe that the Great Pyramid of Cheops contains the golden ratio. The ancient Greeks knew about the golden ratio from their works in geometry, but they never truly believed it was as important as numbers such as pi (π). Many works of art in the Renaissance are thought to have used the golden ratio within paintings and sculptures, although it may have been subconsciously incorporated into their compositions. In 1509 Luca Pacioli published the work *Divina Proportione,* which explored the mathematics of the golden ratio, along with its use in architectural design.

Of course, humans aren't the only ones who "practice" the golden ratio. It is also seen in nature as the result of the dynamics of some systems. For example, the spacing of sunflower seeds—and even the shape of the chambered nautilus shell—are often claimed to be related to the golden ratio.

In what way do some historians **link mathematics** to the **pyramids**?

The pyramids in Egypt were built as royal tombs for the pharaohs—first along the edges of cliffs as low rectangular structures called mastabas, then as tall, four-sided

pyramids. The three structures most of us associate with pyramids were built at Giza, near Cairo, about 2500 BCE. The largest of these, the great pyramid of the pharaoh Cheops, measures 481 feet (147 meters) high. But there are no intricate pathways in this pyramid; it is merely a pile of limestone blocks weighing between 3 and 15 tons.

Why did the Egyptians choose the pyramid form? Historians know that the Egyptian Sun god, with its rays reaching for the Sun, was represented by pyramid-shaped stones, or ben-bens. And because the Egyptians worshipped the Sun as their chief god, the pyramids are thought to be huge renditions of the ben-bens. After a pharaoh died, the pyramids would be symbolic of the pharaoh ascending the Sun's rays to join his Sun god.

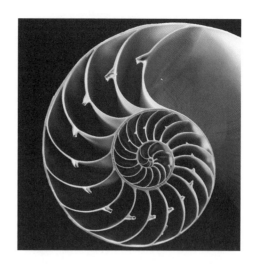

Cutting a nautilus shell in half reveals chambers that follow the rules of the golden ratio. It is a beautiful example of how math can be found all around us in nature. *The Image Bank/Getty Images*.

Some historians also believe that the pyramids may have had some (still hidden) numerical significance. In particular, some believe the ratio of the pyramid's perimeter to its height times two, or $P/(2 \times H)$, gives a close approximation to the value of pi (π); another claim is that the slopes of the pyramids' sides were also expressions of pi.

How did people in the **Renaissance** approach **architecture**?

During the Renaissance, not only mathematics but also architecture made great strides. In particular, church buildings were no longer based on the shape of the cross, but rather on the circle. This is because Renaissance architects believed that ancient mathematicians equated circles with geometric perfection, and that the circle must then represent the perfection of God.

What was the **original use** of **Stonehenge**?

One of the most famous ancient stone complexes is England's Stonehenge, an impressive engineering and mathematical feat. Several groups of local inhabitants constructed this collection of large and small stones grouped in four concentric circles—two of which are created by paired upright stones bearing huge capstones—between 2950 and 1600 BCE. Based on how the stones line up with various astronomical events, historians believe the entire structure represents a huge (literally monumental) calendar. For example, various stones align with the Moon (the full moon's extreme positions on the horizon are marked

Who was Vitruvius?

Marcus Vitruvius Pollio (c. first century BCE) was a Roman writer, architect, and engineer. He was the author of *De Architectura libri decem,* known today as *The Ten Books of Architecture.* This Latin treatise, dedicated to Octavian, the heir and adopted son of Julius Caesar, was written around 27 BCE. (For more about Octavian, see "Mathematics throughout History.")

Possibly the first works published about architecture, the books entailed a compilation of architectural ideas of Vitruvius's day and covered the following ten subjects: principles of architecture; history of architecture and architectural materials; Ionic temples; Doric and Corinthian temples; public buildings, theaters, music, baths, and harbors; town and country houses; interior decoration; water supply; dials and clocks; and mechanical engineering with military applications. Specifically, the topics included such forward-thinking ideas as the manufacture of building materials and dyes (material science), machines for heating water for public baths (chemical engineering), amplification in amphitheaters (acoustical engineering), and the design of roads and bridges (civil engineering).

De architectura was wildly successful, and Vitruvius's architectural advice was followed for centuries. But because Vitruvius's books were passed down through the ages, they were copied by various people, especially throughout the middle ages. Many medieval engineers added information to the texts, treating the books as handbooks, not documents to be preserved. In the end, historians have had to winnow away the added sections to find the true writings of Vitruvius.

at Stonehenge) and with the Sun (including the summer solstice). It is also thought that the complex served as a place in which ancient rites and rituals were performed on significant days of the year. And it was all built without the help of computers.

What are the **modern connections** between **architecture and mathematics**?

The foundations of modern architecture began with mathematics. Mathematical planning goes into creating almost every working, free-standing structure, from the smallest monument to the tallest buildings and bridges. For example, in order to build a structure, the area where the building will be erected must be measured to see if it will fit (measurements and surveys); then the building plans need to be drawn up in scale drawings made proportionally smaller to the structure's real size (for more information, see below); and the amount of building material must be estimated (mathematics to figure out the budget). Finally, the actual structure must be built to specifications so the building will stand without collapsing (geometry and measurement).

336

What **famous structures** were built using **mathematics**?

Actually, *all* famous structures needed mathematics, especially in the initial phases of design and construction. Some of the more famous—and exceptionally challenging—buildings include the Chrysler Building in New York (a steel frame skyscraper built around 1930 that was the tallest building in the world before the Empire State Building); the Empire State Building in New York (a steel-framed, stone-clad commercial office skyscraper built in 1931 that rises 1,252 feet [381 meters] high); the Eiffel Tower in Paris, France (designed by architect Gustave Eiffel and built between 1887 and 1889 as a 985-foot- [300-meter-] tall exposition iron observation tower); and the Sears Tower in Chicago (built between 1974 to 1976, it is a steel-frame with glass struc-

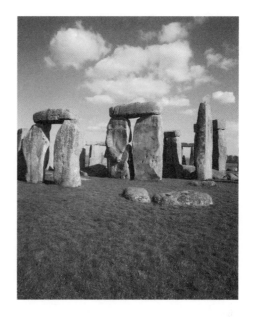

The famous ancient structure of Stonehenge in England is now believed to have been designed to measure astronomical events as part of a religious tradition. *The Image Bank/Getty Images.*

ture standing at 1,450 feet [442 meters] tall and is, to date, the tallest building in the United States). Places such as the Monterey Aquarium (built around 1980 with reinforced concrete and made compatible with surrounding waterfront structures) also needed mathematics in order to be constructed. Of course, when one gets down to it, *all* types of construction require some math knowledge for them to be built, even a modest piece of cabinetry.

What are **scale drawings**?

Scale drawings are drawings or illustrations that are proportional in scale to the real structures they represent. In order for a new building to be designed, an architect must convert his or her ideas to drawings. But since the drawings can't be as large as the building, the architect uses scale drawings to depict the structure. These miniature versions of the actual structure show the sizes, shapes, and arrangements of rooms, along with structural parts, windows, doors, closets, and other important details of construction. The scale drawings of these buildings must be in exact proportion to the actual structure, with various scales used for this purpose. For example, 1/8 inch might be used to represent one foot; thus, an eight-foot-long building feature would be drawn as an inch long on paper. One of the most common scales used by architects is 1/4 inch = 1 foot. (These measurements can also be translated into the metric scale.)

Gustave Eiffel used mathematical concepts to design his famous Eiffel Tower in France in 1889. *National Geographic/Getty Images.*

Scale drawings are also used in other engineering fields, such as surveying. For example, distances measured in the field can be translated to a smaller scale (such as a drawing) in order to accurately depict what was measured. The ratio between the real distance and the drawn distance is called the drawing scale. If the measurement is 200 feet in the field, and on paper the desired line is 8 inches long, then 8 inches on the paper would equal 200 feet on the ground, and 1 inch would be equal to 25 feet on the ground. This is translated as a diagram with a scale of 1" = 25' (1 inch equals 25 feet), or 1:25. There is another way of approaching such an illustration: If the longest distance measured in the field was 300 feet and the desired drawing scale is 1 inch = 25 feet, then the minimum length of paper needed would be 12 inches, or 300/25.

How are the **principles of ratio, proportion,** and **symmetry** applied to architecture?

The definition of a *ratio* is a comparison by division of two quantities expressed as the same unit measurement. For example, a building that is 200 feet wide and 100 feet tall has a ratio of 2:1 (200:100) between its width and height; it is also seen as the fraction 1/2. Such a relationship was understood as far back as ancient Greece and Rome, when people used mathematics to give structure and aesthetics to buildings. This is especially important in architecture, in which building design is based on complex mathematical ratios.

Proportion is an equation stating that two ratios are equal. Every proportion has four terms, with the first and fourth terms being the extremes; the second and third terms are called the means. In each proportion, the product of the means equals the product of the extremes. The Greeks and Romans often used proportions in their buildings and other structural designs. (The Roman architect Vitruvius was also instrumental in praising the virtues of proportion and symmetry in architecture; for more about Vitruvius, see above.) During the Renaissance, architects applied proportion (and other mathematical formulas) to produce aesthetically pleasing buildings—beauty that still holds true today.

Although there are other types of *symmetry,* the most common is line symmetry, in which a line divides an object, line, or other structure into two equal halves (an

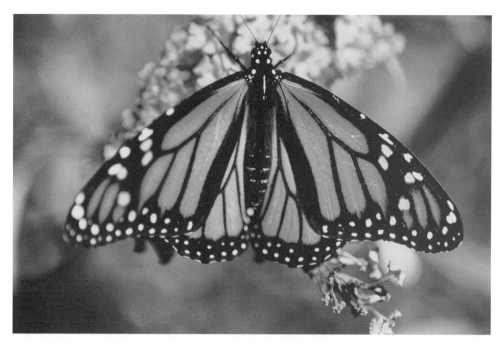

The patterns on a butterfly's wings demonstrate the concept of symmetry in nature. *The Image Bank/Getty Images.*

example in nature would be the wings of a butterfly). If a line of symmetry is drawn, each point on one side of the line has a corresponding point on the opposite side of the line. If you connect these two points, the line is perpendicular to the line of symmetry. There is a more mathematical way of defining symmetry: Two points are symmetric about a line if the line is the perpendicular bisector of the segment joining the two points. Symmetry was used by ancient (and modern) architects to maintain visual and sometimes structural balance of a building or structure.

ELECTRICAL ENGINEERING AND MATERIALS SCIENCE

How is mathematics important to **electrical engineering**?

There are many branches of mathematics that are important to electrical engineering. For example, abstract math is used in communication and signal processing. Complex differential equations—solving equations involving derivatives—are used in circuit theory and systems design; also in circuit theory, engineers need to know algebra and trigonometry. Engineers who deal with electromagnetism need to know calculus, especially Maxwell's equations. (For more about Maxwell, see "Math in the Physical Sciences.")

339

How are **imaginary numbers** used in **electrical engineering**?

Imaginary numbers are used in electrical engineering because complex numbers are an integral part of electrical problems. In fact, there are often more imaginary numbers in electrical engineering problems than there are real numbers. This is because a complex number is a pair of numbers in which one number is real, the other imaginary (or a real number multiplied by the value i, defined as the square root of -1; for more information about imaginary numbers, see "Math Basics").

For instance, we know electricity flows through an electrical circuit component such as a light bulb. The bulb actually resists the flow of some electricity by doing work—or shining—thus, the current is real and measured by a current meter. But if the current can't flow through a device, the current becomes imaginary. For example, a capacitor is two pieces of metal that do not touch; therefore, if one adds a voltage, no real current can flow through it.

Is math used to describe the **strength of materials**?

Materials science is also a major part of engineering, and includes a great deal of mathematics. For example, engineers need to know how materials stand up to stress and strain from the pressure of either a structure or overlying materials. A basic understanding of how structures respond to the action of forces and how these forces

affect the performance of various building materials, such as wood, steel, concrete, and so on, is essential.

Are **advances in technology** fundamental to our **understanding** of **materials**?

Yes, advances in technology are truly fundamental to our understanding of materials, and vice versa. For example, transistors and superconductors were developed by understanding the mathematics of the materials that make up these objects. Math and materials science together have also advanced technology used in infrastructure (such as highways and overpasses), aerospace (such as satellites and shuttles), and micro-electronics (such as those found in automobiles). Today, the need to describe material behavior resulting from physical interactions is forcing

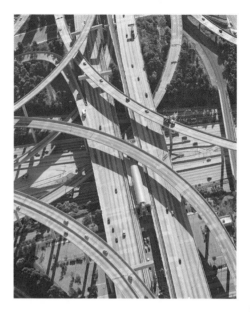

The interweaving, bow tie-like structures of a freeway overpass in Los Angeles are made possible only through the knowledge of mathematics and materials science. *Stone/Getty Images.*

scientists to develop new mathematics. In fact, materials science has evolved to the point in which researchers from physics, engineering, and applied mathematics are working together on common problems.

CHEMICAL ENGINEERING

How is mathematics used in **chemical engineering**?

Mathematics is used a great deal in chemical engineering, especially since chemical engineers design materials and the processes by which those materials are made. To solve chemical problems, many types of mathematics are used, not least of which is calculus (including partial differential equations). Even simple calculations, such as working on chemical formulas and equations, involve mathematics. (For more information about chemical formulas and equations, see "Math in the Physical Sciences.")

Traditionally, chemical engineers worked in the petroleum and large-scale chemical industries. More recently, they have spread out to the pharmaceutical, foodstuff, polymer and material, microelectronics, and biotechnology industries. Using mathematics, they are involved in such studies as thermodynamics, chemical reaction processes, and process dynamics, design, and control. They help to develop new chem-

There are numerous examples of how mathematical models are used by chemical engineers—too many to mention them all here. One good example is modeling crystal growth: Liquids—from water to molten metals—become crystalline solids as they are cooled. Engineers can mathematically design software that helps in the manufacture of superior crystalline growth, especially for electronics and other industries. These improved crystal forms advance the quality of electronic hardware, including computers, and help engineers design better alloys for a wide range of applications.

ical products and processes, test processing equipment and instrumentation, gather data, and monitor quality.

Chemical engineers also build mathematical models and analyze the results, mostly to help understand the performance of a process. In fact, the "solution" to a math problem is often in the understanding of the behavior of the process described by the mathematics, rather than the specific numerical result.

How is math used to understand **chemical reactions**?

One of the simplest examples of mathematics used in understanding chemical reactions is based on two chemicals, A and B (they can be molecules or ions). If A and B encounter one another, they can rearrange themselves into molecules or ions of two other substances: in this example, C and D. The reaction that takes place can give off or absorb energy, making the molecules move faster or slower. Although this is a simple example of what can happen in a chemical reaction, it can still be analyzed using mathematical modeling. For example, given starting amounts of A, B, C, and D molecules at time $t = 0$, what would the molecules be like at time t_1 (or after a specific amount of time)? These and more complex chemical engineering questions can be answered using mathematical modeling.

INDUSTRIAL AND AERONAUTICAL ENGINEERING

How are **statistics** used in **industrial engineering**?

Industrial engineers study the efficient use of personnel, materials, and machines in factories, stores, repair shops, and offices. They prepare layouts of machinery and equip-

ment, plan the flow of work, make statistical studies, and analyze production costs. In particular, they rely heavily on a branch of mathematics called statistics and probability. (For more about statistics and probability, see "Applied Mathematics.")

How is math used in **statistical process control**?

Statistical process control (SPC) involves using statistical techniques to measure and analyze the variations within a process. With SPC, industrial engineers monitor, control, and, ideally, improve a process through statistical analysis. The four basic steps include measuring the process, eliminating variables within the process to make it consistent, monitoring the process, and, finally, improving the process to make the (usually better) intended product. But it is not the answer to everything. All the SPC does is ensure that the product is being manufactured and designed as intended. Thus, SPC does not tell whether the design is good or bad, just if it's made according to plan.

What is **statistical quality control**?

Quality control has been around for a while in a crude sense. When a certain product was manufactured, and consumers chose that product, the makers would try to improve the quality of the product or lower its price. The improvement of the quality did not stop with the product, but also included the process for making the product. But the use of mathematics was minor in early quality control. It was not until the 1920s that statistics was applied to industry and quality control, mainly because of the development of sampling theory. (For more about sampling, see "Applied Mathematics.")

Modern statistical quality control refers to using statistical techniques for measuring and improving the quality of processes; it is often broken down into statistical process control (SPC, see above) and statistical quality control (SQC). Both terms are usually used interchangeably, although SQC has a broader focus than SPC. To compare, SPC is the application of statistical techniques to control a process, reducing variation so that performance remains within specific limits; SQC is the application of statistical techniques to control quality and includes acceptance sampling (inspection of a sample from a lot to decide whether to accept that lot) as well as SPC.

What is **reliability**?

Industrial engineers use another type of statistical technique called reliability, a system that always produces the same results and that hopefully meets or exceeds its specifications. A product is analyzed using the reliability function (or survivor function): the probability of a unit in a system that does not fail in a certain specified time interval. If the unit does fail in a system, it means the end of the unit's ability to perform the required function. This is determined by the failure distribution function, or the probability of an item failing in a specific time interval.

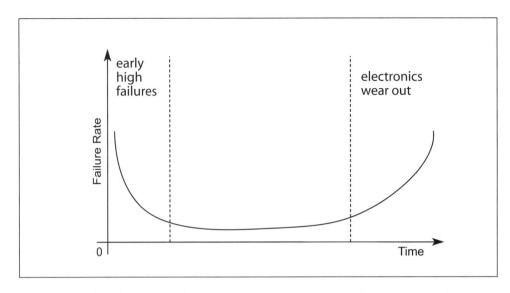

Calculations involving the common failure rates of various mechanical devices often results in a graph with a bathtub-shaped curve.

What is the **bathtub curve**?

Industrial engineers usually know about the bathtub curve, especially in reference to an operating or failing unit. In other words, if enough units from a given population are observed operating and failing over time, it is relatively easy to compute week-by-week (or month-by-month or year-by-year) estimates of the failure rate. The results of the calculated population failure rates over time produces a graph. Because the shape of this failure rate curve resembles the end-to-end section of an antique bathtub, it is widely known as the "bathtub curve." This type of analysis is usually used in industrial settings. For example, it can describe the expected failure rate of certain electronics over time: initially high; then dropping to 0 failures for most of the system's lifetime; then rising again to the other end of the "tub" as the electronics "tire out."

What is an **aerospace engineer**?

Aerospace engineers are directly involved with putting objects—from airplanes and the space shuttle to deep-space craft—into the sky and beyond. Using sundry mathematical models and techniques, they install, construct, maintain, and test systems used to launch, diagnose, or track aircraft and space vehicles. They may calibrate test equipment and determine causes of equipment malfunctions. Using computers and communications systems, aerospace engineers often record and interpret test data.

What is **orbital mechanics**?

Orbital mechanics, also called flight mechanics, is the study of the motions of artificial

satellites and space vehicles moving under the influence of forces such as gravity, atmospheric drag, thrust, and so on. It is a modern spin-off of celestial mechanics, or the study of the motions of planetary and celestial bodies. One of the main scientists who built the foundations of orbital mechanics was mathematician Isaac Newton (1642–1727), who put forth his laws of motion and formulated the law of universal gravitation. (For more about Newton, see "History of Mathematics" and "Mathematical Analysis"; for more about Newton's laws, see "Mathematics in the Physical Sciences.") Today's aerospace engineers apply orbital mechanics to such problems as rocket and spacecraft trajectories, reentry and landing of space vehicles, rendezvous computations (such as the Space Shuttle to the International Space Station), and lunar and interplanetary trajectories for manned and unmanned vehicles.

Without knowing the mathematics involved in orbital mechanics, the international space station would plunge back into the Earth instead of circling it in a stable orbit. *Stone/Getty Images.*

How do engineers determine the **escape velocity** of a **rocket**?

A ball thrown into the air will rise and then return, thanks to the Earth's gravity. If the ball is given a larger initial velocity, it will rise even higher and then return. With even more velocity, the ball will reach a certain escape velocity, in which the ball "escapes" the gravitational pull of the planet. If the ball is launched with an initial velocity greater than the escape velocity, it will rise and not return. In this case, physicists say that the ball was given enough kinetic energy to overcome all of the negative gravitational potential energy—or, it launches into space. Thus, if m is the mass of the ball, M is the mass of the Earth, G is the gravitational constant, v is the velocity, and R is the radius of the Earth, then the potential energy is equal to GmM/R. The kinetic energy of the launched ball is equal to $mv^2/2$. That means the escape velocity is equal to:

$$\sqrt{\frac{2GM}{R}}$$

This is independent of the mass of the ball. To see how this works to an aerospace engineer, just replace the word "ball" with "space vehicle."

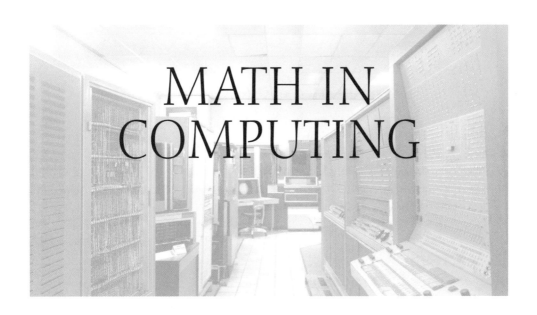

MATH IN COMPUTING

EARLY COUNTING AND CALCULATING DEVICES

Why were **counting devices developed**?

Early counting devices were developed for a logical reason: to allow people to count items in order to trade or to keep track of stock, such as cattle. They also used simple counting devices to keep track of the seasons (mostly for agriculture—in other words, to know when to plant), and for religious reasons, such as marking days for certain feasts. (For more about counting in ancient times, see "History of Mathematics.")

What were some **early counting devices**?

The very earliest counting devices were human hands, with the fingers used as digits. There were limitations to this device, though, especially since each hand only has five fingers. To count more items, some cultures assigned even larger counts to other parts of the body. Such counting methods became tedious, so merchants and others who needed to keep track of assorted items turned to nature, using sticks, stones, and bones to count.

Eventually, devices called counting boards were developed. At first, the counting "boards" were simple, usually entailing drawing lines with fingers or a stylus in the sand or dirt. After all, merchants at outdoor markets needed to count items and calculate the cost of the goods in order to sell, and there was always plenty of sand and dirt at hand.

Portable boards made of wood, stone, or metal soon became more popular, with carved (or even painted) grooves or lines indicating units. These counting boards soon became more sophisticated, with beads, pebbles, or metal discs moved between the

347

What are the oldest surviving counting boards?

To date, the oldest surviving counting board is the Salamis tablet. Discovered in 1846 on the island of Salamis, it was once thought to be a gaming board, but historians have since determined that the white marble slab was actually used to count items. The tablet, which measures 59 inches (149 centimeters) in length, 30 inches (75 centimeters) in width, and is 1.8 inches (4.5 centimeters) thick, was used by the Babylonians around 300 BCE. It contains five groups of markings, with a set of five parallel lines equally divided by a vertical line in the center; below that is a group of 11 parallel lines, all divided by a perpendicular line.

But this was not the only counting board of that time. After the Salamis tablet was developed, the Romans brought out the Calculi and the hand-abacus around 300 BCE to 500 CE. These counting boards were made of stone and metal. One example of a Roman abacus had eight long and eight short grooves arranged in a row; beads would slide into the grooves, indicating the counted units. The longer grooves were marked I to indicate single units, X to indicate tens, and so on up to millions; the shorter grooves were used to indicate multiples of five (five units, five tens, and so on). There were also shorter grooves on the right side of the abacus, which were probably used to indicate Roman ounces and for certain weight measurements.

grooves or lines, allowing for an even larger number of items to be counted. Over even more time, they grew into what is called an abacus, a device with a frame holding rods with free-moving beads attached.

What is an **abacus**?

An abacus (the plural being either abacuses or abaci) is one of the earliest counting devices. The term comes from a Latin word with origins in the Greek words *abax* or *abakon,* meaning "tablet" or "table"; these words probably originated with the Semitic word *abq,* or "sand." The devices—originally made from wood but now usually including plastic—perform arithmetic functions by manually sliding counters (usually beads or discs) on rods or wires.

Contrary to popular belief, abaci were not truly calculators in the sense of the word today. They were used only as mechanical aids for counting. The calculations were done inside the user's head, with the abacus helping the person keep track of sums, subtractions, and carrying and borrowing numbers. (For more about carrying and borrowing in arithmetic, see "Math Basics.")

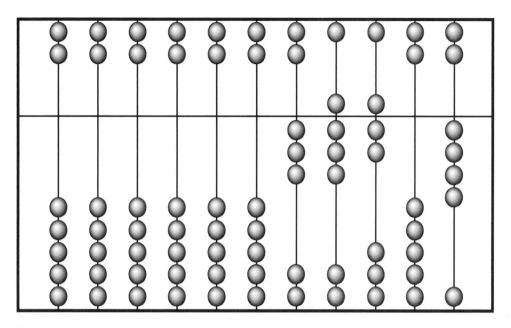

The beads are arranged in this illustration of an abacus to represent the number 38,704.

Have there been **different types of abaci** over the centuries?

Yes, there have been many different types of abaci over the centuries, including the Roman abaci mentioned above. The first type of abacus came into use in China about 1300 and was called a suanpan. Historians do not agree as to whether it was a Chinese invention or not; some say it came from Japan via Korea. Although merchants used this type of abacus for standard addition and subtraction operations, it could also be used to determine square and cube roots of numbers.

The Japanese abacus, or soroban, was similar to the Chinese abacus, but it eliminated one bead each from the upper and lower deck in each column. Thus, it is more similar to the Roman abacus. The Russians also have their own version of an abacus; it uses ten beads on each wire, and a single deck. The separation in the wires is created by one wire with fewer beads.

How are **modern abaci used**?

Today's standard abacus is typically constructed of wood or plastic and varies in size. Most are about the size of a small laptop computer. The frame of the device has a series of vertical rods or wires on which a number of wooden beads slide freely. A horizontal beam separates the frame into two sections called the lower and upper decks.

For example, in a Chinese abacus, the lower and upper decks each have 13 columns; the lower deck has five beads per column, while the upper deck has two 349

beads. Each bead on the upper deck has a value of five, while each bead on the lower deck has a value of one (thus, it is called a 1/5 abacus). To use the abacus, users place the abacus flat on a table or their laps; they then push all the beads on the upper and lower decks away from the horizontal beam. From there, the beads are manipulated, usually with the index finger or thumb of one hand, to calculate a problem. For example, if you wanted to express the number 7, you would move two beads in the lower deck and one bead in the upper deck: $(1 + 1) + 5 = 7$.

This modern abacus is still used by shopkeepers in Asia and many so-called "Chinatowns" in North America. Students continue to be taught how to use the abacus in Asian schools, especially to teach children simple mathematics and multiplication. In fact, it is an excellent way to remember multiplication tables and is useful for teaching other base numbering systems, because it can adapt itself to any base. (For more about base numbers, see "History of Mathematics" and "Math Basics.")

What is a **khipu**?

Khipus (or quipu, in Spanish) were used by the Incas of South America. A khipu is a collection of knotted strings that record certain information. The approximately 600 surviving khipus use an arrangement of knotted strings hanging from horizontal cords. But these knots are nothing like those made by other cultures: They include long knots with four turns, single knots, figure-eight knots, and a whole host of other knot types. Historians believe these strings and knots represent numbers once used for accounting, inventory, and population census purposes.

There are also researchers who believe the khipus may contain certain messages in some sort of code—a kind of language used by the Incas—based on the strings, knots, and even a khipu string's type (usually alpaca wool or cotton) and color. But it may turn

The people of the ancient Inca civilization of South America used knotted strings called khipus to make mathematical calculations. *The Image Bank/Getty Images.*

out that historians will never know the real story behind the khipus. When the Spanish conquered the Inca Empire starting in 1532, they destroyed most of the strings, believing they might be idolatrous items containing accounts of Incan history and religion.

What are **Napier's Bones**?

A tool called Napier's Bones (also called Napier's Rods) was invented by Scottish mathematician John Napier (1550–1617). These were multiplication tables inscribed on strips (also called rods) of bone (not Napier's, but animal bone), ivory, or wood. He published the idea in his book *Rabdologia,* which contained a description of the rods that aided in multiplication, division, and the extraction of square roots. (For more about Napier, see "Algebra.")

Each bone is a multiplication table for a single digit, with the digit appearing at the top of its bone. As seen below, consecutive, non-zero products of this digit are carved in the rod, with each product occupying a single cell. For example, to multiply 63 by 6, the two bones or rods corresponding to 6 and 3 would be put alongside each other and would look like the illustration below.

The first number would be the number in the diagonal at the sixth position (3); then the product (or solution) is evaluated diagonally, or by adding the two numbers

351

Scottish mathematician John Napier designed a system for representing multiplication tables called "Napier's Bones." *Library of Congress.*

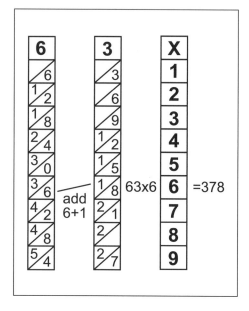

In this example of how to use Napier's Bones, 63 is multiplied by 6 to get the correct result of 378.

diagonal from each other, or 7 (6 + 1); and the next number in the separate diagonal, or 8. In other words, 63 × 6 = 378.

Initially, the tables were used by merchants to speed up calculations. German astronomer and mathematician Wilhelm Schickard (1592–1635) would eventually build the first calculating machine based on Napier's Bones in 1623. His device could add, subtract, and, with help, multiply or divide. This is why he is often called the "father of the computing era" (see below).

MECHANICAL AND ELECTRONIC CALCULATING DEVICES

Who built the **first** known **adding machine**?

No one knows who built the first adding machine, although many historians believe it was German mathematician Wilhelm Schickard (1592–1635) who first invented a mechanical calculator in 1623 based on Napier's Bones (see above). Schickard and his family perished from the Bubonic plague. It was not until the mid-20th century that his notes and letters were discovered. They showed diagrams of how to construct his machine. Schickard apparently built two prototypes: One was destroyed in a fire and the other one's location is unknown, if it survived at all. His device, which he called

the "calculating clock," was able to add and subtract up to six-digit numbers using a mechanism of gears and wheels.

But not all historians credit Schickard. Some believe that there was an even earlier attempt at mechanical computing by Leonardo da Vinci, who also apparently designed an adding machine. Some of his notes were found in the National Museum of Spain in 1967 and describe a machine bearing a certain resemblance to Pascal's machine (see below).

How did Gottfried Wilhelm von **Leibniz** advance **calculating devices**?

German mathematician and philosopher Gottfried Wilhelm von Leibniz (1646–1716) not only described the binary number system—a central concept of all modern computers—but also co-invented dif-

Some people believe that the famous Renaissance inventor and artist Leonardo da Vinci was also the creator of an early type of mechanical computer. *Library of Congress.*

ferential calculus and designed a machine that would perform the four basic arithmetic functions. By 1674 he had completed his design and commissioned the building of the Leibniz Stepped Drum, or the Stepped Reckoner, as he called his machine.

The device used a special type of gear named the Leibniz wheel (or stepped drum), a cylinder with nine bar-shaped teeth along a length parallel to the cylinder's axis. As the cylinder was rotated with a crank, a ten-toothed wheel would rotate from zero to nine positions, depending on its position from the drum. The movements of the various mechanisms would be translated into multiplication or division, depending on what direction the stepped drum was rotated.

Although there were apparently only two prototypes of the device (both still exist), Leibniz's design—along with Pascal's—were the basis for most mechanical calculators in the 18th century. As with most such machines that could not be mass produced—much less understood by the masses—they were more curiosities for display than machines put to actual use.

How did **Joseph-Marie Jacquard's invention** benefit calculating devices?

In the late 18th century, French weaver and inventor Joseph-Marie Jacquard (1752–1834) developed a practical, automatic loom that wove patterns into fabric; it was controlled by a linked sequence of punched cards. This in itself was a major advance in the production of textiles, but it would also prove to be a boon to calculating devices. Bor-

rowing Jacquard's idea, both Charles Babbage and Herman Hollerith (see below) would use such cards on their own computing machines. The company that Hollerith formed eventually became International Business Machines (IBM), a company that for 30 years promoted and benefited from mechanical punched card processing.

What was the **difference engine**?

Because of its automatic sequential approach, the difference engine is thought of by most mathematical historians to be the precursor to modern computers. Johann H. Müller, an engineer in the Hessian army, first developed the concept in 1786. His idea was to have a special machine that would evaluate and print mathematical tables by adding sequentially the difference between certain polynomial values. But he could not get the funds to build the machine.

Müller's idea was soon lost before it was resurrected in 1822, when Charles Babbage obtained government funds to build a programmable, steam-powered prototype of Müller's device (for more information about Babbage, see below). Because of technical limitations, funding cuts, and Babbage's interest in a more advanced device of his own design, Müller's difference engine was only partially completed. Eventually, Swedish inventors George Scheutz (1785–1873) and his son Edvard (1821–1881) would in 1853 build the difference engine, the first calculator with the ability to print.

Who was **Charles Babbage**?

English inventor and mathematician Charles Babbage (1792–1871) is considered by

When was the calculating machine first mass produced?

Between the 1624 invention of Blaise Pascal's calculating machine and 1820, there were about 25 manufacturers of such devices. Because there were so many—most had little funding and only one person involved—very few machines were actually manufactured in any quantity.

By 1820 the first calculating machine to be commercially successful and produced in large numbers was the "Arithmometer." Invented by Frenchman Charles Xavier Thomas de Colmar (1785–1870) while he was serving in the French army, it was based on a Leibniz's "stepped drum" mechanism (see above). Colmar's machine used a simple system of counting gears and an automatic carry (automatically shifting a 1 to the left when the sum of a certain column was greater than 9). The technology of the times also helped catapult Colmar's success. Because it included springs and other machinery that offset the momentum of moving parts, the Arithmometer stopped at a specific, intended point, unlike what often happened with the older calculating machines.

some historians to be the "father of computing." The main reason was his Analytical Engine, which is thought to be the true precursor of the modern computer.

One of his first ventures into calculating machines was the difference engine, which was based on Johann Müller's design and some of Thomas de Colmar's Arithmometer features (for more information about both, see above). The idea was sound, but the execution eventually lacked government funding, not to mention suffering from disputes with the artisan who was making the parts for the machine. Not only that, but Babbage's ambitions may have caused the difference engine prototype to come to a halt. Initially, he wanted the device to go to six decimal places and a second-order difference; then he began planning for 20 decimal places and a sixth-order difference. This much-larger machine was an overwhelming concept for its time.

The abandonment of the difference engine did not stop Babbage, however. Again approaching the government for funding, he promised to build what he called the Analytical Engine, an improved device capable of any mathematical operation, effectively making it a general purpose, programmable computer that used punch cards for input. This new device would use a steam engine for power, and its gears would function like the beads of an abacus, with the main tasks of calculating and printing mathematical tables. For eight years, he attempted to get more money from the government, but to no avail. He would never build his Analytical Engine.

Although the Analytical Engine was never completed in Babbage's lifetime, his son Henry Provost Babbage built the "mill" portion of the machine from his father's

Recognized for his connection to the famous difference engine, English mathematician Charles Babbage had to abandon his elaborate plans for a mechanical computer because the device was simply too expensive. *Library of Congress.*

drawings, and in 1888 he computed multiples of pi (π) to prove the acceptability of the design. This is often thought to represent the first successful test of a "modern" computer part.

What is a **troncet**?

The troncet (or addiator) is credited to J. L. Troncet of France, who invented the device in 1889. He called it his *Arithmographe*. (In actuality, his work was based on earlier designs that were first started by Claude Perrot [1613–1688].) It was used principally for addition and subtraction.

A troncet's flat, mechanical, palm-held calculator had three main components: the part for the calculation, a stylus, and a handle to reset the addiator. By inserting the tip of the stylus into notches along a metal plate, numbers could be added by sliding either up or down strips of metal with numbers marked on them. No gears or interlinked parts were involved. To "carry one" when the sum of two digits was greater than ten, the stylus was moved up to and around the top of the device.

What is a **slide rule**?

The slide rule is a ruler-like device with logarithmic scales that allows the user to do mathematical calculations. It is portable, with the most common slide rules using three interlocking calibrated strips; the central strip can be moved back and forth relative to the other two. Calculations are performed by aligning marks on the central strip with marks on the fixed strips, then reading marks on the strips. There is also a "see through" sliding cursor with a hairline mark perpendicular to the scales, allowing the user to line up numbers on all the scales.

Sadly for mathematical traditionalists, the use of the slide rule was eventually overtaken by the pocket calculator by the mid-1970s. But in other ways, this development was welcome. The slide rule had two major drawbacks, especially for calculations in mathematics, engineering, and the sciences: It was not easy to add with the device and it was only accurate to three digits.

Who is sometimes called the "first programmer"?

One of the first "programmers"—in this case, of a calculating machine—was Ada Augusta Byron (1815–1852; also known as Ada King, Countess of Lovelace), the daughter of Lord George Gordon Noel Byron (1788–1824), the famous English poet. Inventor and mathematician Charles Babbage met Ada Byron around 1833, while still working on his difference engine. Her interest was reportedly more in his mathematical genius, not his machines.

Besides her admiration for him, Ada Byron also put Babbage's name on the computing map, writing up most of the information about his work, which was something Babbage supposedly could not do as well. For example, she translated an 1842 account of his Analytical Engine (written by French-born Italian engineer and mathematician Luigi Federico Menabrea [1809–1896]) from French into English. Babbage was so impressed that he suggested she add her own notes and interpretations of the machine. With his encouragement, she added copious notes, describing how the Analytical Engine could be programmed, and wrote what many consider to be the first-ever computer program. Her account was published in 1843. She was also responsible for the term "do loop" in computer language (a part of a program she called "a snake biting its tail") and for developing the "MNEMONIC" technique that eventually helped simplify assembler commands.

Ada Byron's life deteriorated after writing her notes because of family difficulties, gambling debts (though not her own), the lack of a scientific project to work on, and probably the fact that none of her friends were as deeply—and intuitively—involved in mathematics or the sciences as she was. Babbage was no help, either, having his own difficulties, including his ongoing attempts to obtain governmental funding for his Analytical Engine. In 1852, at only 37 years of age, Ada Byron died of cancer, but she was not forgotten. She was remembered and honored in 1980 when the ADA programming language was named after her.

How did the **slide rule evolve**?

In 1620 English astronomer Edmund Gunter (1581–1626) was responsible for constructing a scale rule that could be used to multiply. He divided his scale according to Napier's principle of logarithms, meaning that multiplication could be done by measuring and adding lengths on the scale. (It is also often considered the first analog computer.)

But there is disagreement as to the true inventor of the slide rule. Many historians give the credit to English reverend William Oughtred (c. 1574–1660), who improved upon Gunter's idea. About 1630 (although that date is highly debated), Oughtred placed two of Gunter's scales directly opposite each other and demonstrated that one could do calculations by simply sliding them back and forth.

The slide rule was not immediately embraced by scientists, mathematicians, or the public. It took until about 1850, when French artillery officer Victor Mayer Amédée Mannheim (1831–1906) standardized the modern version of the slide rule, adding the movable double-sided cursor that gives the slide rule its familiar appearance. Slide rules were used for many decades as the major calculator for the sciences and mathematics, and ranged in shapes from straight rules to rounded.

When was a **mechanical calculating device** first used for the **American census?**

When government officials estimated that the 1890 census would have to handle the data from more than 62 million Americans, there was a slight panic. After all, the existing system was slow and expensive, using tally marks in small squares on rolls of paper, which were then added together by hand. One estimate determined that such an endeavor would take about a decade to complete, which would be just in time to start the process all over again for the 1900 census. In desperation, a competition was set up to invent a device that could easily count the 1890 U.S. census.

Thus, in the 1880s, American inventor Herman Hollerith (1860–1929), who is also known as the father of modern automatic computation, presented his competition-winning idea. He used Jacquard's punched cards to represent the population data, then read and collated the information with an automatic machine. With his Automatic Tabulating Machine—an automatic electrical tabulating device—Hollerith would put each individual's data on a card. With a large number of clocklike counters, he would then accumulate the results. From there, he would use switches so the operators could instruct the machine to examine each card based on a certain characteristic, such as marital status, number of children, profession, and so on. It became the first such machine to read, process, and store information.

The machine's usefulness did not end there, though. Eventually, Hollerith's device became useful for a wide variety of statistical applications. Certain techniques used in the Automatic Tabulating Machine were also significant, helping in the eventual

Has there ever been a competition between a calculator and an abacus?

Yes, there was once a competition between someone using a calculator and another person using an abacus. Although the abacus is often considered a "crude" device to do simple calculations, in expert hands it can work just about as fast as a calculator.

The contest took place in Tokyo, Japan, on November 12, 1946, between the Japanese abacus and an electric calculating machine. The event was sponsored by the U.S. Army newspaper *Stars and Stripes.* The American working the calculating machine was Private Thomas Nathan Wood of the 20th Finance Disbursing Section (from General MacArthur's headquarters), who was considered an expert calculator operator. The Japanese chose Kiyoshi Matsuzaki, himself an expert operator of the abacus, from the Savings Bureau of the Ministry of Postal Administration. In the end, the 2,000-year-old abacus beat the electric calculating machine in adding, subtracting, dividing, and a problem including all three with multiplication thrown in. The machine won only when it came to multiplication.

development of the digital computer. Hollerith's company would also eventually become well-known, becoming International Business Machines, or IBM, in 1924.

What were some of the **first motor-driven calculating devices**?

Many historians believe that the first motor-driven calculating machine was the *Autarigh,* a device designed by Czechoslovakian inventor Alexander Rechnitzer in 1902. The next step occurred in 1907, when Samuel Jacob Herzstark (1867–1937) produced a motor-driven version of his Thomas-based calculators in Vienna. In 1920 a prolific Spanish inventor named Leonardo Torres Quevedo (1852–1936) presented an electromechanical machine wired to a typewriter at the Paris Calculating Machine Exhibition. His invention performed addition, subtraction, multiplication, and division, and then used typewriters as input/output devices. Interestingly enough, even though the machine made a hit at the exhibition, it was never produced commercially.

More and more such calculating devices with electric motors were invented. By the 1940s, the electric-motor-driven mechanical calculator had become a common desktop tool in business, science, and engineering.

What is an **electronic calculator**?

Most people nowadays are familiar with electronic calculators: small, battery-powered digital electronic devices that perform simple arithmetic operations and are limited to

handling numerical data. Data are entered using a small keypad on the face of the calculator; the output (or result) is most commonly a single number on an LCD (Liquid Crystal Display) or other display. It took a long time to go from the electronic motor-driven mechanical calculator to the electronic calculator. In 1961, the company Sumlock Comptometer of England introduced the *ANITA* (A New Inspiration To Arithmetic), the first electronic calculator.

MODERN COMPUTERS
AND MATHEMATICS

What is the definition of a **modern computer** in today's sense of the word?

In general, and simply put, a computer is a machine that performs a series of mathematical calculations or logic operations automatically. Computer specialists often divide computers into two types: An *analog computer* operates on continuously varying data; a *digital computer* performs operations on discrete data.

The majority of today's computers easily process information much faster than any human. The response (output) of a computer depends on the data (input) of the user, usually controlled by a computer program. A computer also can perform a large number of complex operations, and can process, store, and retrieve data without human interference.

There is a great deal of overlap for the word "computer." In a more archaic sense, a computer is called an electronic computer, or computing machine or device; the more common expressions include data processors or information processing systems. But remember, there is a definite difference between a computer and a calculating machine: a computer is able to store a computer program that allows the machine to repeat operations and make logic decisions.

What **problems** were **computers** originally invented to **solve**?

Originally, computers were invented to solve numerical problems. Now, very few mathematicians, scientists, computer scientists, and engineers can imagine a world without computers. Advances in technology have also led to an increase in accuracy, the number of problems that can be solved, and the speed in solving those problems—a far cry from the earliest computers.

What **type of number system** is used by **modern computers**?

Modern computers use the *binary system,* a system that represents information using sequences of 0s and 1s. It is based on powers of 2, unlike our decimal system based on powers of 10. This is because in the binary system, another number place is added

every time another power of two is reached, for example, 2, 4, 8, and so on; in the decimal system, another place is added every time a power of 10 is reached, for example, 10, 100, 1000, and so on.

Computers use this simple number system primarily because binary information is easy to store. A computer's CPU (Central Processing Unit) and memory are made up of millions of "switches" that are either off or on—the symbols 0 and 1 represent those switches, respectively—and are used in the calculations and programs. The two numbers are simple to work with mathematically within the computer. When a person enters a calculation in decimal form, the computer converts it to binary, solves it, and then translates that answer back to decimal form. This conversion is easy to see in the following table:

Decimal	Binary
0	0
1	1
2	10
3	11
4	100
5	101
6	110
7	111
8	1000
9	1001
10	1010
11	1011
12	1100
13	1101
14	1110
15	1111
16	10000
17	10001
18	10010
19	10011

What was the **Turing machine**?

In 1937, while working at Cambridge University, English mathematician Alan Mathison Turing (1912–1954) proposed the idea of a universal machine that could perform mathematical operations and solve equations. This machine would use a combination

of symbolic logic, numerical analysis, electrical engineering, and a mechanical version of human thought processes.

His idea became known as the Turing machine, a simple computer that performed one small, deterministic step at a time. It is often thought of as the precursor to the modern electronic digital computer, and its principles have been used for application in the study of artificial intelligence, the structure of languages, and pattern recognition. (For more on Alan Turing, see "History of Mathematics.")

Who built the **first mechanical binary computer**?

German civil engineer Konrad Zuse (1910–1995) built the Z1—often thought of as the first mechanical binary computer—in his parent's living room around 1938. His goal was to build a machine that would perform the lengthy and tedious calculations needed to design building structures. His computer's design stored intermediate results in its memory and performed sequences of arithmetic operations that he programmed on punched paper tape (he initially used old movie film). This machine led to the Z3 in 1941. Considered by some to be the first large-scale, fully functional automatic digital computer, the computer used a binary number system.

What were some **highlights** in the **development** of **modern computers**?

The first general-purpose analog computer was designed in 1930 by American scientist Vannevar Bush (1890–1974), who built a mechanically operated device called a differential analyzer. The first semi-electronic digital computing device was built by mathematician and physicist John Vincent Atanassoff (1903–1995) and one of his graduate students, Clifford E. Berry (1918–1963), between 1937 and 1942. It was created primarily to solve large systems of simultaneous linear equations. It is interesting to note that Atanassoff's computer was overshadowed by the Electronic Numerical Integrator and Computer (ENIAC; see below), which was once credited as the first

computer. In 1973, however, a federal judge recognized Atanassoff's work and voided Sperry Rand's patent on the ENIAC, saying it had been derived from Atanassoff's invention. Today, Atanassoff and Berry get the credit.

The *Harvard Mark 1,* or the Automatic Sequence Controlled Calculator, was built between 1939 and 1944 by American computer scientist Howard H. Aiken (1900–1973) and his team. It is thought of as the first large-scale automatic digital computer. But there are disagreements about this, with some historians believing that German engineer Konrad Zuse's Z3 (see above) was the first such machine.

Other early computers were the *ENIAC 1* and the *UNIVAC.* The ENIAC (Electronic Numerical Integrator And Calculator) was completed in 1946 at the University of Pennsylvania; it used thou-

Vannevar Bush was an American scientist who invented the differential analyzer, the first general-purpose analog computer. *Library of Congress.*

sands of vacuum tubes. Until 1973, it was thought of as the first semi-electronic digital computer. That credit was subsequently given to Atanassoff and Berry (see above). The UNIVAC (UNIVersal Automatic Computer) was built in 1951 and was the first computer to handle both numeric and alphabetic data. It also was the first commercially available computer.

The third-generation integrated-circuit machines were used primarily during the mid-1960s and 1970s, making the computers smaller, faster (close to a million operations per second), and far more reliable. The first commercial microprocessor was the Intel 4004, which appeared in 1971. It could only add and subtract, and it was used to power one of the first portable electronic calculators. The real push in microprocessors came during the late 1970s to 1990s, allowing for increasingly smaller and more powerful computers. For example, in 1974 the Intel 8080 processor had a clock speed of 2 megahertz (MHz); by 2004 the Pentium 4 ("Prescott") had a clock speed of 3.6 gigahertz (GHz). Computers that use microprocessors include the personal computer and personal digital assistant (PDA). The computer industry has continued rapid growth, mainly thanks to the increased performance of advanced microprocessors.

What is a **microprocessor**?

A microprocessor is a silicon chip that contains a CPU, or central processing unit, which is normally located on the main circuit board in a computer. (In the world of personal

A personal digital assistant (PDA) is just one of many conveniently small and handy computer devices made possible by advances in microprocessors. *Taxi/Getty Images.*

computers, the terms microprocessor and CPU are often used interchangeably.) These chips, or integrated circuits, are small, thin pieces of silicon onto which the transistors making up the microprocessor have been etched. The microprocessor is the heart of any normal computer, from desktops and laptop machines to a larger server. They have many uses. For example, they control the logic of almost all familiar digital devices—from microwaves and clock radios to fuel-injection systems for automobiles.

What is the **difference** between a **minicomputer** and a **microcomputer**?

The term minicomputer is not used much today. It is considered to be the type of computer built mainly from about 1963 to 1987, and refers to the "mini" mainframe computers that were not large enough to be called mainframes but were large enough to take up the space of a small closet. These computers were once popular in small businesses that could not afford the money or space for a mainframe computer. They were much less powerful than a mainframe and were limited in hardware and software, and they were built using what was called low-integration logic integrated circuits. Eventually, they were overtaken by microcomputers built around the microprocessor.

The microcomputer was a later development in computing. It was developed as a general-purpose computer designed to be operated by one person at a time. The single-chip microcomputer (complete with microprocessor) was, in many respects, a landmark development in computer technology, resulting in the commercialization of the personal computer. This is because computers became smaller and less expensive, and the design made parts easier to replace.

What are the **main parts** and **types** of **computers** in use today?

The basic parts of a computer are the central processing unit (CPU), memory, a keyboard and other optional input devices such as a scanner, and output devices such as a screen, a printer, and audio speakers. The biggest differences between the various types of computers are the amount of memory and speed of the machines.

It's also easy to see why the word "computer" has so many connotations. They vary greatly, but include the following basic types, which are based mostly on the size and number of people who can simultaneously use the machines. The *mainframe* is

Mainframe computers like these can fill a room in a company's office, but smaller businesses that did not have the money or space for mainframes once used smaller mainframe units called minicomputers. By the late 1980s, microcomputers had become powerful enough to replace minicomputers. *Stone/Getty Images.*

considered the largest and most powerful general-purpose computer system. It is usually used to fill the needs of a large agency, company, or organization, because it uses hundreds of computer terminals at the same time. *Supercomputers* are sophisticated machines designed to perform complex calculations at maximum speed. Because of their speed—and the great amount of data they can work through—they are most often used to model huge dynamic systems with many variables, such as weather patterns and groundwater flow.

Microcomputers are usually subdivided into personal computers (or desktop computers) and workstations. Oftentimes, microcomputers are linked together in a local area network (LAN) or by joining the microprocessors in a parallel-processing system. This allows smaller computers to work in tandem, often giving them comparable power and computational abilities to mainframes.

Familiar to many people today are *notebooks* and *laptops,* which are very similar. Laptops are small enough to fit on a person's lap; notebooks are usually a bit smaller and lighter than a laptop. The newest ones often have the same capabilities as a desktop computer.

365

What is **computer science**?

Computer science is, of course, the science of studying computers. It is the study of computation and information processing, involving hardware, software, and even mathematics. More specifically, it is the systematic study of computing systems and the computations that go behind making the computer function. Computer scientists need to know computing systems and methods; how to design computer programs, including the use of algorithms, programming languages, and other tools; and how software and hardware work together. They also need to understand the analysis and verification of the input and output.

What are **computer codes** and **programs**?

Mathematics is an important part of computers, because math is used to write computer codes and programs. The codes are the symbolic arrangement of data (or the instructions) in a computer program, a term often used interchangeably with "software." The code (also called the source code, or just source) is any series of statements written in some programming language understandable to the user. This source code within a software program is usually contained in several text files.

The program is the sequence of instructions (or computations) that a computer can interpret and execute. In other words, most programs consist of a loadable set of instructions that will determine how the computer reacts to user input when the program is running. The connection between codes and programs is often heard by students studying computer science or working professionals—and even in action movies and television programs, as in, "I need to add more lines of code to the program!"

APPLICATIONS

How have computers been used to **factor large composite numbers**?

Computers have often been used to factor large numbers—and not just by number theorists having some fun. In fact, factoring such numbers has helped to test the world's most powerful computer systems, to promote designs of new algorithms, and in cryptography used by people who need to protect sensitive information on their computers. For example, in 1978 several computer experts proposed using the reconstruction of the prime numbers from the product of two large prime numbers as an encryption technique. This method of encrypting sensitive data soon blossomed, especially because of the needs of the military and banking industry. The public also reaped the benefit of this idea as it eventually led to encryption methods such as the public-key encryption for banking and personal pages on the Internet.

Have **computers** been used to **solve mathematical proofs**?

Yes, there have been many mathematical proofs solved with the help of computers. One example is the *four color theorem,* which stated that it is possible to have a geographic map colored with only four colors so that no adjacent regions will have the same color. Another way of looking at the problem is: What is the smallest number of colors needed to color any flat map so that any two neighboring regions always have different colors? This idea was first presented in 1852, when Francis Guthrie (1831–1899) colored a map of English counties using only four colors. The idea of only four colors took on a mathematical bend and ended up

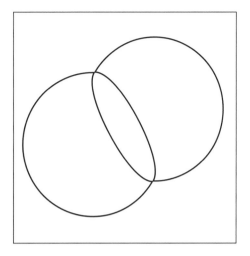

In the problem of the double bubble, ancient Greek mathematicians worked on a proof that would demonstrate the efficient use of space created by two joined bubbles.

being a theorem to be proved. It took until 1976, with the help of modern computers, before the four-color conjecture was finally proven to be true. But some mathematicians are troubled by this computer proof, feeling that the theorem is so easy to understand that it should have been proven by hand. Thus, anyone who can prove the theorem without using a computer may win the Fields Medal, the math equivalent of the Nobel Prize.

Another proof solved with computers is the *double bubble.* The double bubble refers to a pair of bubbles that intersect; they are also separated by a membrane bounded by the intersection of the two bubbles. This is similar to two bubbles stuck together when a child blows bubbles using a water and soap mixture. Since the ancient Greeks, mathematicians have worked on the problem of finding a mathematical proof of the efficiency of a single round bubble. The problem became even more rigorous when considering enclosing two bubbles—or two separate volumes. The problem was solved around 1995 by mathematicians Joel Hass, Michael Hutchings, and Roger Schlafly. They used a computer to calculate the surface areas of the bubbles and found that the double bubble has a smaller area than any other when the enclosed volumes are the same. But this isn't the last word: Scientists are currently working on triple bubbles.

How are **algorithms** connected to **computers**?

Algorithms are essentially the way computers process information. In particular, a computer program is actually an algorithm that tells the computer what particular steps to perform—and in what order—so a specific task is carried out. This can include

Yes, computers have been used to determine the value of pi, but no computer has yet found the "final" number in the long progression of numbers. They probably never will, because pi is considered to be an infinite number. But for the sake of just trying, larger and faster computers are often used for this task. To date, pi has been found to more than six billion places. (For more about pi, see "Mathematics throughout History.")

anything from working out a company's payroll to determining the grades of students in a certain class. (For more about algorithms, see "Foundations of Mathematics.")

What is **cryptography**?

Because of the extensive connections between computers across the Internet, it has become necessary to find ways to protect data and messages from tampering or reading. This includes protecting people's personal information when they buy things over the Internet and those who need to keep banking data secure. One of the major techniques for ensuring such privacy of files and communications is called cryptography.

Cryptography is a mathematical science used to secure the confidentiality (and authentication) of data sent by a user to a certain site. It secures the data by replacing it with a transformed version that can then be reconverted to show the original data, but only by someone with the correct cryptographic algorithm and key. This is why, when ordering over the Internet, it is important to see the "lock" icon at the bottom of the screen. This is your way of knowing that cryptography is working and the data is secure, thus preventing the data's unauthorized use.

Have **computers** had an **effect** on the field of **statistics**?

Yes, computers have had a definite effect on the field of statistics. In particular, personal computers, software such as spreadsheets and professional statistical packages, and other information technologies are now an integral part of statistical data analysis. These tools have enabled statisticians to perform realistic statistical data analysis on large amounts of data faster and cheaper than ever before.

Statistical software systems are most often used to determine examples, understand existing concepts in the statistics, and to find new trends in statistical data. Most of the packages allow the data to be entered into the computer program, but the emphasis then switches to the statistician's ability to interpret the data.

There are two well-known programs often used in statistics: the SAS and SPSS, both of which are commercial statistical packages. The SAS system is a statistics,

graphics, and data management software package available for personal computers. It allows the desktop computer user to get the quality of results once reserved for users of mainframe computers. The SPSS (Statistical Package for the Social Sciences) is also a popular software package for performing statistical analyses. It enables the user to summarize data, determine if there are significant differences between groups, examine relationships among all the variables, and even graph the results.

To date, what **computer** has the **top speed** in operations per second?

In the 2004 TOP500 list—a ranking of supercomputers by speed, with the results announced early each summer—the top three supercomputers were as follows (note: a teraflop is a measure of a computer's performance, in which one teraflop is 10^{12} operations per second):

- IBM's BlueGene/L—70.72 teraflops
- NASA's Columbia—51.87 teraflops
- NEC's Earth Simulator—35.86 teraflops

One of the next contenders is Sandia Lab's Red Storm, with 41.5 teraflops. This list will continue to change each year as more advanced technology becomes available.

In actuality, the fastest computer in the world is the human brain, an amazing computing device with the best processor on Earth. To compare, the fastest computers measure speed in trillions of operations per second, but scientists speculate the brain can handle 10 quadrillion operations per second. The actual numbers are probably even higher than that.

MATH ALL AROUND US

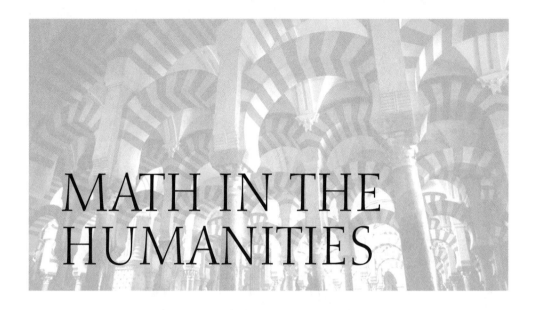

MATH IN THE HUMANITIES

MATH AND THE FINE ARTS

What are the **humanities**?

Humanities are those studies dealing with the fine arts (painting, drawing, and so on), literature, philosophy, and cultural science. This field focuses on the idea of expanding human thought, intellectual skills, and accomplishment through the study of these branches. Although the humanities seem far removed from mathematics, there are actually many connections.

Has **mathematics** been **used in art**?

Yes, mathematics has been used in art—either consciously or not—over the centuries. Many of the fine arts deal with trying to comprehend the reality around us. And because both art and math try to explain reality using some concrete or abstract elements, it's easy to see the connection.

For example, there are some obvious relationships between art—especially painting—and mathematics. An observer can use geometry to analyze a painting in terms of shapes, such as lines, points, circles, or other geometric shapes. Perspective can also be thought of as a mathematical "quality" of many realistic paintings, allowing the viewer to see reality (in the case of a painting) in two dimensions. And, in a way, Jackson Pollock's method of literally throwing paint on a canvas can be thought of in terms of fractal geometry.

Mathematicians may work with quantifying and counting, but most of them will tell you that math is really more than a technical exercise and juggling numbers. Creativity is involved in discovering nuances in certain mathematics, with the mathematician creatively deciding what to concentrate on in order to solve certain complex

Giotto di Bondone developed the concept of linear perspective in his art to depict three-dimensional space realistically in his two-dimensional paintings. *Library of Congress.*

problems. Just like a picture, mathematical concepts need to be created, invented, and discovered, much like the French painter Claude Monet, and others, invented Impressionism.

How did **early painters** integrate mathematics and painting?

Early painters had a problem: how to represent the three-dimensional world on a two-dimensional canvas; or, how to give paintings depth and perspective. Although perspective does follow mathematical guidelines, many early painters were able to incorporate perspective (or depth perception) intuitively.

The idea of perspective flourished around the time of the ancient Greeks, who used a form of perspective to design certain architectural structures and even theatrical stage settings. But it is unclear whether they actually understood the mathematics behind perspective. One of the first to create the impression of depth by using certain rules was Italian painter, sculptor, and architect Giotto di Bondone (c. 1267–1337). But those rules were of his own devising and were probably not based on mathematics. Whatever the case, he clearly worked out a way to represent depth in space, and he came close to understanding linear perspective.

But it took until the Renaissance before painters explored the science behind perspective. In the early 1400s, sculptor Filippo Brunelleschi (1377–1446) made the first correct formulation of linear perspective using mirrors. He understood that there was a single vanishing point to which all parallel lines in a plane (or on his canvas) converge; he also understood scales, calculating the relationship between the actual length of an object and its length in the picture, depending on its distance in the painting. Writer and mathematician Leone Battista Alberti (1404–1472) was the first person to write the rules of perspective in 1435, using the principles of geometry and the science of optics.

By 1450 artist and mathematician Piero della Francesca (1412–1492) had written an even more extensive mathematical work on perspective, including concepts of art, arithmetic, algebra, and geometry. The writings of mathematician Luca Pacioli (c. 1447–c. 1517; or Friar Luca dal Borgo; who was also known as the father of accounting, see below) delved more into perspective, including in his book *De Divina propor-*

tione, which relied heavily on Francesca's work. The illustrations in Pacioli's book were done by none other than the famous scientist, painter, and Renaissance figure Leonardo da Vinci (1452–1519), who was himself a great contributor to the study of perspective. Pacioli was one of da Vinci's teachers, and he instructed the eventual great master a good deal about proportion and perspective.

What is **perspective**?

In painting and photography, perspective gives an image depth, allowing a person to perceive three dimensions in a two-dimensional picture. In art, this can be achieved using certain "tools of the trade." For example, the horizon line is the eye level

This tulip field in Nordwijkerhout, Holland, is one example of how lines lead to a vanishing point. Artists employ this concept to create the feeling of depth in their paintings. *Robert Harding World Imagery/Getty Images.*

of the artist (and viewer); by raising and lowering the viewpoint, it enables the viewer to see more or less of the horizontal plane that lies between the viewer and the horizon. The vanishing point is a point on the horizon line at which parallel lines seem to meet—the classic example is how train tracks seem to come together in the distance. The horizontal baseline is parallel to the horizon line and represents the base of the picture plane. The vanishing point is on the horizon line and is directly opposite (or in front) of the viewer. All of these points, lines, and angles—and other ways of bringing depth into a painting—are based on concepts in mathematics.

How is **geometry** used to create **Islamic patterns**?

Many of the intricate patterns on Islamic architecture, structures, walkways, and fabric incorporate geometry in the construction of the design. From simple shapes, Islamic designs evolved into complex geometric shapes involving a high degree of mathematical symmetry. One of the best places to see such work is in the Alhambra Palace, a Moorish structure in Granada, Spain, built in the 15th century.

Some Islamic designs were drawn by eye to be aesthetically pleasing; others involved collaborations between mathematicians and artists. For example, mathematician and astronomer Abū al-Wafā (940–998 CE; born in today's Iran) worked with artisans in the 10th century, helping to design the ornamental patterns in wood, tile, fabric, and other materials. He discussed such mathematical concepts as constructing a perpendicular at the endpoint of a line segment, constructing a regular polygon, dividing segments into equal parts, and even the bisection of angles. Abū al-Wafā's

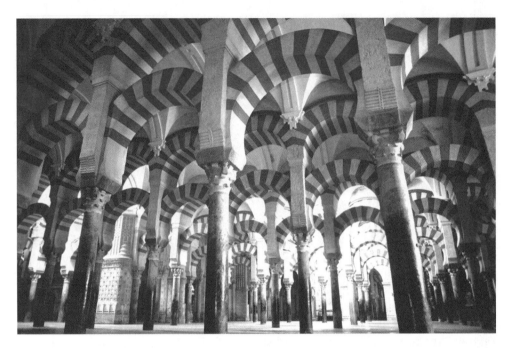

Moorish architecture, as seen in this example of the La Mezquita Cathedral in Andalucia, Cordoba, Spain, features beautiful designs that emphasize geometric shapes and symmetry. *Photographer's Choice/Getty Images.*

teachings, and those of others like him, eventually led artisans to design the many intricate patterns seen in Islamic art today.

What is a **Vedic square**?

Historically, in order for Islamic artisans to create a variety of geometric patterns, a Vedic square was often used. (The Vedic is an ancient religion associated with India; for more information, see below.) These squares generate patterns by the selection of number sequences and angles. Each square is made by taking a multiplication table and reducing all the numbers to single digits. (For example, if a number in the table was 81, it would be reduced by adding 8 + 1, or 9.) Then, to get a design, for example, a certain line of numbers would be chosen, such as connecting all the number 5s; and by using a fixed angle of rotation with the lines, various designs were drawn. These number patterns are distinctively Islamic and helped create many of the famous traditional Islamic geometric designs seen today.

Who was **M. C. Escher**?

Maurits Cornelius Escher (1898–1972) was a Dutch painter famous for his mixing of art and mathematical concepts. He had no formal math training past secondary

school, but he was, and still is, admired by mathematicians for his amazing visualizations of mathematical principles.

Escher's interest in mathematical patterns began around 1936, after viewing the Alhambra Palace's collection of Islamic art (see above). He was fascinated with structures represented in plane and projective geometry—both Euclidean and non-Euclidean geometry (for more information, see "Geometry and Trigonometry")—not only using the geometry of space, but also the logic of space. For example, he used the idea of tessellations, or arrangements of closed shapes (usually polygons or similar regular shapes, like those used on a tiled walkway) that cover the entire plane without overlaps or gaps. Not only did he use regular tessellation shapes, but also irregular ones, making the shapes change and interact with each other. He also made many of his amazing patterns by taking a basic

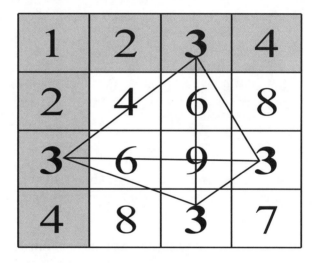

Using Vedic squares, Islamic artisans could create designs based on geometric shapes formed by the arrangements of certain numbers.

pattern and either distorting it and/or applying what geometers call reflections, translations, and rotations. For many other mathematical patterns, he used regular solids, Platonic solids, and visual aspects of topology in his work. (For more about Platonic solids, see "Geometry and Trigonometry.")

What are **mathematical sculptures**?

Mathematical sculptures are sculptures that represent some mathematical design. They are usually a geometrical figure and can be made of materials such as metal, wood, concrete, or stone shaped into squares, triangles, cylinders, specific curves, rec-

What is the "Mozart Effect"?

The Mozart Effect is a term coined in the 1950s by physician and researcher Alfred A. Tomatis (1920–2001). It refers to the supposed increase in brain development that children under age three experience when they listen to music composed by Wolfgang Amadeus Mozart (1756–1791). A more recent interpretation originated in 1993 from physicist Gordon Shaw and Frances Rauscher, a former concert cellist and expert on cognitive development. After a few dozen students listened to the first ten minutes of Mozart's Sonata for Two Pianos in D Major (K.448), the researchers determined that the students experienced a short-term enhancement of their spatial-temporal reasoning (based on a certain IQ test). But many other researchers claim that no one has ever been able to reproduce these results.

Over the years, the Mozart Effect has reached further into the public "psyche," with highly debated claims of better health, memory improvement, and therapeutic uses of music, especially classical music. Proponents also claim the Mozart Effect can be applied to learning such subjects as mathematics. They believe that exposure to certain types of music—especially classical music early in life—leads to higher future scores in spatial visualization, abstract reasoning, and sundry other mathematical concepts. But all these claims remain highly contested.

tangles, spirals, and so on. Many of these sculptures represent different mathematical concepts, from polyhedral geometry and topological knots to fractal designs.

In what ways does **mathematics** apply to **music**?

There are many ways in which mathematics applies to music—from harmonics to the concept of octaves and scales. There is a most interesting connection in that both music and mathematics are experienced as pure objects of the brain—and both have meaning outside of the brain only through artificial connections.

For example, there is a "mathematical" reason why certain note combinations sound harmonious to most people. One good example is a violin string: When it is plucked, the string vibrates back and forth, creating mechanical energy that travels through the air as wave patterns. The number of times the waves reach a person's ear is called the frequency, which is measured in Hertz (Hz, or cycles per second); if more waves are heard within a certain time period, the note's pitch sounds higher.

Chords sound harmonious to us because of waves and ratios. For example, take the C major chord, in which the note middle C is 261.6 Hz, the note E is 329.6 Hz, and

the note G is 392.0 Hz. The ratio of E to C is about 5/4, or every fifth frequency wave of E matches with every fourth wave of C; the ratio of G to E is also about 5/4; and the ratio of G to C is about 3/2. Because the notes' frequencies match with the other notes' frequencies, they all sound harmonious to a person's ear.

It is interesting to note that none of the ratios for the familiar "western-style" scale are truly exact, but rather are approximations. This is because when this scale was put together, the creators wanted to make the notes go up in equal intervals and the ratios to be in tune. The only way to do this was to compromise and use "inexact" ratios.

The music of the brilliant composer Wolfgang Amadeus Mozart is said by some to increase intelligence in babies who are exposed to regular doses of his melodies. This is known as the "Mozart Effect." *Library of Congress.*

What is the **"music of the spheres"**?

Not only was the Greek mathematician and philosopher Pythagoras of Samos (c. 582–c. 507 BCE) credited as the first to prove the Pythagorean theorem, he also discovered the "music of the spheres." He found that the pitch of a musical note depends on the length of the string producing the sound, enabling him to develop intervals of the musical scale with simple numerical ratios. When a stringed instrument is played, if the musician puts pressure halfway along the string's length, he or she produces a note that is one octave above the string's note—in other words, the same quality of sound but at a higher pitch. Octaves increase by one step each time a string vibrates at twice the frequency of the previous note; this is expressed mathematically as a frequency ratio of 1:2 (string:octave). Pythagoras recognized other ratios, too, such as the perfect fifth (ratio 2:3) and the perfect fourth (3:4), thus developing the mathematical basics of musical harmony.

Pythagoras took music and mathematics a bit further, believing that the musical octave was the simplest and most profound expression of the relationship between spirit and matter (for more about Pythagoras, see below and "History of Mathematics"). He also taught that each of the known planets produced a particular note (generated by its motion) according to the planet's distance from the Earth, calling this *Musica mundana,* or the "music of the spheres," it was music no one could really hear. He and his followers, called the Pythagoreans, further used music to heal the body and to elevate the soul, yet they believed earthly music was just a faint echo of the universal notes. Although today it may seem more "magic" than hard science and

379

mathematics, maybe Pythagoras was right: After all, many researchers believe music does have the capacity to heal a person under certain circumstances.

MATH AND THE SOCIAL SCIENCES

What is **sociology**?

Sociology is the study and classification of human societies. It delves into the relationships among peoples—mostly within, but also outside, each culture. Sociology studies are based on the idea that behavior of peoples is influenced by social, political, occupational, and intellectual groupings; it is also based on the immediate and particular settings in which individuals reside.

What is **computational sociology**?

Computational sociology is a branch of sociology that uses computation to understand social phenomena. It takes great advantage of statistics to determine trends in data and uses computer simulation in the construction of social theories. It is also referred to as the study of social complexity.

What are some **mathematical concepts** used in **demography**?

Demography is the statistical study of human populations that reveals the characteristics of the population. This includes such factors as size, density, growth, distribution, and vital statistics. Some common demographic statistics include birth and death rates, life expectancy, and infant mortality rates.

The *birth rate* (also called the crude birth rate) is usually represented in demography as the ratio of live births in an area population annually, most often expressed as the annual number of live births per 1,000 people (or a time division, such as at mid-year). Related to this is the *infant mortality rate,* which is the annual number of deaths of children less than one year old per 1,000 live births.

The *death rate,* or crude death rate, refers to the ratio of deaths in an area population annually. It is often expressed as the annual number of deaths per 1,000 people. *Life expectancy* is related to death rate, and indicates the average lifespan of people within a certain population. It is calculated on the basis of statistical probabilities, usually represented by the average number of years a group of people born in the same year can be expected to live, if mortality at each age remains constant in the future.

Based on its shape, a population pyramid quickly shows whether this crowd of people has a high or low birth rate, as well as a high or low mortality rate. *Lonely Planet Images/Getty Images.*

What is a **population pyramid**?

A population pyramid is actually two back-to-back bar graphs that show the distribution of a population by gender: One shows the number of males and the other shows the number of females (most often in five-year age groupings). These "pyramids" take a variety of shapes. For example, a triangular population distribution (also called a pyramid or exponential distribution) is usually indicative of a high birth and death rate, and a short life expectancy that is typical of less economically developed countries. A rectangular population distribution pyramid shows little change in size between the age groups, with more people reaching old age; this is typical of a more economically developed country.

There are several uses for population pyramids. Besides calculating life expectancies, they are often used to find the number of dependents being supported in a specific population. (Dependents in this case are considered to be children under 15, those attending school full time and "unable" to work, and those over 65 years old or retired, although this is a general definition and not applicable in all cases in all countries.) Thus, governments can use such graphs to determine how much of the working population can support dependents. The graphs are also used to help predict future

changes (in terms of decades) of a population's age structure. Historians use them, too, to find significant information based on past population pyramids.

What is a **life table**?

A life table is a statistical chart of mortality and survivorship of a group of people (also called a cohort) of the same age at each "stage" of life. The results are usually shown as the probability of death and survival, as well as the life expectancy at various ages.

There are two basic life tables, or expectations of life charts. A current (period or cross-sectional) life table is based on age-specific mortality rates for a given period (either a single or many years). It assumes that as a group ages it has a set pattern of age-specific mortality rates that never change from year to year. But this is based on a hypothetical model of mortality, not the true mortality of the cohort.

In a cohort (longitudinal or generation) life table, the people in the group are well-defined, usually of the same age (such as all the people are born in 2000), and have common experiences or exposures. Thus, it is based on age-specific mortality rates that allow for known or projected changes in mortality in later years. (If the group is followed up for the incidence of diseases, for example, such as osteoporosis or heart disease development, it is called a cohort or prospective study.)

To compare, a current life table of life expectancy at age 65 in 2000 would be worked out using the mortality rate for age 65 in 2000, for age 66 in 2000, for age 67 in 2000, and so on. A cohort life table of life expectancy at age 65 in 2000 would be worked out using the mortality rate for age 65 in 2000, for age 66 in 2001, for age 67 in 2002, and so on. But note that not all cohorts are used in mortality studies. Some are also used for other reasons, such as a professor following a cohort of students who study together for the length of a certain degree program.

MATH, RELIGION, AND MYSTICISM

How were **mathematics and religion** intertwined in **ancient times**?

One of the first references to religion and mathematics was around 4000 BCE (although this date is highly debated), through the Vedic religion, which is followed by the Indo-Aryan peoples. Two works written in Vedic Sanskrit were the *Vedas* and *Vedangas,* both of which not only discuss religion, but also include a great deal of astronomical and mathematical knowledge throughout the text.

Another connection between religion and mathematics grew through the practice of astrology, which is thought to have started around the 4th century BCE by the Babylonians. Astrology, the belief that celestial bodies control the affairs and fates of individuals, kings, and nations, was a kind of "religion" in ancient times; it was based on the position of the Moon, planets, and constellations. In order to practice astrology, the astrologer needed an extraordinary knowledge not only of astronomy, but also of mathematics, including the use of algorithms to calculate some of the "predictive" results.

Still another connection is the one between Christianity and math, although it is often debated whether mathematics affected Christianity more or vice versa. What is known is that people who studied math and science in the past were often deeply entrenched in Christianity. For example, in the 16th and 17th centuries, great scientists such as Galileo Galilei (1564–1642), Johannes Kepler (1571–1630), Isaac Newton (1643–1727), and Nicolaus Copernicus (1473–1543) were all deeply religious Christians who saw their scientific works as a religious undertaking.

To actually list all the other connections between religion, science, and mathematics is far beyond the scope of this book. But as English physicist Freeman Dyson (1923–) once said, one of the basic connections came as the result of theological debate, and such arguments nurtured analytical thinking that could be applied to the analysis of natural phenomena.

What was the **Pythagorean Society**?

Greek philosopher and mathematician Pythagoras of Samos (c. 582–c. 507 BCE) not only contributed the Pythagorean theorem to mathematics, but also started a group (some say "cult") called the Pythagorean Society. The school he founded stressed the necessity for people to be well-rounded and also taught reincarnation and mysticism, making it similar to—or perhaps influenced by—the earlier Orphic cult. The fundamental belief of the Pythagoreans was that "all is number," or that the entire universe—even abstract ethical concepts like justice—could be explained in terms of numbers. In fact, Pythagoras was so enamored with the concept of numbers that his beliefs were centered around the premise that the entire universe was based on a mathematical algorithm.

But the society also had some interesting non-mathematical beliefs, too. Those in the "inner circle" of the society were called the *mathematikoi.* The lived permanently

Sangaku—literally, "mathematical tablet" and often seen as "san gaku"—is the name for a form of traditional Japanese temple geometry. From 1639 until 1854, Japan was isolated from the West. Because of this, Japan developed a kind of native mathematics that was used by samurai, merchants, and farmers. They would solve geometry problems, marking their work on delicately inscribed, colored wooden tablets that hung under the roofs of shrines and temples. In general, sangaku problems dealt with Euclidean geometry, but they were much different from Western geometric studies. Although the majority of the sangaku are simple to solve by Western standards, others require the use of calculus and other complex methods.

with the other Pythagoreans, had no personal possessions, and were vegetarians. Those in the "outer circle" of the society were known as *akousmatics.* They lived in their own houses and only came to the society gatherings during the day. Even women were allowed to join, some of them becoming famous philosophers. Although Pythagoras is thought to have written many works, the secrecy of his school and its communalism has made it difficult to distinguish between the works of Pythagoras and those of his followers. (For more information about Pythagoras and the Pythagoreans, see above and "History of Mathematics.")

How did **Jainism** influence mathematics in **India**?

Jainism was a religion and philosophy founded around the 5th century BCE following the decline of the Vedic religion on the Indian subcontinent. Along with Buddhism, it became one of the area's main religions.

Over the next several centuries, Jainism also became a major influence in Indian science and mathematics. It was steeped in cosmological ideas in which time was thought of as eternal and without form. The world was also considered to be infinite and to have always existed. In fact, their cosmology included time periods with numbers larger than our modern guesses about the universe's age. Even their astronomical measurements came close to some of our modern values. For example, the synodic lunar month was thought to be 29 16/31 days (29.516129032 days); the correct value is 29.5305888.

Jainism also produced a plethora of mathematical ideas, some of which were surprisingly advanced for their time. For example, they understood such concepts as the theory of numbers, arithmetic operations, sequences and progressions, the theory of sets, square roots, knowledge of the fundamental laws of indices, geometry, an approximation of pi (they believed π equaled the square root of 10, or 3.162278, which is

close to the actual value of 3.141593), operations with fractions, and simple, cubic, and quartic equations.

What is **numerology**?

Numerology is the study of the (supposed) influence numbers have on human affairs. It is considered an occult art in which numbers are used to reflect the spiritual characteristics of people. In general, numerologists match numbers (such as birth dates) to letters (such as names), then "predict" things such as a person's purpose in life, their areas of talent, and even behavior patterns. Numerology charts use the numbers 1 through 9, 11, and 22. The values from birthdays and names are divided so that they equal one of these numbers, with each number indicative of something about the person, such as the Life Path Number, Expression Number, and so on. Although many people use numerology to make life plans or decisions in their life—and some people claim it does work for them—there is no mathematical or scientific basis for numerology claims.

What are examples of **"unlucky" and "lucky" numbers** in different cultures?

The concept of lucky and unlucky numbers in various cultures abounds. As many people agree, it's actually the "luck of the draw" that determines a lucky or unlucky number. Some cultures have tried to attach good or bad fortunes to numbers. For example, in terms of good luck and fortune, the number seven seems very popular in many cultures. It is the holy number, symbolizing God in all forms, and is considered sacred to many people. The Japanese, for instance, have Seven Gods of Good Fortune; the Egyptian god Hathor can be seven cows at once; and there are the seven days of creation in the Bible (actually, the seventh was a day of rest).

On the other hand, the same number can represent something that is "bad." For example, many mythological monsters have seven heads; in Native American lore, the Sioux's saltwater snake Uncegila could only be killed by a blow to her "seventh spot"; and the "seven deadly sins"—avarice, envy, gluttony, lust, pride, sloth, and wrath—is a Western cliché.

One famous unlucky number is 13, which is thought by most Western people to be bad luck because of the connection between Judas Iscariot, the disciple and betrayer of Jesus who was the 13th man in the room at the Last Supper. But in other places, such as Italy and China, 13 is considered lucky. Even ancient civilizations did not shy away from the number 13. In the Celtic and Native American systems of astrology, there were 13 lunar months in the year and 13 astrological signs.

What **numbers** have **special significance** in various cultures?

There are plenty of numbers that have special significance in various cultures—too many to mention here. But some common examples can be found around the world.

What is the "beast number"?

The so-called "beast number" is 666, a number that is mentioned in Revelations 13:18: "Here is wisdom. Let him that hath understanding count the number of the beast: for it is the number of a man; and his number is 666." It is also thought of as the number of the Antichrist.

There are other related numbers that have gained "cult" aspects, too. For example, a number having exactly 666 digits is called an apocalypse number. A number of the form 2^n that contains the digits 666—in other words, powers of 2 that contain three consecutive sixes in the decimal representation—is called an apocalyptic number; 2^{157} is one such number (it equals 182,687,704,*666*,362, 864,775,460,604,089,535,377,456,991,567,872). The first few such powers are 157, 192, 218, 220, and so on.

There are also some interesting mathematical characteristics of the beast number that mathematicians have found as they played with numbers. For example, the beast number is:

- equal to the sum of the squares of the first seven primes: or $2^2 + 3^2 + 5^2 + 7^2 + 11^2 + 13^2 + 17^2 = 666$

- it's a sum and difference of the first three 6th powers: $1^6 - 2^6 + 3^6 = 666$

- There are exactly two ways to insert "+" signs into the sequence of 123456789 to make the sum 666, and exactly one way for the sequence 987654321:

 $1 + 2 + 3 + 4 + 567 + 89 = 666$
 $123 + 456 + 78 + 9 = 666$
 $9 + 87 + 6 + 543 + 21 = 666$

- Even more amazing is that the Dewey Decimal System classification number used in libraries for "Numerology" is 133.335; reverse the number and add to get $133.335 + 533.331 = 666.666$

- And, just as weird, if you write the first six Roman numerals, in order from largest to smallest, you get 666: DCLXVI = 666.

In Indian numerology, for example, everyone has three relevant numbers that explain and predict human behavior: psychic, destiny, and name numbers—and they mean something different to each person. In Norse numerology, three, nine, and multiples of three and nine are magically potent. For example, the number nine is significant in Norse magical practice and often points to Odin. Considered the supreme god in Germanic and Norse mythology, Odin, legend says, hung himself from a tree for nine days

and nights, a time in which he learned nine magical songs and 18 magical runes. In addition, there were nine realms of existence, 40 Norse Valkyries, and 13 Norse Gods were present when Loki caused the death of Baldur.

In China, the number three is lucky, and eight is even luckier. Six and nine also run close behind. The number six implies that everything will go smoothly; eight was originally favored by the Cantonese, since in their language eight means to make a great fortune in the near future (later, the Chinese favored the number); nine implies something everlasting, especially in friendship and marriage. Numbers ending in these digits are often favored today when Chinese people choose phone numbers, room numbers, and car licenses. As for unlucky numbers, four (unlucky also in Japan) and seven are at the top of the list, the former implying death and the latter meaning "gone." Fourteen is also bad news. In fact, some cities in China ban the number from car licenses, and many buildings do not have fourth and fourteenth floors.

How do **astrologers** perceive **numbers** in relationship to the **solar system**?

Astrologers—people who look at the universe in terms of how it affects human behavior and future events—perceive that each number from 0 to 9 is ruled by a certain celestial body in our solar system. Based on various factors, people are given certain numbers that then can be used to determine a person's dominant personality, and even his or her future. For example, the number two is ruled by the Moon; a person with that number is cooperative, emotive, and has a great deal of feeling; the number seven is ruled by Neptune, with the person being spiritual, and thus it is the number of mystics, visionaries, and seers. Not everyone subscribes to this way of "knowing ourselves," but soothsayers have been around for centuries, finding numbers and sundry other cosmic quantities for people who believe in astrology.

MATH IN BUSINESS AND ECONOMICS

What is **money**?

Actually, money is a mathematical concept. In each culture, money (or currency) is the most common medium of exchange, and through an "agreement" within a community it can be traded and exchanged for goods, services, or obligations. Each type of currency represents specific units that can be subdivided (such as dollar bills and quarters). A good or service is then given a certain value when compared to other goods and services, what is usually referred to as a price. Because of this, people who use money need to know some of the most basic of mathematical concepts, especially addition and subtraction.

Of course, money is actually an abstraction. Now and over the centuries, it has been essentially a token. That is why money has often been represented by such things

How did money evolve?

Before the advent of money, exchanging goods was merely a matter of bartering. Livestock and grains often were used in ancient times as methods of barter, with one item exchanged for another. But there were definite limitations to this. For instance, if a farmer wanted to barter for a horse with wheat in the winter, he would have to know how to store the wheat in order to get the horse. Timing became essential, and sometimes meant the difference between surviving and dying.

Some of the first types of money included limestone coins (ranging in size depending on their value), tobacco, shells, whale's teeth, and even bread in medieval Iraq. The use of gold and silver coins is thought to have started around 650 BCE by the Lydians, who lived in the area between the Black and Mediterranean Seas. Before that, metals were traded as money in other forms, such as nuggets, rings, and bracelets. Paper currency first appeared about 300 years ago and was usually backed by some "standard" materials of natural value (such as gold) that could be converted on demand.

Using currency was a great improvement over the bartering system (although bartering still has its place in many countries, especially throughout smaller communities). For example, when gold became a trade standard, it was easier for the farmer to buy the horse in the winter with gold coins than try to store enough wheat (and keep away mice) to make the trade. In the West, as more and more people participated in the gold standard, the banking industry grew, paralleling the growth of trade and industry.

Today, monetary systems are deeply entrenched, with various types of currencies in a multitude of countries. Type of money keeps changing, too. For example, in 1988 the world's first durable plastic currency was introduced by Australia (plastic bills were seen as a way to frustrate counterfeiting). In 2000 the European Economic community introduced the Euro, a monetary unit that puts much of western Europe under a common currency. All this exchanging and changing of currency involves, of course, mathematics.

as naturally scarce precious metals (such as gold and silver), conch shells, trinkets, and today's entirely artificial representation of money: paper banknotes.

What is **interest** and how is it **calculated**?

In finance, interest can be viewed in two ways: The interest given to a person on

money "loaned to" (deposited in) a bank or financial institution, or as a fee (payment)

for borrowing or lending money, most often based on a percentage of the requested amount. The two most common types of interest are simple and compound. (For more information on interest, see "Everyday Math.")

In terms of a savings account, the interest is usually compounded, which means that any interest earned is reinvested, or compounded, to generate even more money, and thus increase future interest. Take, for example, a person starting with $1,000 in a money market fund earning 5 percent per year with quarterly interest payments (or the person gets 5 percent divided by four, or 1.25 percent per quarter). After one year, the $1,000 has grown to $1,050.95, making the compound interest rate actually 5.095 percent—not 5.00 percent—because interest was also paid on the accumulated interest for each quarter. (An interest of 5 percent only is called the simple interest rate; but most banking institutions pay compound interest on savings.)

In 2000 the Euro was formally introduced as the preferred currency for members of the European Economic Community. This greatly simplified trade between these countries, which no longer had to worry about converting currencies at rates that could change daily. *Photographer's Choice.*

Simple and compound interest are both used in borrowing and lending. With simple interest, the interest is paid strictly on the amount of the initial principal (original amount borrowed or lent), which is usually represented by the formula: $a(t) = a(0)(1 + rt)$, in which $a(t)$ is the sum of the principal and interest at the time t for a constant interest rate r.

Compound interest has a more complex formula. This type of interest is calculated not only on the initial principal, but also the interest accumulated (or accrued) over time. For example, a person purchases a home for $250,000 and pays $50,000 as a down payment. The remaining $200,000 is taken out as a loan at 8 percent interest for 30 years (compounded monthly) with equal monthly payments. The monthly mortgage payment (M), in which P is the principal, i is the interest rate, n is the number of years, and q is the number of pay periods per year, would be:

$$M = Pi / [q(1 - [1 + (i/q)]^{-nq})]$$

$$= (\$200,000)(0.08) / [(12)(1 - [1 + (0.08/12)]^{-(30)(12)}]$$

$$= (\$1333.333333\ldots)/(1 - [1.006666666\ldots]^{-360})$$

$$= \$1,467.53 \text{ per month}$$

What is a **stock market** or **stock exchange**?

The stock market (or stock exchange) is a way of providing companies a means of issuing shares to people who want to invest in various companies; it provides a way of buying and selling those shares. Stocks, or the ownership investment in a corporation, is represented by shares, which are claims on a company's assets and earnings. Common stock means the investor has voting rights in the company; preferred stock means there are no voting privileges. Owners of preferred stock, however, have first claim not only on the assets and earnings of the company over common stock, but also the dividends are paid first to people who own these shares.

How are some **stock market financial indicators** calculated?

There are numerous financial indicators associated with the stock market, and they all have to do with mathematical calculations. *Dividends* are taxable payments put forth by a company's board of directors and distributed to its shareholders. These moneys come from the company's current or retained earnings and are usually paid out every quarter. They can be either in the form of a cash dividend (usually in check form), stock dividend, or other property.

The *dividend yield* is the yield paid out by a company to shareholders in the form of dividends, calculated by the amount of dividends paid per share over a year, then divided by the stock's price. For example, if a stock pays out $2 in dividends over the course of a year and trades at $40, the dividend yield is 5 percent.

The *P/E ratio* stands for "price/earnings ratio," a measure of a stock's amount to its earnings per share. It is a measure of a stock's market capitalization divided by its after-tax earnings over a 12-month period. The higher the P/E ratio, the more the market is willing to pay for each dollar of annual earnings. Of course, companies with negative earnings (or those that are not profitable) have no P/E ratio at all.

What are **market indexes**?

A market index is a statistical measure of the changes in a portfolio of stocks, all of which represent a portion of the overall stock market. For example, the price of the stock market index called the Dow index in the United States was determined by adding up the prices of 12 of the largest public companies and dividing that number by 12 (averaging the prices). Today, the Dow Jones Industrial Average uses 30 of the largest and most influential companies in the United States to determine its stock index, using much more complicated methods that call for the use of computers.

Is math used in **accounting**?

Yes, math is definitely used in accounting. The field includes the classification, analysis, and interpretation of the financial records (or bookkeeping) of an institution, company, or other enterprise, with most of the mathematics being addition and subtraction. Bookkeeping records not only tell the financial health of an organization—in terms of profits or losses—but are also used for auditing and the occasional balancing of a company's ledger from a set period of time.

Bookkeeping has a long history. The idea was probably understood long before mathematician Luca Pacioli's (c. 1447–c. 1517) treatise on bookkeeping was published in 1495 (for more about Pacioli, see p. 374). The idea of a double-entry bookkeeping system—or a debit from an account and a corresponding credit in another account—is popular today, but it was first used in medieval Europe.

What are **economics** and **econometrics**?

Economics is a branch of the social sciences that deals with the production, distribution, and consumption of goods and services, including their management. The application of mathematical and statistical techniques to economics and financial data is called econometrics. It includes the study of problems, the analysis of data, and development and testing of theories and models.

How are some **economic indicators calculated**?

There are many economic indicators that are mathematically calculated, mainly in order to understand a nation's economic health. The gross domestic product (GDP) is the current measure of a nation's economic performance. It is the total market value of all final goods and services produced in a country during any quarter or year. Its calculation includes such variables as total consumer spending, business investments, government spending and investments, and the value of exports minus the value of imports.

The consumer price index (CPI) is a measure of the average change over a period of time of prices paid by consumers for a "market basket" of consumer goods and services. (A market basket is based on information from a sampling of consumers on their spending habits—what they "put in their basket.")

There are also the M1 and M2 indicators. The M1 is a measure of the money supply—including currency in circulation, such as travelers checks, deposits, and checking account balances—that can be spent immediately. M2 is the M1 plus assets that are invested for the short term, such as certain overnight repurchase agreements, savings deposits, time deposits (less than $100,000), and money market mutual funds. There is also an M3 that relates to the big bucks from places such as lending institutions.

MATH IN MEDICINE AND LAW

What is **mathematical medicine**?

Mathematical medicine uses math to understand certain medical problems. For example, some medical researchers are using noises and sounds from a patient's lungs as a diagnostic tool. Based on the quality of the sound—measured by ear or mathematical analysis of a frequency spectrum—it may prove to be a noninvasive way of detecting lung disease. Other examples are the mathematical modeling of how drugs are delivered through the body to target diseases and the mathematical modeling of the progression of age-related macular degeneration. Overall, mathematical modeling and simulation techniques are the most important part of mathematical medicine, and will likely be used in the future to help solve problems with the human body.

What is **epidemiology**?

Epidemiology is a form of statistical study of the incidence, distribution, and reasons for an infection, disease, or other health-related event that occurs in a population. Using statistics, it asks such questions as "Who has the infection or disease?"; "Where are they located geographically and in relation to each other?"; "When is the infection or disease occurring?"; "What is the cause?"; and "Why did it occur?"

How is **mathematics** used in **psychology**?

Like many other fields in the social sciences, psychology uses a great deal of mathematical statistics. Many psychologists, especially those in research, use the scientific method to collect and analyze data, then summarize the data using tables, diagrams, and descriptive statistics. Statistics and general mathematics are everywhere, from the nature of measurement to the different types of variables and forms of scales (ratio, interval, ordinal, and nominal) used to measure a range of psychological phenomena.

Statistics plays a big role in professions besides business and engineering. Psychologists and psychiatrists, for instance, use statistics to assist them with diagnoses. *Taxi/Getty Images.*

In fact, statistics plays a major role in the manual used as a guide for psychiatric diagnosis: Called the *Diagnostic and Statistical Manual of Mental Disorders* (or DSM), it is published by the American Psychiatric Association. The DSM contains standard definitions of psychological disorders, including mental illness, developmental disorders, and some other neurological conditions, as well as plenty of statistics concerning those disorders.

What are some types of **math** used in the **legal profession**?

Lawyers—from civil to criminal—need to know mathematics for a number of reasons. For example, when an attorney writes a brief for a case, he or she structures his or her argument like a geometric proof, which is similar to proving theorems by presenting all the relevant facts and laws. Statistics and accounting are valuable, too, because many times large amounts of money are involved in estate suits or other such cases. Knowledge of percentages, interest, and so on are needed, as well, to determine fair settlements for all the parties involved. Lawyers even have to delve into the field of logic, especially symbolic logic that helps them interpret bad arguments and fallacies in the courtroom.

How do you find **statistics** about **criminal justice**?

One of the best places to find statistics about criminal justice is the Bureau of Justice Statistics. The Bureau offers a multitude of useful crime statistics, including criminal victimization, crime characteristics, victim characteristics, and incident-based statis-

tics, all of which help the police and other crime-fighting agencies to better understand crime and how it works. But the statistics don't stop there: There are also statistics about law enforcement officers, campus law enforcement agencies, and crime labs; a census of law enforcement training academies; and even statistics about the courts and criminal sentencing.

There are also other places to find such statistics. For example, the Federal Bureau of Investigation offers statistical reports, such as the *Uniform Crime Reports* (UCR), a program started in 1929 that collects information about crimes reported to the police. But in the late 1980s, in response to law enforcement's need for more flexible, in-depth data, the UCR Program evolved into the National Incident-Based Reporting System (NIBRS). This more complete system collects detailed statistical data on each reported crime incident and is used by law enforcement, researchers, governmental planners, criminal justice students, and the general public.

Is **math** ever used to **solve crimes**?

Yes, certain types of mathematics are often used to solve crimes. For example, police officers and crime scene experts use geometry to determine the conditions behind the crime. Angles come in handy to determine from what direction a drop of blood originated, determine where a car was coming from when it struck another car, or to otherwise reconstruct a crime scene. Length measurements can be used to determine the stride and shoe size of a robber from the person's footprints. Some higher math comes in handy to see how fast a car was traveling as it slammed on its brakes—a task solved by measuring skid marks. At murder scenes, forensic personnel can measure the splatter pattern of bloodstains on a wall to determine where the murdered person was standing; or even measure the size and angle of a bullet wound to determine the distance from, and position of, the murdered person to the murderer.

Probably the biggest connection between mathematics and solving crimes is how mathematics teaches people to improve their thinking habits. Math is a good way to develop problem-solving skills, and in everyday police work there is always another problem to solve.

EVERYDAY MATH

NUMBERS AND MATH IN EVERYDAY LIFE

(Note: Medical information contained in this chapter should not be a replacement for seeing your doctor. Please consult your doctor for any exercise programs you want to employ or any medical problems.)

What are the various **methods of keeping time**?

There are three basic methods of keeping time in common use today. They include time based on 12-hour intervals (a.m. and p.m.); a 24-hour clock; and Coordinated Universal Time clock (called UTC or Zulu time). All of these have to do with simple mathematics—mainly addition and subtraction.

A 12-hour clock (using the 12-hour interval concept) is familiar to most of us. It represents half a day (24 hours divided by 2). The abbreviations "a.m." and "p.m."— terms that originated from the use of longitudinal meridians—are used to differentiate between the morning and afternoon hours. The term *meridian* (from the Latin *meri,* a variation of *medius,* or "middle," and *diem,* or "day") once meant noon. Thus, the time before noon was called "ante meridiem" and after noon was called "post meridiem." They were eventually shortened to "a.m." and "p.m.," respectively. Whether the terms are capitalized or not is not an issue either, as both are used in texts. The 24-hour clock is, naturally, divided into 24 hour increments. It is commonly used by the U.S. military and other government agencies throughout many countries.

UTC (or Coordinated Universal Time; the letters are not a true abbreviation, but a variant of Universal Time) is equivalent to mean solar time at the prime meridian (0° longitude), formerly expressed as GMT, or Greenwich Mean Time. (The change was done to eliminate using the name of a specific location in an international standard.) Other names for UTC are World time, Zulu time, and Z time. UTC should not be confused with UT, or Universal Time, the basis for the coordinated broadcast of time signals, counted from 0000, or midnight.

A brass strip marks 0° longitude (the prime meridian) in Greenwich, England.

What are the ways to **convert** between **a.m and p.m.** and **24-hour time**?

When converting a.m./p.m. to 24-hour time for the hours between 12:00 a.m. and 12:59 a.m., just subtract 12 from the time. For instance, 12:45 a.m. minus 12 will equal 0045, and 12:59 a.m. minus 12 will equal 0059. Times between 1 a.m. and 12:59 p.m. are a straight conversion, such as 9:00 a.m. equals 0900 and 11:00 a.m. equals 1100. For the conversion between 1 p.m. and 11:59 p.m., add 12, such as 8:34 p.m. plus 12 equals 2034, and 11:59 p.m. plus 12 equals 2359 (also seen as 20:34 and 23:59).

To convert in the opposite direction from the above, add 12 for the hours between 0000 (midnight) and 0059, such as 0034 plus 12 equals 12:34 a.m., and 0059 plus 12 equals 12:59 a.m. To convert from 0100 to 1159 and from 1200 to 12:29, there is a one-to-one conversion, the former to a.m. and the latter to p.m.. For example, 0100 equals 1 a.m., 0345 equals 3:45 a.m., 1235 equals 12:35 p.m., and 1259 equals 12:59 p.m. To convert between 1300 and 2359, subtract 12, such as 1424 minus 12 equals 2:24 p.m., and 2359 minus 12 equals 11:59 p.m.

How are local **a.m./p.m.** and **24-hour times converted** to **Coordinated Universal Time** (UTC)?

When it comes to converting to specific times, Coordinated Universal Time (UTC) is a "standard." It is used for a variety of reasons, including certain scientific studies, such as weather forecasting and astronomical data. The following lists several examples of

converting local standard time (a.m./p.m. and 24-hour) to UTC time, and local daylight savings time (a.m./p.m. and 24-hour) to UTC:

Converting 24-Hour Time to UTC

Local date	a.m./p.m.	24-hour clock	UTC
April 10	9:59 a.m. EST	0959	1459
April 10	5:00 p.m. EST	1700	2200
April 10	9:30 p.m. EST	2130	0200 April 11
May 10	9:59 a.m. EDT	0959	1359
May 10	5:00 p.m. EDT	1700	2100
May 10	9:30 p.m. EDT	2130	0100 May 11

Note: In the United States, for Eastern Standard Time (EST) add five hours to get UTC time, six hours to Central Standard Time, seven hours to Mountain Standard Time, and eight hours to Pacific Standard Time. For Daylight Savings Time, add four hours to Eastern Daylight Time to get UTC time, five hours to Central Daylight Time, six hours to Mountain Daylight Time, and seven hours to Pacific Daylight Time.

What is the **Dewey Decimal System**?

The Dewey Decimal System of Classification is a numerical method libraries use to classify nonfiction publications into groups based on subject. It was invented by American librarian Melville Louis Kossuth Dewey (1851–1931) as a system for small libraries. The subject of a book is classified by a three-digit numeral that represents ten classes of subjects (000–999). In order, these are Generalities, Philosophy and Psychology, Religion, Social Science, Language, Natural Science and Mathematics, Technology (Applied Sciences), Arts, Literature, and Geography and History. For example, *The Handy Math Answer Book* would be found in the Dewey Decimal System under the 500s for Natural Science and Mathematics.

A Dewey Decimal classification number is followed by the Cutter number, or Cutter. This method was invented by Charles Ammi Cutter (1837–1903) and is an alpha-numeric way to represent words or names by using one or more letters followed by one or more Arabic numerals used decimally. Both systems—the Dewey Decimal System and Cutter—are together called the "call numbers," a way of locating every book in a library.

What is a **ruler**?

Usually made of wood, metal, or plastic, a ruler is a measuring stick. Most rulers have a straight edge used for drawing straight lines and measuring lengths. The simplest and most well-known ruler has small scales, measured in terms of inches (or centimeters).

In order to read a ruler, the user needs to know the main divisions. For example, when looking at a foot ruler, the longest increments are represented by inches and are

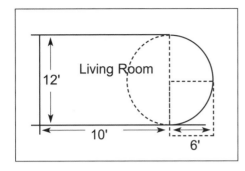

Measuring a room for a circular rug involves some basic knowledge of mathematics.

usually numbered 1 through 12; the measurement starts on the left end of the ruler, which may or may not be marked with a "0." The next divisions, from smallest to largest, are: The distances between the smallest increments represent a sixteenth of an inch, the distances between the next largest increments represent an eighth of an inch, the next represent a fourth of an inch; and, finally, a half of an inch. (For more about measurement, see "Mathematics throughout History.")

How can a person **calculate** the **amount of carpeting** needed to redecorate a room in a house?

Watching remodeling programs on television, visiting someone else's house, or even getting cabin fever in the winter often triggers the need to redecorate from floor to ceiling. But just how much new carpeting does a house need? The main way to discover how to cover up the floors is by using simple geometry, the branch of mathematics dealing with area, distance, volume, and the properties of shapes and lines. (For more about geometry, see "Geometry and Trigonometry.")

For example, if a person needs to buy new carpeting for a bedroom, just use the formula $A = L \times W$, or the area equals the length times the width. For a bedroom that measures 10 feet by 12 feet, the area would equal 120 square feet. Thus, the room would need 120 square feet of carpeting.

If a living room has a circular alcove, there is another measurement one can use: $A = \text{pi } (\pi) \times r^2$, or the area (A) equals pi times the radius squared. If the room measures 12 feet in width at the long end of the alcove (or the diameter of the alcove), the radius of the alcove is 6 feet. Thus, $A = 3.14159 \times (6)^2$, or 113 square feet (rounded to the nearest square foot), would equal the area of an entire circle. Because an alcove is half of a circle, divide the 113 square feet in half, or about 56.5 square feet. Add this area to the rest of the room. For instance, if the rest of the room measures 10 by 12 feet, multiply $10 \times 12 = 120$, then add the alcove area of 56.6 to get the needed amount of carpeting: 176.5 square feet.

Are **ratio** and **proportion** important in cooking?

Yes, ratio and proportion—two major mathematical concepts—are important in cooking (as is addition, subtraction, multiplication, and division). For example, when a recipe calls for 1 cup of flour and 2 eggs, the relationship between these two quantities is called a ratio. In this case, the relationship of cups of flour to eggs is 1 to 2, or 1/2,

or 1:2. Any change in this ratio and the recipe's result will not be the same—and the food might not be edible.

Another way of looking at cooking and mathematics is when changing the amount of ingredients in a recipe. For example, when a recipe calls for a certain amount of each ingredient, and the cook wants to make half a recipe, everything in the list is divided in half. Two cups of sugar becomes one cup, 1/2 teaspoon of vanilla becomes 1/4 teaspoon, and so on. The same logic applies if the cook wants to double the recipe, but in this case everything is multiplied by two. Two cups of sugar becomes 4 cups; 1/2 teaspoon of vanilla becomes 1 teaspoon; and so on.

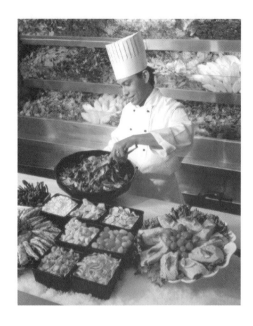

Chefs and other cooks use mathematical ratios all the time when measuring ingredients for recipes. *Stone/Getty Images.*

What does it mean when a **test score** is marked **"on the curve"**?

A test score marked "on the curve" means that the marks will roughly follow what is often called a Gaussian Probability Distribution—or, more commonly, a bell curve. This symmetrically shaped curve is based on the test scores of the exam. In a perfect world, one-sixth of the scores would be on either end of the curve, with more than two-thirds falling in the middle, creating a normal distribution.

But most test results are not ideal. Thus, when a plot of the number of students versus the marks received are viewed, the relative difficulty of the test is known. It is then up to the teacher to decide how to distribute the grades (this is usually done by comparing each student's mark to the distribution curve). The teacher then decides where to cut off the passing and failing marks, which is often referred to as curving the grades. (For more about normal distribution, see "Applied Mathematics.")

What do the numbers on **gas and electric meters** measure?

Gas and electric meters measure the daily gas or power usage, usually of a house or other building. The standard electric meters are clocklike devices that record the amount of usage. As a house or business draws on the electric current, a set of small gears inside the meter move. The number of revolutions are recorded by dials in the meter, with the speed of revolutions determined by the amount of power consumed. Newer digital models record power usage digitally.

How is political polling done?

Although polls seem to be magical predictors of election results or the success of product advertising, they are merely a matter of taking information and applying some simple statistics. Polling is a technique that uncovers the attitudes or opinions of a segment of the population, and is based on certain questions about politics, the economy, and even social conditions. The sample population can be chosen randomly, or by other methods. People can be polled via a telephone interview, questionnaire in the mail, or personal interview, such as an exit poll during an election (polled as a person leaves his or her voting place). Statistics such as averaging and resulting percents are then used to determine the overall "pulse" of the public. Many commercial poll takers not only claim their results help in market research and advertising, but also get the people's concerns out in the open.

Of course, just because statistics are used does not make polling infallible, or even reliable. For example, some questions may be misleading. The media is notorious for asking such polling questions as, "Do you agree that keeping the environment clean is important?" as if the majority of people would say "no." Such questions often make the results of polling somewhat questionable, especially if the results are not presented along with the original questions. Errors and questionable outcomes are also caused by people lying, bad interview techniques, and even the sample of people interviewed. All too often, the results of polling create another consequence: swaying public opinion and creating a "jump on the bandwagon" effect. (For more information about statistics, see "Applied Mathematics.")

A gas meter measures the amount of gas used by a home or business. In this case, the meter measures the force of the moving gas in the pipeline. The dials on the meter turn faster as the flow of gas increases, or slower as the gas flow decreases. To understand how the gas and electric companies charge a customer takes only an understanding of simple mathematics. In both cases, the difference between one month's reading and the next month's reading is the amount charged. For example, if the electric current reading is 3240 and the previous month's reading is 3201, the amount of electricity used is 39 kilowatts for one month. Then the company multiplies the number of kilowatts by the amount per kilowatt hour to arrive at the charge to the customer.

What are some numbers associated with **tire pressure**?

Tire pressure is measured using a tire pressure gauge, with the most common measuring device being about the size of a pen. First, a bit about pressure: The atmos-

phere at the surface of our planet measures about 14.7 pounds (6.67 kilograms) per square inch, or a 1 inch (2.54 centimeters) square column of air weighs 14.7 pounds. This changes depending on the altitude. For example, at an altitude of 10,000 feet (3,048 meters), the air pressure decreases to 10.2 pounds (4.63 kilograms) per square inch.

But a car, truck, SUV, or bike tire needs more pressure in order to inflate. By increasing the number of atoms inside the tire, there are more collisions between the atoms and more pressure exerted on the sides of the tire. In other words, a pump stuffs more air into a constant volume (the confines of the tire), so the air pressure within the tire rises. A car tire's pressure is typically at about 30 pounds (13.61 kilograms) per square inch; a bicycle tire's pressure can be around 90 pounds (40.82 kilograms) per square inch.

What is the purpose of the numbers found in a **mailing address**?

The Zone Improvement Plan (or ZIP) Code is a grouping of numbers assigned by the U.S. Postal Service to designate a local area or entity in order to speedily deliver and distribute mail. ZIP Codes most often refer to a street section, a collection of streets, a structure or building, or a group of post office boxes, but the numbers do not rigidly conform to boundaries of cities, counties, states, and other places.

Depending on the area, a ZIP Code includes 5, 7, 9, or 11 digits. In the most common codes—five-digit ZIP Codes—the first digits divide the country into ten large groups of states numbered from 0 in the Northeast to 9 in the far West. Each state is divided into geographic areas identified by the second and third digits of the ZIP Code. For example, New York and Pennsylvania have ZIP Codes starting with numbers between 090 and 199; Indiana, Kentucky, Michigan, and Ohio begin with numbers between 400 and 499. The fourth and fifth digits of a ZIP Code identify the local delivery area.

MATH AND THE OUTDOORS

How is the **amount of rainfall** measured?

The amount of rainfall—liquid precipitation that falls to the surface—is measured by a rain gauge. The most commonly used free-standing rain gauge is a cylinder with increments (most often in inches) inscribed on the outside of the tube. It is put in an area that is not obstructed by buildings, trees, or other tall structures than can impede the collection of rainfall.

The rain gauge can also measure snow, but added steps are needed to calculate this. In this case, a rain gauge measures the liquid equivalent of snow. This is why meteorologists will often say that a snowstorm that produces 10 inches (25.4 centimeters) of snow will have a liquid equivalent of 1 inch (2.54 centimeters) of rain, or a

What do the numbers on an aneroid barometer signify?

The non-liquid aneroid barometer measures air pressure. As the atmospheric pressure changes because of storm systems (or lack of storms), the instrument records the changes. (Because atmospheric pressure changes with distance above or below sea level, a barometer can also be used to measure altitude.) The aneroid barometer contains a small capsule that acts like a bellows, but with the air removed. When the air pressure increases, the sides of the capsule are pushed in and the connected needle rises (moves clockwise). When the air pressure decreases or falls, the capsule's sides puff out and the needle moves in a counterclockwise direction.

The numbers on a common barometer range from about 26 to 31, with divisions of 10 or more in between each number. A needle (actually a hand similar to a clock hand) points to the numbers on the barometer and moves in response to the changing air pressure. These numbers are based on the principle that atmospheric pressure supports 30 inches (76.2 centimeters) of liquid mercury in a tube with one end sealed; and this information is based on the mercury barometer, the first type ever made.

How is an aneroid barometer read? Well, a falling hand on a barometer indicates a low pressure system is on the way with poor weather (usually a storm with snow or rain); a steady barometer means there will be no changes with the ongoing pressure system; and a rising barometer means high pressure and fair weather. An even higher reading, around 31, means an extremely dry atmosphere. The timing of the barometric change is also telling: A change of a degree either way in a few hours means that the weather will change quickly; a slow change of 0.3 or so a day indicates weather arriving in 12 to 24 hours. A quick rise in the barometer also often indicates high winds and unsettled weather.

All barometers work because of our weather systems: Changes in air pressure are caused by differences in air temperature. And this, in turn, creates the wind and weather patterns that carry the high and low pressure systems around the Earth's lower atmosphere.

ratio of 10:1. But this generalization can be tricky. If the weather system is super-cold, such as an Arctic air mass over Canada and the northern United States, the below-freezing temperatures might create more than 10 inches of snow per inch of rain. Meteorologists often call this the "fluff factor," because the snow seems "fluffier" due to the fact that there is more air between the snow crystals at much colder temperatures. In fact, in very cold air the snow-to-liquid equivalent ratio can be 15, 20, or even 30 to 1. (For more about weather and math, see "Math in the Natural Sciences.")

How is **wind speed measured**?

Wind speed is most often measured using an anemometer. The simplest ones are made with cup-shaped devices that spin and catch the winds. The number of spins translates into an approximate value for the wind speed. The more accurate electronic anemometer has a free-moving turbine suspended in the middle of the instrument that, when held correctly, measures the wind speed. The speed of the turbine sensed by an infrared light relays the signal to an electrical circuit. From there, the wind speed is digitally displayed.

What do **plant hardiness zone numbers** mean?

Plant hardiness zones are another way numbers are used. In this case, they indicate the average annual minimum temperatures for landmasses around the world. For example, one common Plant Hardiness Zone Map is broken down into 20 different zones based on the average annual minimum temperatures. In zone 5a, for example, the average annual minimum temperature ranges from −20 to −15 degrees Fahrenheit (−26.2 to −28.8 degrees Centigrade; for example, Des Moines, Iowa). In zone 11, such temperatures are above 40 degrees Fahrenheit (4.5 degrees Centigrade; for example, Honolulu, Hawaii). And in zone 1, such temperatures are below −50 degrees Fahrenheit (−45.6 degrees Centigrade; for example, Fairbanks, Alaska). There are other maps, too, that break down the zones into even more detail.

MATH, NUMBERS, AND THE BODY

Is a **human's normal body temperature** really 98.6 degrees Fahrenheit?

Although everyone seems to be taught that the normal human body temperature is 98.6 degrees Fahrenheit, in reality, "normal" has a range. In fact, a person's actual measured temperature is rarely exactly 98.6 degrees Fahrenheit. One reason for this has to do with how the old standard was calculated: using an oral mercury thermometer and basing the results on a small human sampling.

Gone are the days of putting a thermometer under one's tongue for five to ten minutes. Today's thermometers are more sophisticated, accurate, and faster. Thus researchers believe that, based on better data—and more people tested—normal body temperatures measured orally range from 97.5 to 98.8, with about 1 in 20 people having a bit higher or lower normal temperatures. These numbers change throughout the day, too, varying from 1 to 2 degrees (on the average, reaching a low at about 2 a.m. to 4 a.m., and a high twelve hours later). An even more accurate representation of a body's temperature is to measure the core temperature, or the actual temperature inside the body, usually by using a rectal or inner-ear (tympanic) thermometer.

What do **blood pressure numbers** mean?

Blood pressure is a measure of how much the blood presses against the walls of the arteries. This creates two forces: The first comes from when the heart pumps blood into the arteries; the second is the force of the arteries to resist the blood flow. When a person "takes a blood pressure," he or she is taking a ratio: The higher number (called the systolic or top number) represents the pressure when the heart contracts to pump blood to the body; the lower number (called the diastolic or bottom number) represents the pressure when the heart relaxes between beats. For example, for a blood pressure of 120/76 (said, "120 over 76"), the systolic reading is 120 and the diastolic reading is 76.

When a nurse or doctor takes your blood pressure, the two numbers they read to you indicate systolic and diastolic pressures in your arteries. *Hulton Archive/Getty Images.*

In actuality, the numbers represent how high the blood's pressure would force a column of mercury to rise in a tube. For example, a systolic reading of 120 means the mercury would rise 120 millimeters (usually labeled mm Hg, with Hg the symbol for mercury) in a tube. Based on the most recent information (and it keeps changing), a blood pressure below 120/80 is considered optimal for adults; 120 to 139 over 80 to 89 is considered "prehypertension." Anything over 140/90 is considered hypertension, which includes three stages, with the highest hypertension reading, stage 3, being anything above 179/109.

How does one calculate **resting heart rate**?

Calculating resting heart rate (RHR) involves easy math. It is the number of heart beats per minute when the body is resting, with the beats per minute representing the number of times the heart contracts. To measure the RHR, just count the number of beats per minute via the pulse—usually taken either on the inside of the wrist or along either side of the neck—for 15 seconds. Then multiply this number by four ($15 \times 4 = 60$ seconds) to get the heart rate per minute. Or count for 10 seconds, then multiply this number times six ($6 \times 10 = 60$ seconds) to get the heart rate per minute. For example, if you count 12 beats in 10 seconds, multiply 12 by 6 to get a resting heart rate of 72.

How does one calculate **heart rate** during **exercise**?

One reason to calculate heart rate during exercise is to know if a person is getting beneficial exercise to keep the heart and body healthy. In order to determine the safe and

effective range of exercise to get cardio-vascular benefits, two measurements are often taken. The first is the *maximal heart rate,* a number related to a person's age (the heart beats slower with age). To estimate the maximal heart rate, subtract a person's age from the number 220. For example, if someone is 40 years old, his or her maximal heart rate is 180.

The next measurement is the *target heart-rate zone*. This number uses the maximal heart rate and represents the number of times per minute at which a heart should be beating during aerobic exercise. For most healthy people, the range is 50 percent at the lower limit to 80 (some charts say 75) percent at the upper limit of their maximal heart rate. When a person's heart rate reaches a value within this zone during exercise, it means he or she has achieved a level of activity that contributes to his or her car-

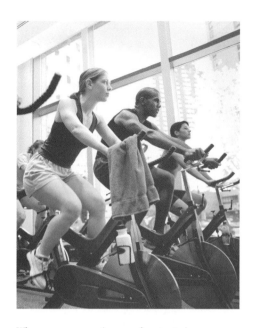

Whenever you exercise, your heart rate increases. Healthy exercise means not exceeding your target heart-rate zone and maximal heart rate, which are based on your age. *The Image Bank/Getty Images.*

diovascular fitness. For instance, from the above example of a maximal heart rate of 180, the beats per minute for the lower range would be 180 multiplied by 50 percent (0.50), or 90 beats per minute; the upper range would be 180 multiplied by 80 percent (0.80), or 144 beats per minute.

If you work out and maintain a lower-than-50- or higher-than-80-percent limit, there are few beneficial effects from the exercise. In terms of the lower limit, the heart is not working hard enough for any cardiovascular benefit; in terms of the upper limit (besides the strain and injuries that can result), the heart is working too fast for any benefit and the body can't replenish oxygen that quickly.

What do **cholesterol numbers** mean?

Cholesterol numbers indicate the amount of cholesterol in the bloodstream (choles-terol is a waxy, fat-like substance manufactured in the liver and found in all tissues). For humans, a total cholesterol number above 200 means there is an increase in the risk of heart disease (between 200 and 239 is considered borderline high cholesterol); for anything below 200, there is less of a risk for heart disease.

But total cholesterol is not the only number to know. There is also Low Density Lipoprotein (LDL), or "bad" cholesterol. LDL is the main source of the buildup and blockage in the arteries (risk levels are above 130, measured in milligrams per

What is a dog's and cat's age equivalent compared to a human?

All animals "age" at different rates, and our domestic dogs and cats are no exception. Unlike humans, these pets seem to age more rapidly in terms of human years, with both animals being of an age equivalent to about 15 human years by the time they are one year old. There are, however, some differences for dogs. Some people have determined an age range depending on the size of the animal. For example, a dog over 90 pounds may only be equivalent to 12 human years by the time it is one year old; a dog 51 to 90 pounds is about 14 human years old when it is one year old; and dogs under 50 pounds are the equivalent of 15 human years by their first year. These age differences are reflected for the entire life of the dog. Below is a simplified table illustrating dog and cat years and their human age equivalents.

Human Years versus Cat/Dog Equivalents

Human Age Equivalent	Cat Age	Dog Age	Human Age Equivalent	Cat Age	Dog Age
10 years		5 months	64 years	12 years	12 years
13 years		8 months	68 years	13 years	13 years
14 years		10 months	72 years	14 years	14 years
15 years	1 year	1 year	76 years	15 years	15 years
24 years	2 years	2 years	80 years	16 years	16 years
28 years		3 years	84 years	17 years	17 years
32 years	3 years	4 years	88 years	18 years	18 years
36 years	5 years	5 years	89 years		19 years
40 years	6 years	6 years	92 years		19 years
44 years	7 years	7 years	93 years		20 years
48 years	8 years	8 years	96 years	20 years	21 years
52 years	9 years	9 years	99 years		22 years
56 years	10 years	10 years	103 years		23 years
60 years	11 years	11 years	106 years		24 years

decaliter). There is also High Density Lipoprotein (HDL), the "good" cholesterol that helps keep the plaque from building up (risk levels are below 40, measured in milligrams per decaliter).

When there is too much cholesterol in the blood, it can build up on the walls of the arteries. Over time, the buildup (often called plaque) causes "hardening of the arteries," or a narrowing of the arteries that restricts or stops blood flow to the heart.

Blood carries oxygen to the heart, and if there is less blood (and thus, less oxygen) reaching the heart, chest pain can result. If there is a near-complete or complete blockage that cuts off the blood supply to a portion of the heart, a heart attack (myocardial infarction) is often the result. This is why most doctors recommend watching your numbers for total cholesterol, HDL, and LDL for signs of any change.

MATH AND THE CONSUMER'S MONEY

How do I **make change**?

"Making change" is an important monetary endeavor no matter where a person travels or lives. It is defined as the change (in coins and/or bills) a customer receives after making a purchase, especially when the customer gives the merchant more money than the amount due. For example, say a person visiting a farmer's market in the United States buys a half pound of garlic for $2.60 and gives the merchant a $20.00 bill. To make change, the merchant would first subtract $20.00 − $2.60 = $17.40; or the customer is to receive $17.40 in change.

The resulting change would start with the highest denomination and work down: a $10 bill, a $5.00 bill, two $1.00 bills, a quarter ($0.25), a dime ($0.10), and a nickel ($0.05). The merchant would then count the change out to the customer, stating each amount in the "opposite" fashion: "That's a total of $2.60 out of $20.00; your change is two sixty-five [$2.65 as the nickel is given], two seventy-five [$2.75 as the dime is given], three dollars [as the quarter is given], four dollars [as the first dollar bill is given], five dollars [as the second dollar bill is given], ten dollars [as the five-dollar bill is given], and twenty [as the ten-dollar bill is given]."

What is a **store discount**?

Many stores—from apparel shops to bookstores—offer a discount on the retail price or, less often, on the sale price. For example, say a sweater that originally costs $50.00 is discounted by 25 percent during a sale at a department store. The customer will actually pay 75 percent (or 75/100 or 0.75) of $50.00. To determine the sale price, multiply $50.00 by 25 percent (50 × 0.25 = 12.50); then subtract the discount from the original price to find the sale price, or $50.00 − $12.50 = $37.50.

How does a person calculate **sales tax** on purchases?

Most states and cities charge sales tax on retail purchases made by customers—from clothes to certain foods. This sales tax is based on a percent of the purchase price, with the percentage of tax called the tax rate. In turn, each state's tax rates vary greatly, often with the counties charging additional taxes: For example, in Vermont, the state

How does a person calculate the amount of a tip?

A tip, or gratuity, is the money given to a person who performs a service for a customer, such as a waiter or waitress at a restaurant. Depending on the service, in the United States a 10 percent to 20 percent tip is usually left, with the most common being 15 percent, although many people have stories about the 0 percent tip they left after a bad experience. The tip is based on the total bill—the meal and the tax—although some people base the gratuity on just the meal. For example, if a meal at a restaurant costs a total of $10.00 (meal and taxes), a 15 percent tip would be $10 \times 0.15 = $1.50. The tip is usually left at the table (or given to the waiter or waitress), or taken out by the establishment and added to a "tip pot" shared by all the wait staff.

There are some mathematical tricks to remember when leaving a tip at a restaurant, to a hairdresser, doorman, or in other appropriate circumstances. A good way to estimate a tip is to round the total bill to the most significant place value. For example, an $18.50 meal would round to $20. Next, move the decimal point of the rounded amount one place to the left ($20 to $2.00), or 10 percent of the total cost. Then divide this amount in half to determine 5 percent (or $2.00/2 = $1.00). Add the two resulting amounts to estimate 15 percent of the total—in this instance, $2.00 + $1.00 = $3.00 tip. (In reality, 15 percent of $18.50 is $2.78, which is close enough to $3.00.)

But remember, not every country tips the same. Tipping is a way of life in Egypt, but taxi drivers don't accept tips; French restaurants must add the tip, usually at 15 percent, to the bill by law; tipping in Australia is almost nonexistent; no one tips in mainland China (mainly because the government tacks on enough charges to visitors); there is no tipping in New Zealand, either, as the price usually includes services; and don't even think about tipping in Japan!

and local sales tax rate stands at 5 percent; Florida has a state sales tax rate of 6 percent. And in various counties of New York, the sales tax rates start as low as 7.25 percent to as high as 8.75 percent.

How is the sales tax calculated at the checkout? Take, for instance, a purchase of a $100.00 ring in New York City, a place with a sales tax rate of 8.265 percent. Multiply $100 times 8.265 percent, or $100 \times 0.08265 = $8.265, which rounds up to $8.27 for sales tax. The merchant would then charge the customer a total of $100.00 + $8.27 = $108.27. If the same $100.00 ring was purchased in Florida, the sales tax would be $100.00 \times 6 percent, or $100 \times 0.06 = $6.00, with a total purchase price of $106.00. Of course, not every purchase will be as straightforward as this, since some cities can also add local taxes and/or surcharges and fees to the bill.

How is **unit price** used to **determine total price**?

The unit price is simply the cost for each item (or unit). The term is often used to compare the cost of the same quantity of items that come in different sizes, or it is used to determine total costs for services. For example, if a person was having a birthday party at a local restaurant for 100 guests, and each meal (unit) cost $7.50, the total cost of the celebration would be $7.50 × 100 = $750.00. Add a tip to the total (see above), and it's easy to see why most people celebrate at home or only invite a few friends.

Be careful to calculate your tips correctly using percentages, or you might end up with a disgruntled waitress. *Stone/Getty Images.*

What is **balancing a checkbook**?

Balancing a checkbook is often a challenge. For some people, forgetting to enter checks written against or deposits made into the account creates the biggest balancing problems. For others, it is not depositing enough money to cover written checks. There really is no "art" to keeping a checkbook. It is just a matter of checks and balances—or debits and credits—and a little bit of simple mathematics.

To keep a healthy checkbook, there are several things a person can do. For example, keep a running balance of distributed checks in a check ledger. Whenever you write a check, write the amount in the ledger booklet most banks give with the checks. In the proper column, list the check number, who the check is made out to (and any other important information), the amount in the negative (−, or debit) column, and subtract the check amount from the last balance.

Along with making out checks (taking out money), keep a record of deposits made in the checkbook register. Deposits are usually written in the positive column (+, or credit). Don't let the money get low—if the account balance goes into negative numbers, the account does not have enough money to cover the checks. If more money is not put into the checking account at this point, the checks will "bounce," or not clear with sufficient funds. (This is *not* good!) Most banks charge substantial fees to the account owner for bounced checks, not the person to whom the check is made out.

Whenever you receive your bank statement, check to see if the balance agrees with your checkbook. This is called "balancing the checkbook." As you compare the

checks that have cleared with the listing in the register, check each off with an "X" or check mark. Also subtract any bank charges, such as ATM (automated teller machine) fees. If all of the checks have cleared, and all charges have been accounted for, the balance of the checkbook and statement should agree (unless the bank gives interest on checking accounts; if so, add the interest to the checkbook under the "+" or "credit" column). If not all the items have cleared, check the bank statement and note the ones not marked; total all these outstanding transactions. Subtract the total of the outstanding transactions from the end balance on the bank statement; then add any deposits that are not on the bank statement to this new balance. The numbers should match the balance in the check register. If they don't, go back over the addition and subtraction in the checkbook register to catch any inaccuracies, which is often the reason why a checkbook doesn't balance.

What is the **annual percentage rate,** or APR?

The annual percentage rate (APR) is an expression of the yearly interest rate that will be paid on a loan (which includes credit cards). It differs from the advertised interest rate, as it includes one-time fees in an attempt to calculate the "total cost" of borrowing the money. Therefore, it is wise to always look for a low APR. Because lenders are required to disclose the APR before a loan or credit application is finalized, it makes it easier to compare lenders. For example, if a person borrows $100 for one year at 5 percent simple interest (they will owe $105 at the end of the year), and the lender charges a $5.00 fee, the total cost to borrow the money will actually be $10, meaning the APR is 10 percent.

What are some **terms** to know when getting a **credit card**?

There are several terms—many that include a bit of mathematics—one should know when getting a credit card. An important one is the *annual fee,* which is often charged by the credit card company and is a flat, yearly charge similar to a membership fee. A *finance charge* is the dollar amount you pay to use credit; it should be listed on your credit card statement. It usually includes the interest on the borrowed money and other charges associated with transactions, such as cash advance fees or exchange rate calculation fees when paying for an item in a foreign country.

Many store items these days are labeled with bar codes, which use lines of varying widths to indicate numbers that can be read by laser scanners. *Taxi/Getty Images.*

As stated in the question above, the *annual percentage rate* (APR) is a measure of the cost (or relative cost) of credit on a yearly basis. With credit cards, APR commonly includes interest and other charges, such as a yearly rate. Credit cards often offer two types of *interest rates,* too. In the variable-rate plan, as the name implies, the interest is variable; it is usually tied to other interest rates, such as the Treasury Bill or prime rates. A fixed-rate plan is a rate not tied to changes in other interest rates; it remains steady, unless the credit company raises or lowers rates for everyone, which they can periodically do.

What are the ways **credit card companies** calculate **finance charges**?

When the credit card issuer calculates the finance charge on a card, it applies a periodic rate to a balance. In order to calculate that balance, the company uses various methods. The most common is the average daily balance method, in which the balance is calculated by taking the amount of debt in the account each day during a specific period and averaging it. The previous balance method uses the outstanding balance at the end of the period to compute the finance charges. The adjusted balance method derives the balance by subtracting any payments made during the cycle from the previous balance, with new purchases not being counted.

What is a **mortgage**?

A mortgage is a method of using property as security for the repayment of a loan. It is based on a 14th-century coinage of a Latin word meaning "dead pledge." The interpretation was that the property was "dead" to the borrower if he defaulted on the debt,

413

and the pledge was "dead" to the lender after the loan was repaid. For centuries, it's been figured out using mathematics.

What **mathematical concept** is used to calculate **mortgage payments**?

In most cases, a mortgage is based on amortization, which is the gradual elimination of a liability (a financial obligation or debt, such as a mortgage) in regular, fixed, systematic payments (such as monthly) over a specific period of time. These payments must be enough to cover both the principal borrowed and the interest. Although it is usually written in a complex set of mathematical calculations, simply put amortization means a part of the payment goes toward the interest cost and the remainder goes toward the principal (or the amount borrowed). The interest is then recomputed on the amount owed, and therefore it gets smaller and smaller as the ending balance

of the loan becomes less and less. That is why the homeowner pays a great deal toward interest and not the principal for the first several years of a home mortgage.

For example, if a mortgage is taken out for $100,000 at 6.5 percent for 30 years, the fixed monthly principal and interest payment is $632.07. For the first month, the homeowner pays interest on the $100,000 (or $541.67), with the remainder of the payment ($90.40) going toward principal. In other words, the debt on the principal is reduced by $90.40. By the next month, the homeowner owes interest on a lesser amount of money—on $99,909.60 (or $100,000 − $90.40), not the $100,000, with $541.18 going toward interest and $90.89 going toward principal. As payments are made month after month, the interest decreases and the principal reduction increases. By the 360th payment (or 30 years later), the payment contributes $3.41 to interest and $628.66 to principal.

MATH AND TRAVELING

How are **positions on Earth** determined?

Positions on Earth are determined using two numbers that represent latitude and longitude. These numbers are actually two angles, measured in degrees (°), minutes of arc ('), and seconds of arc ("). On a globe of the Earth, latitude lines circle parallel to the equator, and differ in length depending on their location. The longest line is at the equator (latitude 0 degrees); the shortest lines—actually pinpoints—are at the poles (90 degrees north at the North Pole; 90 degrees [or −90 degrees] south at the South Pole). In the Northern Hemisphere, latitude degrees increase as you move north away from the equator; in the Southern Hemisphere, latitude degrees increase as you move south away from the equator.

Longitude lines, or meridians (once called "meridian lines" and eventually shortened to "meridians"), are those that extend from pole to pole, slicing the Earth like segments of an orange, with each meridian crossing the equator. In the Western Hemisphere, longitude increases as you move west from Greenwich, England (0 to 180 degrees). In the Eastern Hemisphere, longitude also increases as you move east from Greenwich, England (again, 0 to 180 degrees). All points on the same line of longitude experience true noon (and any other hour) at the same time. But note: Longitude lines are not to be confused with time zones, most of which follow a more erratic demarcation. (For more information on time zones, see below; for more information about latitude and longitude with regard to polar coordinates, see "Geometry and Trigonometry.")

Why is **Greenwich, England,** called the **Prime Meridian**?

The reason for Greenwich, England, being the Prime Meridian is historical. An imaginary line passes through the old Royal Astronomical Observatory, which was chosen

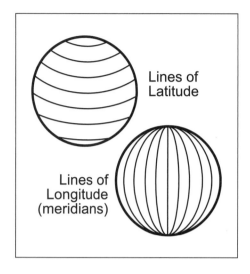

Lines of Latitude

Lines of Longitude (meridians)

by astronomers of the day as zero longitude. The observatory is now a public museum located at the eastern edge of London. It is a great spot for tourists, who can find there a long strip of brass that stretches across the yard marking the "prime meridian." Here it is possible to straddle the line with one foot in the Earth's Eastern Hemisphere and the other in the Western Hemisphere.

How does one **convert latitude and longitude** to degrees from readings containing degrees, minutes, and seconds?

It takes simple mathematics to convert the degrees, minutes, and seconds of latitudes and longitudes into degrees only. It helps to know that there are 60 seconds in one minute, and 60 minutes in one degree. Therefore, to translate 65° 45' 36" (also written as 65:45:36) south latitude into degrees you would do the following calculation: -65 degrees (south makes the number "negative") + 45 minutes × (1° /60') + 36 seconds × (1'/60') × (1°/60') = $-65.76°$ latitude.

What are **time zones**?

Time zones are any of the 24 regions on the globe (loosely divided by longitude but more erratic in their demarcations), which is divided according to the number of hours in a day. Within each zone, all clocks are set to the same time. Time zones include the international date line, an imaginary line of longitude generally 180 degrees east or west of the prime meridian; the date becomes one day earlier to the east of this line.

Time zones are truly a product of mathematics. Initially, people used local solar time, resulting in slightly different times between towns. With technology—especially trains and telecommunications—more accurate timekeeping became a necessity. Thus, time zones helped solve many of the problems by setting the clocks of a region to the same mean solar time. The time zones made it easy for neighboring time zones to be labeled "one hour apart." Not that the system is perfect. The hour separation is not universal, and because they often follow political boundaries, the shapes of the time zones can be extremely irregular.

How does one determine the **distance** and **amount of time** a trip will take?

There are several ways to determine the distance and amount of time a trip will take. For example, if the traveler knows that his or her car travels 20 miles for every gallon

Our planet has been divided into 24 time zones—indicated by somewhat erratic lines that run north and south—which are used by most people to set their clocks and coordinate schedules. *Taxi/Getty Images*.

of gas in the tank, he or she could represent the total number of miles it is possible to drive based on how much gas is put into the tank. This can be expressed as 20*g*, in which *g* stands for the number of gallons in the tank. For example, if someone bought 10 gallons of gas to reach a destination (10 gallons × 20 miles/gallon), the traveler can go about 200 miles (the gallons in the top and bottom of the equation cancel each other out, leaving miles).

Another example involves cars and mileage. If the traveler drives down a highway at a steady rate of 65 miles per hour, the driver can determine how long it will take to get to a destination, especially if he or she knows how many miles it is to that destination. The equation is mileage divided by 65 miles per hour (m/65). For example, if the traveler is going 650 miles, the (650 miles)/(65 miles per hour) equals 10 hours (the miles in the top and bottom of the equation cancel each other out, leaving hours).

What is the **scale of a map**?

Most travel, street, or highway maps show a measurement scale, usually in terms of miles and kilometers. To determine a straight line (horizontal) distance on a map, take a piece of paper and mark the origin and destination as tick marks on the paper. Then measure the "distance" between the tick marks based on the map scale to find the distance in miles or kilometers.

417

Hindu-Arabic	Greek	Hebrew	Japanese
1	α′	א	一
2	β′	ב	二
3	γ′	ג	三
4	δ′	ד	四
5	ε′	ה	五
6	ϛ′	ו	六
7	ζ′	ז	七
8	η′	ח	八
9	θ′	ט	九
10	ι′	י	十

Besides our familiar Hindu-Arabic system, numbers are expressed using different characters in such languages as Greek, Hebrew, and Japanese. (For Chinese characters, see table in the chapter "Math Basics.")

Topological maps also have scales, but in this case the scale is a ratio representing the measure on a map to some number of the actual units of measure on the Earth's surface. For example, a map with a scale of 1:25,000 means that one inch on the map is equal to 25,000 inches on the ground. Because both numbers have the same units, it can also be interpreted as any unit measure. For example, the same map could also be interpreted as 1 centimeter equals 25,000 centimeters on the ground, or 1 meter equals 25,000 meters, and so on. For those who prefer to measure in miles and kilometers, most topographic maps also offer a graphic scale in the legend. (For more about scales, see "Math in Engineering.")

What is a **currency exchange rate**?

When traveling to another country, it is important to know the currency exchange rate, which is the value of a traveler's home currency compared to the currency of the country being visited. For example, like all currency, the U.S. dollar fluctuates daily when compared to other countries' currencies. If you travel to Canada, and the U.S. dollar can buy $1.40 Canadian dollars, then the exchange rate is 1.40 to 1. If you go to New Zealand, and the U.S. dollar exchange rate is 0.5477, then 1 New Zealand dollar is worth 54.77 U.S. cents.

What are some **non-Hindu-Arabic numerals** encountered around the world?

Although Hindu-Arabic numerals are the dominant numerals used around the world, there are some places in which other number symbols are used. For example, there are Chinese, Japanese (kanji), Greek, Thai, and Hebrew numerals. The illustration lists some of the ones encountered by world travelers (for more information about numbers, see "Math Basics.")

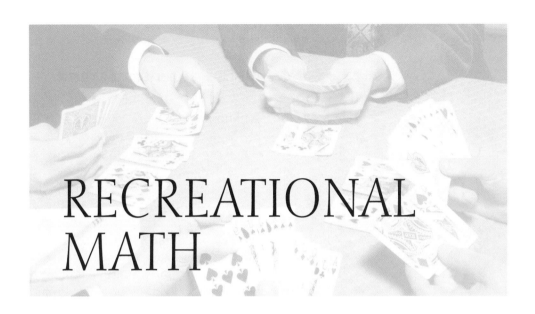

RECREATIONAL MATH

MATH PUZZLES

What is a **puzzle**?

A puzzle is a mathematical problem that produces a solution often in the form of rearranging pieces (often geometric) or filling in the blanks (such as a crossword puzzle). Puzzles do not typically require superior mathematical knowledge, but many originate from more advanced mathematical or logistical problems. They also include board games (such as chess) and brain teasers.

What are some **types of puzzles**?

There are numerous categories of puzzles. The following lists the most common:

- Puzzles derived from board games include chess-type problems (chess and eight queens puzzle)
- Logic puzzles (paint by numbers)
- Mechanical puzzles (Rubik's Cube)
- Computer puzzle games
- Cube games
- Solitaire-type puzzles
- Brain teasers
- Riddles
- Tiling puzzles (including jigsaw, polysquare, and Tangram puzzles)
- Whodunits
- Word puzzles (anagrams, cracking a code, crossword puzzles, find-the-word puzzles, and verbal arithmetic puzzles

What is a **cryptarithmetic puzzle**?

Some of the most challenging verbal arithmetic puzzles are called cryptarithmetic puzzles. These number puzzles (often called cryptarithms) are made up of mathematical equations whose digits are represented by letters or symbols; the goal is to identify the numerical value of each letter. In such a puzzle, each letter represents a unique digit, but there are rules. As in ordinary arithmetic notation, the leading digit of a multi-digit number must not be zero; also, the puzzle usually has only one solution.

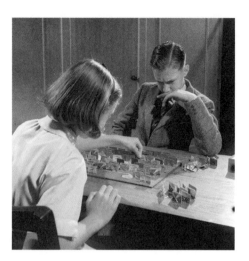

Playing games with puzzles that often demand mathematical skills has been a popular pastime for generations. *Hulton Archive/Getty Images.*

Cryptarithmetic puzzles are most often divided into two types. An alphametic cryptarithm is one in which the letters are used to represent distinct digits. These are derived from related words or meaningful phrases in the puzzles. A digimetic cryptarithm is one in which the digits are used to represent other digits.

What are **tiling** and **dissection puzzles**?

Tiling puzzles are two-dimensional shapes that are reassembled into a larger given shape without overlaps. The best examples of these are dissection puzzles; the most common ones are those in which an object is converted to another by making a finite number of cuts, then reassembling the pieces. (Most of the cuts are represented by straight lines, but not always; in addition, sometimes the cut object can be reassembled into two or more shapes.)

What is a **Tangram**?

Some puzzle forms go back thousands of years, such as the Tangram, which is also known as a dissection puzzle. The Tangram is of Chinese origin—literally called the *Seven-Board of Cunning*—but the word itself is of English origin, and is built from the words "tang" (thought to be a synonym for "Chinese" in the Cantonese dialect) and "gram." It is thought that the Pythagorean Theorem was discovered in Asia before Pythagoras's time with the help of Tangram pieces (for more about the Pythagorean Theorem, see "Geometry and Trigonometry").

The Tangram consists of a square divided into seven pieces called *Tans,* all of which must be arranged to match particular designs, usually a square. The seven

What is a famous cryptarithmetic puzzle?

One of the most famous cryptarithmetic puzzles was developed by Englishman Henry Ernest Dudeney (1857–1930) and published in a 1924 issue of the *Strand Magazine*:

$$\begin{array}{r} \text{SEND} \\ + \text{MORE} \\ \hline \text{MONEY} \end{array}$$

In this puzzle, the addition sum of each letter represents a digit, with all the letters being different digits. These letters represent the numbers as follows: O = 0, M = 1, Y = 2, E = 5, N = 6, D = 7, R = 8, and S = 9. Or

$$\begin{array}{r} 9567 \\ + 1085 \\ \hline 10652 \end{array}$$

pieces include five triangles of various sizes, one square, and one parallelogram. All the shapes must be used in the final form and none can overlap.

What is a **stomachion**?

A stomachion is a dissection puzzle similar to a Tangram. It uses 14 pieces in the puzzle, each of varying polygonal shapes and arranged into a 12 by 12 square grid. Each of the pieces has an area that is an integral fraction: for example, 24, 12, 9, 6, or 3, of the total area of the square, which is 144 units. The object of a stomachion is to arrange the pieces into interesting, and often recognizable shapes, such as people, animals, and objects.

This puzzle was also an ancient game known to the Greek mathematician Archimedes (c. 287–212 BCE), and thus is also called the *Loculus of Archimedes* ("Archimedes Box"). It is doubtful Archimedes invented the puzzle, but he did explore its geometric aspects. It also now appears that in contemplating the different solutions to this puzzle, Archimedes actually was anticipating the branch of mathematics we now call combinatorics.

What are some other **examples** of **dissection puzzles**?

There are several other example of dissection puzzles, including the following:

- The haberdasher problem
- Pythagorean square puzzle
- T-puzzle

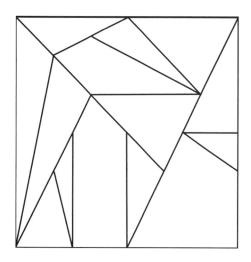

This is an example of a stomachion, a type of puzzle in which a geometric shape is dissected into several smaller shapes that can then be rearranged.

Puzzle inventor Henry Ernest Dudeney (1857–1930) of England was instrumental in developing many more puzzles than the cryptarithmetic puzzle described above. One of the most famous of his geometrical puzzles is the "haberdasher's problem," which asks how an equilateral triangle can be cut into four pieces and reassembled to form a square. (His model used hinges that would move the pieces into place.)

In the *Pythagorean square puzzle* the two squares on the left are combined to form a single large square on the right.

The T-puzzle is a dissection puzzle that forms the letter T. Four pieces are used to create the capital letter, as seen on p. 425.

What is the **15 Puzzle**?

The 15 Puzzle was introduced in 1878 by American amateur mathematician Samuel Loyd (1841–1911). He called it the "Boss Puzzle" and later the "15-16 Puzzle." It is one of the most famous puzzles in his book *Sam Loyd's Cyclopaedia of 5,000 Puzzles, Tricks and Conundrums,* published in 1914 after his death by his son, Sam Loyd. This puzzle has 16 squares; 15 of them are numbered from 1 to 15 and placed in a 4 by 4 configuration, with one position, the 16th, left open. The idea was to reposition the squares from a given arbitrary arrangement by sliding them from place to place until they were in numerical order (1, 2, 3, and so on). For some initial starting points, the rearrangement was possible; for others, it was not.

But Loyd offered a twist to the puzzle—he switched the positions of the squares numbered 14 and 15—and offered $1,000 to anyone who could solve the puzzle. Working out the puzzle became a craze in America, with reports of companies prohibiting employees from playing during office hours—it was as popular as playing computer solitaire is today. Even in Europe, the craze grew. Deputies in Germany's Reichstag played the puzzle, and in France it was claimed to be a greater curse than alcohol or tobacco. But Loyd knew no one could solve the puzzle, much less remember all the steps taken to try and get to a solution, because there was no solution!

What are **logic puzzles**?

In general, a logic puzzle involves the description of an event or contest. It is derived from the mathematical field of deduction: Clues are provided and the puzzle player

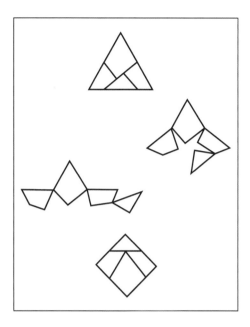

In the haberdasher problem, an equilateral triangle is divided and rearranged into a square.

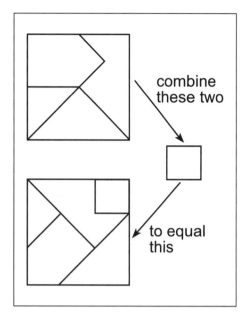

combine these two

to equal this

The Pythagorean square puzzle tests your geometry skills by asking you to take two square shapes and reconfigure them into one larger square.

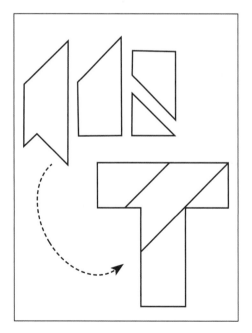

In the T-puzzle, several shapes must be assembled into the letter T.

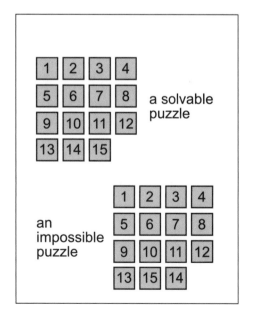

a solvable puzzle

an impossible puzzle

A 15 Puzzle might not be a fair challenge of a person's skills, if the numbers don't begin in the correct order.

425

The popular Rubik's Cube puzzle requires a player to spin sections of the cube in order to make the colors on all six sides match. *Hulton Archive/Getty Images.*

has to piece together what actually happened by using clear and logical thinking (thus the name).

One of the most famous logic puzzle pioneers was writer, photographer, mathematician, and illustrator Charles Lutwidge Dodgson (1832–1898), otherwise known as *Alice's Adventures in Wonderland* author Lewis Carroll. In his book *The Game of Symbolic Logic* he introduced several games, asking the reader to solve a puzzle, such as: "some games are fun," "every puzzle is a game," thus, "are all puzzles fun?" Puzzles such as this are known as syllogisms, in which the reader is given a list of premises and asked what can be deduced from the list. (For more information about syllogisms, see "Foundations of Mathematics.")

What are the **number of possible positions** for a **Rubik's Cube**?

Rubik's Cube was invented in the 1970s by the Hungarian architect, inventor, and mathematician Ernö Rubik (1944–), who also invented a number of other puzzles, including Rubik's Clock. The cube measures 3 by 3 by 3, with a total of 26 subcubes on the outside. All the subcubes are hinged, making them easy to turn (by a quarter turn in either direction) in any of the planes on the cube. Initially, each of the six sides are painted a certain color; the object is to move the cube planes in a random way, then return the cube so that each side has a single color again.

What are the possible number of positions of a Rubik's cube? Mathematicians need to use factorials (symbolized with the ! sign; for more information about factorials, see "Algebra") in order to find out the many iterations, as seen in the following equation:

$$\frac{8! \times 3^8 \times 12! \times 2^{12}}{2 \times 2 \times 3}$$

The number of positions turns out to be 43,252,003,274,489,856,000, or more than 43 quintillion turns.

What is the **St. Ives problem**?

The St. Ives problem is one of deduction and reasoning. The centuries-old original poem states:

What is an example of a logic puzzle?

Most of us have encountered logic puzzles before, usually on mathematics tests given in grade and high school. They often contain numbers (the mathematics connection) and a (often seemingly convoluted) sequence of events to which the reader has to determine the outcome.

For example, the following will jar the memory banks of everyone who has ever seen such logic puzzles on an exam or elsewhere: One weekend, three people check into a bed and breakfast. They pay $30 ($10 each) to the B&B owner and go to their room. The owner remembers that there is a special deal that weekend, and the actual room rates come to $25. He gives $5 to his brother (who works with him) and tells him to return the money to the guests. On the way to the room, the brother realizes that $5 would be difficult to share between three people—5/3 = 1.666 ..., an uneven number—so he pockets $2 and gives $1 to each person. Therefore, each person paid $10 and got back $1; this means they paid $9 each, totaling $27. The brother has $2, so the entire total is $29. Where is the remaining dollar?

The solution? Deduction comes in handy here. It's the owner who's important to pay attention to, not the brother. Overall, each guest paid $9 (a total of $27), the owner now has $25, and the brother has $2; thus, the brother's amount should be either added to the owner's money (25 [owner] + 2 [brother] + 3 [guests] = 30), or subtracted from the guests' amount of $27 (27 [guests] − 2 [brother] + 5 [refund from owner] = 30)—not added to the guests' amount. This proves that not only is the owner's brother a bit dishonest, but also how easy it is to con some people out of a few dollars if they aren't used to solving logic puzzles!

While on my way to St. Ives,

I met a man with seven wives.

Each wife had seven sacks;

Each sack had seven cats;

Each cat had seven kits.

Kits, cats, sacks, wives;

How many were going to St. Ives?

By this time, most people start adding and multiplying, trying to come up with the answer. But in reality, it's a trick question: The narrator is on the way to St. Ives; the group he or she met along the way were leaving, not going to, St. Ives. Therefore, the number "going to St. Ives" equals (at least) one: the narrator.

Of course, there are some mathematicians who can't leave well enough alone, and have calculated the total number of cats, kits, sacks, and wives, based on a geometric series. According to this equation, 2,801 were going to St. Ives, if the man, his wives, their cats, etc. had turned around.

MATHEMATICAL GAMES

What is a game?

A game is a recreational activity that involves a "conflict" resulting in gains and losses between two or more opponents (although some games can involve one player acting alone). In general, all games must have a goal that the players are trying to reach, and the opponents must follow strict, formal rules that determine what the players can or can't do within the game. If any of the rules are broken during the game, it is often referred to as a foul—or, at its worst, cheating.

The study of games is also a branch of mathematics and logic that is called game theory. In game theory, games can be simple and solved with mathematics that result in a complete "solution" (result of the game). It also includes the analysis of more complex games, such as cards, chess, and checkers, and can even be applied to real-world situations in economics, politics, and warfare.

What is gambling?

Gambling is the act of playing a game for stakes—it is thought of as the art of taking chances. It is also often called betting. A bet is the amount of money, or other object of value, that is risked in a wager. Most people gamble with the hope of winning a certain stake, usually a cash payment. But in order to get such a payoff, the gambler must risk money or valuables, betting these items on the outcome of a game, contest, or other event. All of this depends on the outcome of activities that are partially or wholly dependent upon chance. What is chance? Mathematically speaking, chance is a measure of how likely it is that an event will occur—a probability. (For more about chance and probability, see "Applied Mathematics.")

What are betting odds and how are they determined?

Betting odds are usually written in the form r:s, in which r is the "chances for" and s is the "chances against." This can be stated as: "r to s" or "chances for to chances against"; or "s to r" or "chances against to chances for." But note that odds of 1:1 are said as "one to one" *not* "one out of one." Odds are usually calculated as follows:

total chances = chances for + chances against;

or chances against = total chances − chances for

Why is it difficult to win a lottery?

According to one state lottery site, a lottery "is a plan that provides for the distribution of money, property, or other reward or benefit to persons selected by chance from among participants some or all of whom have given a consideration for the chance of being selected." In other words, a person buys a chance at winning a certain sum of money. But in reality—as with many games of chance—the game is not in the participant's favor. With most lotteries, such as a "lotto-type" lottery, a person has a better chance of being in a car or plane accident—or even being hit by lightning—than winning. But that doesn't stop many people. One recent statistic shows that, in the United States, an average of more than $96 million is spent on lotteries every day, or more than $35 billion per year.

The reason for this "dream of winning" is simple: It's how this game of chance is perceived. Many people believe that if they just keep the same number, it will eventually be chosen. What they often don't understand when playing a lottery is the idea of replacement. Take a 52-card deck to represent a lottery, with the participant asked to chose a card as the "winning" card, such as the queen of hearts. In the first choice, the king of diamonds is picked, and not reshuffled back into the deck. After each choice, if the cards are not put back into the deck, eventually, the participant's chances of picking the queen of hearts gets better and better. After all, the choices of cards in the deck become less; if one card is left, the participant knows he or she will win.

But a regular lottery does not "reshuffle" the numbers. Instead, lotteries chose from the same "group" of numbers each week, which makes it even more difficult to win. There may be repetitions in winning numbers, but the odds of winning are the same each time the lottery is played. For example, the odds of winning a recent California Super Lotto game were 1 in 18 million. Thus, if a person bought 50 lottery tickets a week, his or her chances of winning would be once every 6,923 years.

Using a 52-card deck as an example, the odds of drawing a king from the deck are 1 in 12—or 4:(52 − 4) = 4:48, which equals 1:12.

What is the difference between **probability** and **odds**?

Probability is usually expressed as a fraction (sometimes as a percentage). For example, if there are ten pieces of fruit in a jar—three apples and seven oranges—then the probability of taking out an orange is 7/10 (or seven chances of an orange out of a total of ten chances).

On the other hand, odds are expressed as the number of chances for (or against) versus the number of chances against (or for). Thus, if there are three chances of picking an apple and seven chances of picking an orange, the odds are 7 to 3 against you picking an apple. Just reverse this to find the odds in favor; or, in this example, the odds would be 3 to 7 in favor of picking an apple.

In order to convert the odds to probability, just add the chances. Thus, if the odds against a horse winning the Kentucky Derby are 4 to 1, that means that, out of 5 (or 4 + 1) chances, the horse has one chance of winning. That makes the probability of the horse winning 1/5, or 20 percent.

What are the **odds** of winning the **powerball lottery**?

It was eventually going to happen: A number of lottery-offering states got together to have lotteries with huge amounts of prize money. The resulting powerball lotteries have been very lucrative—not for the players, but for the states. Such gigantic sums of money tempt quite a few people to take the risk, with many buying hundreds of tickets in an attempt to better their odds.

But does it work? Not really. There is a way to determine the odds of such "lotto-type" lotteries in which numbered balls (or numbers) are randomly chosen to represent a winning number. This is usually expressed as: n! / (n − r)! r!, in which n is the highest numbered ball and r is the number of balls chosen. (The n! is "n-factorial"; for more information about factorials, see "Algebra.")

In math, this type of equation is called a combination. For example, if there are 50 balls and 5 are chosen, there are 50 possible numbers that can come up first, leaving 49 that can come up second, and so on. The equation becomes:

$$\frac{50!}{(50-5)! \times 5!} = 2,118,760$$

or the chances of winning are about 2 million to 1.

But don't think that powerball lotteries give a person an edge. For example, a powerball lottery can be one in which 5 out of 50 balls (or numbers) are drawn, with an extra powerball pulled out of a different number of balls. This is *not* like figuring the odds for a "drawing 6 out of 50 balls" contest, but is actually two separate lotteries: one with 5 out of 50 balls and one with the powerball group.

The probability of matching the first 5 balls is determined as above. But the powerball is separate: For instance, say the powerball is taken from a group of 36, making the powerball group's odds 36:1. The probability of winning the entire jackpot can be determined by adding this to the results of the 5-ball draw. Now the odds become even higher: The 50/5 drawing + a powerball of 36 = (2,118,760 × 36):1 = 76,275,360:1 or about 76 million to one. There are even worse odds if more balls are in the powerball group.

**What are the probability and odds
of drawing a certain card from a deck?**

The probability of an event is usually described as the chances for the event to occur over the total chances for the event to occur (chances for / total chances). In the case of a deck of 52 cards, the probability of drawing a king from the deck is 4/52 = 1/13 = 0.077, or 7.7 percent. As seen above, the odds, or the ratio of chances for to chances against (chances for : chances against) can be found by the formula: total chances equals the chances for plus chances against (or total chances = chances for + chances against, or chances against = total chances − chances for). Thus, the odds of picking a king from a deck of 52 cards is 4:(52 − 4) = 4:48 = 1:12.

CARD AND DICE GAMES

What are **card games**?

A collection of cards (or a deck) is a set of *n* rectangular pieces, usually made of heavy coated paper or cardboard, that hold special varying markings on one side and a uniform, identical pattern on the other. The special markings make each card unique, with each marking representing something playable in a certain card game.

The most common cards for games is a 52-card deck represented by four specifically colored suits (spades and clubs in black, diamonds and hearts in red), with 13 cards of each suit numbered 1 through 10, followed by several face cards—jack (J), queen (Q), and king (K). Card "1" is usually an "ace"; card 11 is represented by a "jack," 12 by a "queen," and 13 by a "king." The value of the ace often changes depending on the game. For example, it can either hold a value of 1 or 11 (in blackjack) or 14 (in bridge). Such cards are also used for many gambling games, such as poker and baccarat. Interestingly enough, the investigation of the probabilities of various outcomes in card games was one of the original motivations for the development of modern probability theory. (For more about probability, see "Applied Mathematics.")

What are the **probabilities** and **odds** of being initially dealt certain hands in five-card **poker** and **bridge**?

Because there are a certain number of cards in a playing deck, mathematicians have worked out the probabilities and odds of being dealt certain hands for certain games. The following explains the odds in terms of "chances against : chances for."

431

What are the possible numbers of distinct five-card hands in five-card poker and thirteen-card hands in bridge?

There is, of course, a mathematical way of determining the possible number of distinct five-card hands in poker—or the various combinations of five cards from a 52-card deck. The formula, in which N is the number of combinations, is as follows (note: the 52 written over 5 is called a binomial coefficient; for more about coefficients, see "Algebra"):

$$N = \begin{bmatrix} 52 \\ 5 \end{bmatrix} = 2,598,760 \text{ combinations}$$

Not to be outdone, such a number can be determined for bridge games, too. In the following, N again represents the combinations:

$$N = \begin{bmatrix} 52 \\ 13 \end{bmatrix} = 635,013,559,600 \text{ combinations}$$

In *five-card poker* the probability of drawing a royal flush (a poker hand with the ace, king, queen, jack, and 10 all in the same suit) is 1.54×10^{-6}, with odds of 649,739:1. For a straight flush (a poker hand with consecutive cards in the same suit, but not a royal flush), the probability is 1.39×10^{-5}, with odds of 72,192.3:1. Three of a kind (three cards with the same value) has a probability of 0.0211, with odds of 46.3:1, while one pair (two cards of the same value) has a probability of 0.423, with odds of 1.366:1.

In *bridge* 13 top honors has a probability of 6.3×10^{-12} and odds of 158,753,389, 899:1. A 12-card suit, ace high, has a probability of 2.72×10^{-9}, and odds of 367,484,697.8:1. Getting four aces has a probability of 2.64×10^{-3}, with odds of 377.6:1.

What is the **casino's or house edge**?

A casino is a gambling facility that normally includes all or a combination of the following: slot machines, video games, card games, and other games such as keno, craps, and bingo. In order to make the casino the major money-maker—not the gambler—there are certain "rules" of the casino, including the casino's edge, otherwise known as the house edge.

The house edge is the ratio of the expected player loss to the initial amount bet; it is an exact measurement, usually expressed as a percent, of the casino's advantage in a game. A casino earns money by paying winners at "house odds," an amount that is

slightly less than the true odds of winning the game. There are definitely exceptions to this rule, which is why there are professional gamblers, but the majority of the time, the casino has the advantage.

For example, if the house edge for blackjack or slots is 5 percent, then for every $100 bet initially made, the player can expect to lose $5. This not only helps the casino, but it is sometimes a way for patrons to compare one game to another, and even estimate how much they may lose.

Of course, this is not always so straightforward. Some people win and others lose in the short run, but the casino always wins in the long run. There are also exceptions because of the way a game is played, such as in craps (see below). The best scenario would be to have no house edge, or an edge against the house.

This woman may have hit it big on the slot machines, but statistics show that if she keeps playing she will eventually lose money to the casino, which has the "house edge." *Taxi/Getty Images.*

What are **dice**?

Dice are small cubes usually used in games of chance. Each die (dice is the plural) has six sides numbered with dots from one to six. The dots are placed on the cubes so that the sum of dots on opposite sides equals seven; the total number of dots on each die equals 21. They are mostly associated with certain types of games, with the simplest involving a player, or many players, who throw (or toss or roll) the dice for the highest sum.

Dice have been around for more than 3,000 years, with evidence found in ancient Egyptian tombs, Chinese burial chambers, and the ruins of Babylon. The Greeks and Romans were avid users of dice and associated games, and they have been popular from the Middle Ages on. Dice have been made from many materials, including ivory, bone, wood, metal, and eventually plastic.

Not all dice are square. The cube (or hexahedron) belongs to the group of five Platonic solids, or solids formed by regular polygons. Thus, other Platonic solids have also been made into dice, including shapes (polyhedra) such as tetrahedrons, octahedrons, and dodecahedrons, which are used for certain types of games. (For more information about Platonic solids and polyhedra, see "Geometry and Trigonometry.")

433

What are the odds when playing craps?

Craps is probably the most popular game of chance in the world; it is also illegal to play in many places. But it has a long history: It was played in ancient Greece and Rome and was even a mainstay of some old 1930s and '40s movies. Craps can be played using a wall and a pair of dice. It is a popular casino game in places such as Las Vegas—and even on the Internet—with betting on craps involving a complex equation.

Its popularity no doubt comes from its simplicity. In craps, a player throws two dice; their number (roll) is the total of the dots on the top faces of the dice. If the initial roll is a 7 or 11 (called a natural), the player wins. If the number 2, 3, or 12 comes up—called craps—the player loses, but keeps the dice. If the sum of the dice adds up to the number 4, 5, 6, 8, 9, or 10, that number becomes the thrower's "point." The player then continues to shoot until he or she throws the point number again, in which case the gambler wins and retains the dice. But if the player shoots a sum of 7 before he or she can roll the point value, he or she loses and gives the dice to the next player.

Craps is truly a game of chance, with the probability mathematics of craps fairly straightforward. For example, take the probability of winning on a roll-by-roll basis, in which $P(p = n)$ is the probability of rolling a point n. The resulting numbers show that the probability of winning is 244/495, or the shooter wins about 49.2929 percent of the time.

SPORTS NUMBERS

What is **sabermetrics**?

Sabermetrics is the study of baseball using objective evidence, such as baseball statistics. It uses scientifically based data and various interpretation methods to explain why teams win and lose. Sabermetrics was taken from the acronym SABR, or the Society for American Baseball Research, and was coined by baseball historian, statistician, and writer Bill James (1949–).

What are some of the **statistics** used in **baseball**?

There are numerous statistics used in baseball, including batting and pitching statistics. *Batting statistics* can be divided into several numbers. The *batting average* (AVG) is the number of hits a player makes divided by the number of times at bat. It does not include walks or sacrifice hits. The *runs batted in* (RBI) is the number of runners who scored on a player's hit, base on balls, or sacrifice. The *on-base plus sluggage* (OPS) is

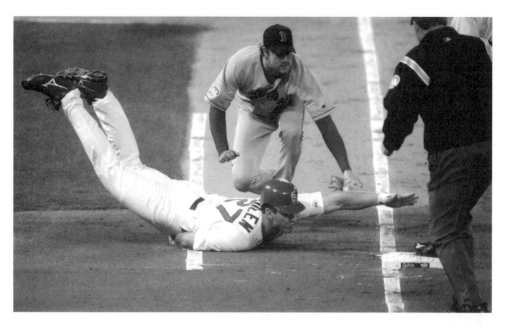

The mathematical study of the American pastime of baseball is called "sabermetrics." *Reportage/Getty Images.*

a good measure of a hitter's ability. This statistic combines getting on-base (on-base percentage, or OBP) and advancing runners (slugging percentage, or SLG). It is also more accurate thanks to the adjustment $1.2 \times \text{OBP} + \text{SLG}$, which compensates for the fact that SLG has a wider range than OBP.

Pitching statistics include the ERA and WHIP. The *earned run average* (ERA) is the earned runs times the innings in a game (most commonly nine) divided by the innings pitched. The *walks plus hits per inning pitched* (WHIP) records the bases on balls (walks) plus hits divided by innings pitched. It's a good way to measure the approximate number of walks and hits a pitcher allows in each inning that he pitches. It then compares that amount to other pitchers to formulate a pitcher's index.

What are some **statistics** used in **football**?

There are several mathematical statistics used in football. The *touchdown percentage* is the touchdown passes divided by the pass attempts. The *passer (quarterback) rating* is determined by four elements and their statistical calculations: percent of completions, average yards gained per attempt, percentage of touchdown passes, and the percent of interceptions. The average is 1.0, the bottom is 0, and the maximum anyone can receive in any category is 2.375 (this is difficult; for a passer to gain 2.375 in completion percents, he would have to complete 77.5 percent of his passes). Those passers who earn a 2.0 rating or better are exceptional. For example, the National Football

League ranks this at 70 percent in completions, 10 percent in touchdowns, 1.5 percent in interceptions, and 11 yards average gain per pass attempt.

Rushing is a common statistic commonly heard after any football game on the after-game special. It is the average yards per carry (AVG), a number measured by the total yards divided by the attempts. In *punting* the net punting average (NET) is the gross punting yards, minus the return yards, minus 20 yards for every touchback, divided by the total punts.

What are some **basketball statistics**?

One of the main ways a National Basketball Association coach evaluates a player's game performance is by *efficiency,* which is determined by using the formula: (points + rebounds + assists + steals + blocks) − (field goals attempted − field goals made) + (free throws attempted − free throws made) + turnovers.

Field goal percentage (FG percent) is determined by the field goals made (FGM) divided by the field goals attempted (FGA). Finally, the *free throw percentage* (FT percent) is determined by the number of free throws made (FTM) divided by the free throws attempted (FTA).

What is a **point spread**?

A point spread is the predicted scoring difference between a game's two opponents; it is used by bookmakers to equalize two teams for betting purposes. For example, if a person chooses the team that is favored, the gambler has to win by more than the point spread in order to get credit for a correct pick. If the underdog team is chosen, the bettor must lose by less than the spread number in order to get credit for a correct pick. If a point spread is listed as "off," this means there is no official point spread for the game.

For example, if a person picks Army to win, and the team is favored to beat Navy by five points (called a spread of five points), Army must win the game by six or more

points in order for the bettor to win. If Army wins by exactly five points, the game is called a "push," and no one wins.

How does a person determine the **odds** in **horse racing**?

The track makes its money in a certain mathematical way: When the odds of the horse race are converted to probabilities, they usually add up to more than 1, giving the track the advantage. For example, say a race track has 12 races with four horses, and the odds of winning each race are as follows (note: race tracks don't use the same horse in each race—this example is for illustrative purposes only):

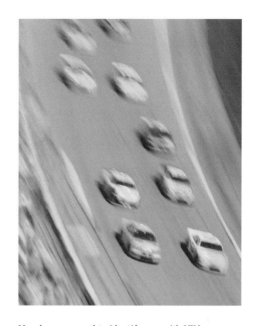

Numbers are used to identify cars with VIN numbers, and to help people figure out which car contains their favorite driver in a NASCAR race. *The Image Bank/Getty Images.*

- Horse 1—1:1; probability = chances for (1) / total chances (1 + 1) = 1/2 or 6/12
- Horse 2—2:1; 1 / (1 + 2) = 1/3 or 4/12
- Horse 3—3:1; 1 / (1 + 3) = 1/4 or 3/12
- Horse 4—5:1; 1 / (1 + 5) = 1/6 or 2/12

If a person puts $1 bets on horse number 2, he or she would have to win 4 out of 12 races to break even. But note: All these numbers add up to 15/12, or 1.25, a higher number than 1, so as long as no horse wins more than its probability, the house wins.

There's another way of looking at this type of betting. In order for the gambler to do "better" than the track, he or she has to win 15 times in 12 races—a physical and mathematical impossibility—which is why the track always makes money. Of course, a gambler may bet on certain low-probability horses that win more races than expected, earning a few more dollars along the way, but don't bet on it.

JUST FOR FUN

What is **number guessing**?

Number guessing is a game (some say trick). Some people can do number guessing in their heads, as long as the numbers are kept small. Have a person think of any positive integer n (not zero or any negative numbers) and apply the following steps:

437

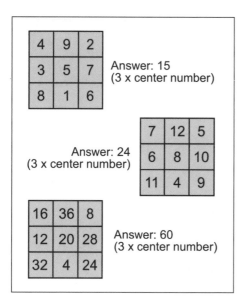

In these 3-by-3 Chinese magic squares, the numbers are arranged so that when added vertically, horizontally, or diagonally, they are always equal to 3 times the center number.

1. For example, a person chooses the number 25 and keeps that number to him- or herself. Ask the person to compute $3 \times n$. ($3 \times 25 = 75$)

2. Ask if the number is odd or even. (odd)

3. If the number is even, tell him or her to divide the number by 2; if the number is odd, tell him or her to add 1 to the number, then divide that number by 2. ($75 + 1 = 76/2 = 38$)

4. Tell him or her to take that number and multiply by 3, then divide by 9. ($38 \times 3 = 114/9 = 12.666 \ldots$)

5. Take that number and multiply by 2 ($12.666 \ldots \times 2 =$ approximately 25). The answer should be the original number—or close to it.

What is a **magic square**?

A magic square is an array of numbers in an n by n square that contains positive integers—from 1 to n^2—with each number occurring only once. The numbers in the squares are arranged so that the sum in any horizontal, vertical, or main diagonal direction is always the same. This is shown in the following formula: $n(n^2 + 1)/2$.

Magic squares are often divided into orders; for example, a three-order magic square means three boxes per row and three boxes per column. In reality, "magic" squares are actually matrices (for more information about a matrix, see "Algebra"); they can be odd-order (such as a 3-by-3, or 5-by-5 matrix) or even-order (such as a 4-by-4 or 6-by-6 matrix) magic squares. Perhaps the simplest magic square is the 1-by-1 square, whose only entry is the number 1.

Such magic squares have been known for centuries. For example, the Chinese knew about the three unique normal squares of order three. In Chinese literature dating from as early as 2800 BCE, a magic square known as the *Loh-Shu,* or "scroll of the river Loh," was invented by Fuh-Hi, who is thought of as the mythical founder of the Chinese civilization.

What is **Pascal's triangle**?

Pascal's triangle, as the name implies, is a collection of numbers in the shape of a triangle. Each number in the triangle is the sum of the two directly above. For example, in the

accompanying illustration, the 6 on line 5 is the sum of the pair of 3's above; the next line is 1, 10 (1 + 9), 45 (9 + 36), 120 (36 + 84), and so on. Although the triangle was known to both the Chinese and the Arab cultures for several hundred years before, it was named after the person who brought it to the forefront of mathematics: French mathematician Blaise Pascal (1623–1662). (For more information about Pascal, see "History of Mathematics.")

```
              Pascal's Triangle
                      1
                   1     1
                 1    2    1
               1   3    3    1
             1   4    6    4    1
           1   5   10   10   5    1
         1   6   15   20   15   6    1
       1   7   21   35   35   21   7   1
                      .
                      .
                      .
                 and so on
```

In Pascal's triangle, each number is equal to the two numbers directly above it.

How can one visualize **Pascal's triangle** using **algebra**?

One way of looking at Pascal's triangle is through its connection to algebra. For example, expand (or remove the brackets around) the expression $(1 + x)^2 = (1 + x)(1 + x) = 1 + 2x + x^2$. The same can be done with a cube; for example, $(1 + x)^3 = (1 + x)(1 + x)(1 + x) = (1 + x)(1 + 2x + x^2) = 1 + 3x + 3x^2 + x^3$; and even the expression $(1 + x)^4$, which equals $1 + 4x + 6x^2 + 4x^3 + x^4$.

The coefficients (the numbers in front of the x's) in the results are the connection. For the first example, the coefficients are 1, 2, 1; for the second one, 1, 3, 3, 1; and for the last expression, the coefficients are 1, 4, 6, 4, 1. These, of course, are the third, fourth, and fifth lines from Pascal's triangle.

What are some **"life questions"** you can figure out **using math**?

There are many questions you can explore about your own body and age with mathematics. For example, approximately how many Sunday nights can you expect to sleep until you are 100 years old? Just take 100 years, minus your current age, and multiply that result by 52 (weeks in a year with a Sunday). For example, if you are 25, the answer would be $(100 - 25) \times 52 = 3{,}900$. How many of those will be good night's sleep is up to you.

To calculate the number of times your heart has beaten since you were born, you need the help of a watch or clock. First, time your heartbeats per minute (to find out how to count your pulse, see "Everyday Math"); then multiply beats per minute × 60 minutes (in an hour) × 24 hours (in a day) × 365.25 days (in a year) × your age. For example, 72 heart beats × 60 × 24 × 365.25 × a person who is 30 = 1,136,073,600 beats since the person was born. Of course, this is an approximation, since the heart usually beats slower at night, and it speeds up when you see the bill for your latest car repair.

This astronaut on the Moon weighs only 17 percent of what he would weigh if he were standing on the Earth. *Photographer's Choice/Getty Images.*

Figuring out how much air you breathe during a lifetime is another fun mathematical calculation. If you optimistically decide you want to eventually be 100 years old, and the average person inhales about one pint (or 0.47 liters) of air per breath, you can do the math. First take the number of breaths you take while at rest per minute (say about 21 per minute) \times 0.47 liters \times 60 minutes \times 24 hours \times 365.25 days \times 100 years old = 519,122,520 liters. Again, this is only an approximation.

What is the **Monty Hall Problem**?

Let's Make a Deal, a once-popular television game show with host Monty Hall (who was the master of ceremonies from 1963 to 1986), is the origin of the Monty Hall Problem. The same type of problem is also represented in a card game called "three-card monte."

The Monty Hall Problem is similar to the show. A contestant is given a choice of three doors. Behind one is a car; behind the other two there is nothing. The host asks the contestant to pick a door. After the pick, the host opens one of the unpicked doors—one the host knows is empty. The host then suggests a switch. The big question, and the problem, becomes does the contestant switch his or her choice or keep the originally chosen door?

The mathematical, statistical answer is yes, switch the doors. Why? Because the original probability that the contestant picked the correct door does not change—it is still 1/3. But if the contestant does switch, the probability becomes 2/3. Most people believe that once one of the doors is eliminated, the probability between the remaining doors becomes 50-50, but it does not.

How can people calculate their **age** and **weight on other planets**?

It is certainly possible to find a person's age and weight on another planet without actually having to travel there. To figure out weight, one just needs to know the gravitational pull on another planet, moon, or other space body. For example, based on the chart below, if a person weighed 100 pounds on Earth, he or she would weigh 38 pounds on Mercury (100 \times 0.38).

A person's age on Earth depends on how many times the planet has orbited the Sun during his or her lifetime. For example, a person 30 years old has traveled around

How can I prove that 1 = 0?

There is a way to show that one equals zero, and it includes an interesting "proof":

Consider two non-zero numbers x and y such that x = y. If that is so, then x^2 = xy. Subtracting y^2 from both sides gives: $x^2 - y^2 = xy - y^2$. Then dividing by (x − y) gives x + y = y; and since x = y, then 2y = y. Thus 2 = 1; the proof started with y as a non-zero, so subtracting 1 from both sides gives 1 = 0.

The problem with this proof? If x = y, then x − y = 0. Notice that halfway through the "proof," the equation was divided by (x − y), which makes the proof erroneous.

our star 30 times. In order to work out how old a person is on another planet, the person's age has to be divided by the period of revolution (in terms of Earth years). Thus, if the same 30-Earth-year-old person lived on Saturn, he or she would be just over one Saturn year old (30/29.5 = 1.02 Saturn years). On Mercury, the person would be 124.5 Mercury years old (30/0.241 = 124.5 Mercury years).

Calculating Your Age and Weight on Other Space Bodies

Space Body	Time to Orbit around the Sun*	Gravity (Weight) Compared to Earth
Mercury	87.9 days	38%
Venus	224.7 days	91%
Earth	1 year	100%
Moon	1**	17%
Mars	1.88 years	38%
Jupiter	11.9 years	254%
Saturn	29.5 years	93%
Uranus	84 years	80%
Neptune	164.8 years	120%
Pluto	248.5 years	unknown
Sun	N/A	2800%

* In Earth days or years.

**Of course, the Moon travels with the Earth around the Sun, so it orbits the Sun once per Earth year. However, the Moon also orbits the Earth 13.37 times per year.

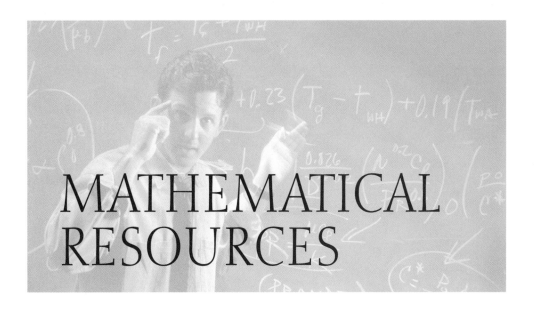

MATHEMATICAL RESOURCES

(Note: The authors have diligently sought and researched the following Web site addresses, mailing addresses, and phone numbers in order to present this mathematically oriented information. Please realize that some of these sites, addresses, and numbers change or are eliminated over time. We apologize for any closed or modified Web site listings, mailing addresses, or phone numbers.)

EDUCATIONAL RESOURCES

What types of **careers** are available to **mathematicians**?

The number of careers available to mathematicians is much too long to list in this text. Some of the classic jobs include architect, statistician, bookkeeper, systems engineer, research scientist (in many fields, such as geology, physics, astronomy, chemistry, and biology), engineering, and even rocket science; some of the more modern applications include mathematics in materials science, computer animation, neuroscience (in a subfield called biomedical mathematics), and nanotechnology.

To find out more about mathematical careers, check out the following Web sites:

American Mathematical Society (http://www.ams.org/careers/)—On this site just link to the "archives" to read about mathematicians in various careers.

Society for Industrial and Applied Mathematics (http://www.siam.org/students/career.htm)—This site lists not only careers but also interviews many mathematicians about their work. It also has a list of questions to consider if you are thinking about a career in mathematics.

Mathematical Association of America (http://www.maa.org/students/undergrad/career.html)—This site looks closely at several job specialties in mathematics, pro-

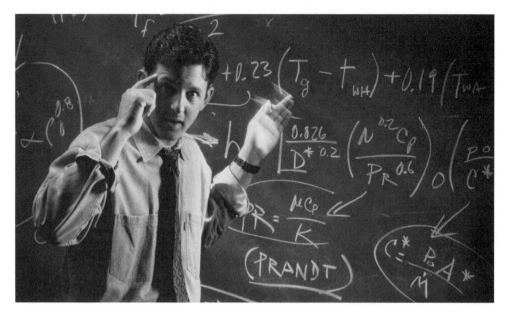

A mathematician is just one of many possible careers involving mathematics. There are a number of resources you can explore to learn about math career options. *Stone/Getty Images.*

vides profiles of many working mathematicians, and lists several books about careers in mathematics.

Where can I get an **undergraduate degree** in mathematics?

Finding a college or university offering an undergraduate degree in mathematics is not difficult. Almost every college in the United States offers such a degree, even at liberal-arts-oriented institutions. The major dilemma is easy to see: Which college or university does one choose to obtain a degree?

Although the choice is up to the individual, there are ways to winnow down the vast numbers of schools. One way is to study the curriculum offered by a college's mathematics department. For example, if the student wishes to go into statistics, explore those colleges with a good reputation in statistics by looking up information about the department and by talking to other mathematicians—or even students—in the field. Also, check out *The Princeton Review,* which offers a good Web site listing undergraduate schools that offer mathematics. To find out more information, go to: http://www.princetonreview.com/college/research/majors/Schools.asp?majorID=168.

What are some well-known **mathematical institutions** around the world?

There are many mathematical institutions around the world, which are often thought of as "think tanks" for mathematics. For example, the Fields Institute for Research in

When is Mathematics Awareness Month?

Mathematics Awareness Month occurs every April under the auspices of the Joint Policy Board for Mathematics (JPBM). This organization is composed of the Mathematical Association of America, the American Mathematical Society, the Society for Industrial and Applied Mathematics, and the American Statistical Association. The observance originated via a proclamation in 1986 by President Ronald Reagan for a Mathematics Week, and was expanded to a full month in 1999. Originally, the week was celebrated nationally by such organizations as the Smithsonian Institution. More recently, the activities are more local and regional, with emphasis on the importance, value, and even the beauty of mathematics. To peruse the MAM's special Web site, log on to http://www.mathaware.org.

Mathematical Sciences is located at the University of Waterloo in Toronto, Canada. This is a center for mathematical research activity that offers, in their own words, "a place where mathematicians from Canada and abroad, from business, industry and financial institutions, can come together to carry out research and formulate problems of mutual interest."

Another example is the Max Planck Institute for Mathematics (MPIM) in Bonn, Germany, which is one of about 80 research facilities that are part of the Max Planck Society—all internationally recognized for their basic research in the sciences, mathematics, and humanities. Mathematicians from all over the world visit the MPIM, which offers visitors the ability to discuss mathematical problems or exchange ideas with colleagues. For a long list of such mathematical institutions, visit http://www.ams.org/mathweb/mi-inst.html.

What are some **clubs and honorary organizations** for students interested in math?

There are many student math organizations, including the following:

Mu Alpha Theta—This organization is sponsored by the MAA, the National Council of Teachers of Mathematics, and the Society for Industrial and Applied Mathematics. It is a mathematics club for high school and two-year college students who enjoy mathematical problems, articles, and puzzles. Mu Alpha Theta publishes a journal, *The Mathematical Log,* and holds regional and national meetings. Web address: http://www.mualphatheta.org/.

Pi Mu Epsilon—This group is an Honorary National Mathematics Society. Its purpose is the promotion of scholarly activities in mathematics among the students in academic institutions, and it has more than 300 chapters at colleges and universities throughout the United States. Web address: http://www.pme-math.org/.

445

Kappa Mu Epsilon—This group is also a national mathematics honor society, but it is more specialized. It was founded in 1931 to promote the interest of mathematics among undergraduate students; it has around 118 chapters in the United States. Web address: http://www.kme.eku.edu/.

ORGANIZATIONS AND SOCIETIES

What are some **mathematics societies** and organizations in the **United States**?

The following lists some of these groups, along with contact information:

American Mathematical Society
201 Charles St.
Providence, RI 02904-2294
Phone: 401-455-4000 worldwide; 800-321-4AMS in the United States and Canada
Fax: 401-331-3842
E-mail: ams@ams.org
Web site: http://www.ams.org

American Statistical Association
1429 Duke St.
Alexandria, VA 22314-3415
Phone: 703-684-1221; 888-231-3473
Fax: 703-684-2037
E-mail: asainfo@amstat.org
Web site: http://www.amstat.org

Association for Symbolic Logic
Box 742, Vassar College
124 Raymond Ave.
Poughkeepsie, NY 12604
Phone: 845-437-7080
Fax: 845-437-7830
E-mail: asl@vassar.edu
Web site: http://www.aslonline.org

Mathematical Association of America
1529 Eighteenth St. NW
Washington, DC 20036-1358
Phone: 202-387-5200; 1-800-741-9415
Fax: 202-265-2384
E-mail: maahq@maa.org
Web site: http://www.maa.org

Society for Industrial and Applied Mathematics (SIAM)
3600 University City Science Center
Philadelphia, PA 19104
Phone: 215-382-9800; 1-800-447-SIAM (in the United States and Canada)
Fax: 215-386-7999
E-mail: service@siam.org
Web site: http://www.siam.org

Are there any **international societies** devoted to mathematics?

As befits a major branch of science, there are many international mathematical societies. A few are listed below:

Canadian Mathematical Society
577 King Edward, Suite 109
Ottawa, ON
Canada K1N 6N5
Phone: 613-562-5702
Fax: 613-565-1539
E-mail: office@cms.math.ca
Web site: http://www.cms.math.ca

European Mathematical Society (EMS)
EMS Secretariat
Department of Mathematics & Statistics
P.O. Box 68 (Gustaf Hällströmink, 2b)
00014 University Helsinki
Finland
Phone: (+358) 9 1915 1426
Fax: (+358) 9 1915 1400
E-mail: tuulikki.makelainen@helsinki.fi
Web site: http://www.emis.de

London Mathematical Society
De Morgan House
57-58 Russell Square
London WC1B 4HS
England
Phone 020 7637 3686
Fax: 020 7323 3655
E-mail: lms@lms.ac.uk
Web site: http://www.lms.ac.uk

New Zealand Mathematical Society
c/o Dr. Winston Sweatman (NZMS Secretary)
Institute of Information and Mathematical Sciences

Massey University
Private Bag 102 904
North Shore Mail Centre
Auckland
New Zealand
E-mail: w.sweatman@massey.ac.nz
Web site: http://www.math.waikato.ac.nz/NZMS/NZMS.html

St. Petersburg Mathematical Society
Fontanka 27,
St. Petersburg, 191023, Russia
Phones: 7 (812) 312 8829, 312 4058
Fax: 7 (812) 310 5377
E-mail: matob@pdmi.ras.ru
Web site: http://www.mathsoc.spb.ru/index-e.html

Swiss Mathematical Society
P.O. Box 300
CH-1723 Marly
Switzerland
Phone: ++ 41 / 26 / 436 13 13
E-mail: louise.wolf@bluewin.ch
Web site: http://www.math.ch

MUSEUMS

What **museum** is **devoted** exclusively to **mathematics**?

The Goudreau Museum of Mathematics in Art and Science, located in New Hyde Park, New York, is devoted exclusively to math. Mathematics teacher and engineer Bernard Goudreau founded the museum in 1980. It has since grown into a unique learning and resource center, offering exhibits, workshops, programs, and special events for the whole family. At the museum, visitors can explore games and puzzles, view exhibits dealing with math in science and the arts, build a math model, take part in math workshops and programs, and utilize a math resource library. The Museum Store has mathematical games, puzzles, and books in stock. For more information, contact:

The Goudreau Museum of Mathematics in Art and Science
Herricks Community Center
999 Herricks Rd., Room 202
New Hyde Park, NY 11040-1353
Phone: 516-747-0777
E-mail: info@mathmuseum.org
Web site: http://www.mathmuseum.org

The Library of Congress created the Vatican Exhibit Mathematics Room, a Web site that contains fascinating resources about Greek and Latin mathematics. *The Image Bank/Getty Images.*

What **major museums** around the world feature online exhibits dealing with math?

There are several museums around the world that offer mathematically oriented online exhibits. The following lists just a few:

The Institute and Museum of the History of Science—This museum is located in Florence, Italy, and its Web site offers a look at the mathematics behind Galileo Galilei. For example, one part examines Galileo's compass, which includes lessons on angles and proportions. The Web site is at http://www.imss.fi.it/museo/index.html.

Library of Congress Vatican Exhibit Mathematics Room—The Web site for this museum contains annotated Greek and Latin manuscripts of mathematics and astronomy; it also has special images, including (under "Greek Mathematics") a 9th century version of *Euclid's Elements* (showing the Pythagorean Theorem), and 13th- and 15th-century versions of Archimedes' works. Its Web address is http://sunsite.unc.edu/expo/vatican.exhibit/exhibit/d-mathematics/Mathematics.html.

The Museum of the History of Science—Information about the history of mathematics and mathematical applications in history are located in two online exhibits: *The Measurers: A Flemish Image of Mathematics in the Sixteenth Century* and *The*

449

Geometry of War, 1500–1750. The museum is in Oxford, England, and its Web site is at http://www.mhs.ox.ac.uk/exhibits/index.htm.

POPULAR RESOURCES

Are there any **magazines** devoted to mathematics?

The following lists a few print magazines about math:

Function
The Business Manager
Function
Department of Mathematics & Statistics
Monash University
Victoria 3800
Australia
E-mail: c.varsavsky@monash.edu.au
Web site: http://www.maths.monash.edu.au/function/index.shtml

This periodical is directed towards students in the upper years of secondary school, as well as anyone with an interest in mathematics.

Mathematical Gazette
The Mathematical Association
259 London Rd.
Leicester LE2 3BE
England
Phone: 0116 221 0013
Fax: 0116 212 2835
E-mail: office@m-a.org.uk
Web site: http://www.m-a.org.uk

The primary focus of this gazette is teaching and learning mathematics. It is free to members of the association, and also available on a subscription basis.

Mathematical Spectrum
The Applied Probability Trust
School of Mathematics and Statistics
University of Sheffield, Sheffield S3 7RH
England
Phone: +44 114 222 3922
Fax: +44 114 272 9782
E-mail: s.c.boyles@sheffield.ac.uk
Web site: http://www.shef.ac.uk/uni/companies/apt/ms.html

This is a magazine for teachers, students, and anyone interested in mathematics as a hobby.

Mathematics Magazine
The MAA Service Center
P.O. Box 91112
Washington, DC 20090-1112
Phone: 800-331-1622; 301-617-7800
Fax: 301-206-9789
E-mail: rchapman@maa.org
Web site: http://www.maa.org/pubs/mathmag.html

Published by the Mathematical Association of America for members, it offers readable and lively expositions on a wide range of mathematical topics.

Math Horizons
Math Horizons Subscriptions
Mathematical Association of America
1529 18th St. NW
Washington, DC 20036-1385
E-mail: maaservice@maa.org
Web site: http://www.maa.org/Mathhorizons/

This publication is mainly intended for undergraduate students interested in mathematics.

What **other magazines** often contain mathematical content?

There are many magazines that often contain news about math or mathematical content. The following lists just a few (notice most of these magazines are science-oriented):

Astronomy
21027 Crossroads Circle
P.O. Box 1612
Waukesha, WI 53187
Phone: 800-533-6644
Web site: http://www.astronomy.com

Discover
114 Fifth Ave.
New York, New York 10011
Phone: 212-633-4400
Web site: http://www.discover.com

New Scientist
Lacon House
84 Theobald's Rd.

Who is Martin Gardner?

Martin Gardner (1914–) is an American recreational mathematician and author. For decades, he was the "Mathematical Games" columnist for *Scientific American.* He is the author of more than 65 books and countless articles. Some of his titles include: *My Best Mathematical and Logic Puzzles* (Dover Publications, 1994; ISBN: 0486281523), *The Colossal Book of Mathematics: Classic Puzzles, Paradoxes, and Problems* (W.W. Norton & Company, 2001; ISBN: 0393020231), *Entertaining Mathematical Puzzles* (Dover Publications, 1986; ISBN: 0486252116), and *Mathematics, Magic and Mystery* (Dover Publications, 1977; ISBN: 0486203352).

London WC1X 8NS
England
Web site: http://www.newscientist.com/home.ns

Popular Science
2 Park Ave., 9th Floor
New York, NY 10016
Phone: 212-779-5000
Fax: 212-779-5108
Web site: http://www.popsci.com/popsci/

Science News
1719 N Street NW
Washington, DC 20036
Phone: 202-785-2255
Web site: http://www.sciencenews.org

Scientific American
Scientific American, Inc.
415 Madison Ave.
New York, NY 10017
Phone: 212-754-0550
Web site: http://www.sciam.com

What are some good **biographical books** about **mathematicians**?

The following books are just a sampling of those available about the lives of mathematicians:

A Beautiful Mind: The Life of Mathematical Genius and Nobel Laureate John Nash by Sylvia Nasar (Simon & Schuster, 2001; ISBN: 0743224574)—The fascinating biography of John Nash, the mathematical genius who descended into schizophrenia for decades, losing his sanity, career, and wife. He ultimately recovered from this illness, and was subsequently awarded the Nobel Prize for his early work. There was also a movie adaptation of this book (see below).

The Man Who Loved Only Numbers: The Story of Paul Erdos and the Search for Mathematical Truth by Paul Hoffman (Hyperion, 1999; ISBN: 0786884061)—The story of a mathematician who had no home, no wife, and no life other than numbers. Living out of two suitcases for more than 60 years, Erdos passionately chased mathematical problems over four continents, thinking and working for 19 hours per day, while interacting with the leading scientists of his day.

Incompleteness: The Proof and Paradox of Kurt Gödel by Rebecca Goldstein (W. W. Norton & Company, 2005; ISBN: 0393051692)—This book explores the remarkable theorem of incompleteness and the eccentric genius behind its discovery, Kurt Gödel.

What are some **nonfiction books** about **specific numbers**?

Although most people wouldn't think that a book about a number would be interesting, the following shows that's not always true:

The Golden Ratio: The Story of Phi, the World's Most Astonishing Number by Mario Livio (Broadway Books, 2003; ISBN: 0767908163)—A history of the number phi (1.6180339887), also known as the golden ratio or divine proportion. Livio discusses examples from nature, as well as phi's use in architecture and art throughout human history.

Pi: A Biography of the World's Most Mysterious Number by Alfred S. Posamentier and Ingmar Lehmann (Prometheus Books, 2004; ISBN: 1591022002)—The story of the number pi throughout history, from the Old Testament to modern politics. An epilogue has pi expressed to 100,000 decimal places.

e: The Story of a Number by Eli Maor (Princeton University Press, 1998; ISBN: 0691058547)—The tale of the development of "e" from both a mathematical and human perspective.

Zero: The Biography of a Dangerous Idea by Charles Seife and Matt Zimet (Penguin Books, 2000; ISBN: 0140296476)—An entertaining story about (literally) nothing. The development and use of nothing, or zero, is covered in detail from ancient times to the present.

What are some **nonfiction books** about **mathematical problems**?

Tales of epic quests to solve some of the most difficult mathematical problems ever considered are served up in the following books:

Fermat's Enigma: The Epic Quest to Solve the World's Greatest Mathematical Problem by Simon Singh (Anchor Books, 1998; ISBN: 0385493622)—The epic quest to solve Fermat's Last Theorem is recounted, replete with human drama and tragedy.

Prime Obsession: Bernhard Riemann and the Greatest Unsolved Problem in Mathematics by John Derbyshire (Plume Books, 2004; ISBN: 0452285259). Math, history, and biography are intertwined in this story of a mathematical mystery that remains unsolved.

Four Colors Suffice: How the Map Problem Was Solved by Robin Wilson (Princeton University Press, 2004; ISBN: 0691120234). A seemingly simple problem that perplexed amateur and professional mathematicians for more than a hundred years is recounted in this interesting work.

What are some **other interesting nonfiction books** about mathematics?

There are hundreds, if not thousands, of nonfiction books devoted to mathematics that prove that math can be both fun and interesting:

Gödel, Escher, Bach: An Eternal Golden Braid by Douglas R. Hofstadter (Basic Books, Inc., 1999; ISBN: 0465026567)—The classic work on human creativity and thought, bringing together the mathematics of Gödel, the art of Escher, and the music of Bach.

The Lady Tasting Tea: How Statistics Revolutionized Science in the Twentieth Century by David Salsburg (Henry Holt & Company, 2002; ISBN: 0805071342)—As the title suggests, this is the story of how statistics changed the way science was done in the 20th century. The methods of statistics are covered in easily understood terms, and there are short biographies of the major contributors to this field.

The Cartoon Guide to Statistics by Larry Gonick and Woollcott Smith (Harper-Collins Publishers, 1993; ISBN: 0062731025)—Inspired cartoons make the learning of statistics fun and (relatively) easy.

The Mathematical Tourist: New and Updated Snapshots of Modern Mathematics by Ivars Peterson (Owl Books, 1998; ISBN: 0805071598)—The second edition of this book updates his 1988 book, including mathematical stories of crystal structure, string theory, mathematicians' uses of computers, chaos theory, and Fermat's Last Theorem. This is only one of many entertaining and fascinating mathematical books written by Peterson.

How has **mathematics** been used in **fiction**?

There are, of course, hundreds of fiction books in literature that use mathematics as a theme, a mathematician as protagonist, or have a mathematical solution. More recent

novels seem to be mostly science fiction. Below is merely a small taste of such titles—past and recent:

1 to 999—This book is by famous science fiction (and nonfiction) writer Isaac Asimov, who once earned a living as a chemist. In this book, cryptologists try to break a simple code, with one of the key clues being the frequency with which letters appear.

Sixty Million Trillion Combinations—Another book by Asimov, in which one of his recurring characters in his "Black Widower" mystery series, Tom Trumbull, tries to convince an eccentric mathematician that his secret password is not safe. Asimov had several more books with mathematical connections, which are a strong emphasis in many of his short story collections and over 500 published books.

Jules Verne wrote the classic science fiction tale *Round the Moon*. As well as being an entertaining space travel story, it includes mathematical discussions about algebra, parabolas, and hyperbolas. *Library of Congress.*

Kepler: A Novel—This John Banville book gives a fictionalized, yet somewhat accurate, portrayal of the Renaissance mathematician and astronomer—from his work to determine the orbits of the planets to some more eccentric ideas, such as why there are only six planets in the solar system in terms of Platonic solids.

The Difference Engine—In this science fiction, alternate reality tale by William Gibson and Bruce Sterling, mathematicians Charles Babbage and Ada Lovelace (Byron) actually succeed in making the Difference Engine (for more about the Difference Engine, see "Math in Computing").

The Nine Billion Names of God—Arthur C. Clarke's classic story in which two programmers hired by a Buddhist sect seek to find all true names of God by exhausting a combinatorial library of possibilities—a story that combines mathematics, computers, and religion.

Round the Moon—Written in 1870 by Jules Verne, this classic book about space travel comes complete with two chapters—chapter 4 "A Little Algebra" and chapter 15 "Hyperbola or Parabola"—containing detailed mathematics as discussed by the spacefaring crew.

Adventure of the Final Problem—And of course, one can't forget Sir Arthur Conan Doyle's Sherlock Holmes, his sidekick Dr. Watson, and Holmes's major adversary, Professor Moriarty. This is the first story that mentions Moriarty, introducing

him as a professor of mathematics who wins fame as a young man for his extension of the binomial theorem.

Are there any **math books** aimed at **children and young adults**?

Yes, there are a plethora of math books intended for children and young adults. The following lists only a *very* few that enlist stories, riddles, or other methods to explain mathematics:

The Adventures of Penrose the Mathematical Cat by Theoni Pappas (Wide World Publishing, 1997; ISBN: 1884550142)—This story tells of Penrose the cat as he explores and experiences a variety of mathematical concepts, including infinity, the golden rectangle, and impossible figures. For ages 9 to 12.

The Number Devil: A Mathematical Adventure by Hans Magnus Enzensberger, Rotraut Susanne Berner, and Michael Henry Heim (Metropolitan Books; Reprint edition, 2000; ISBN: 0805062998)—This great humorous book for kids is about mathematics. It begins when young Robert's dreams take a decided turn for the weird. Instead of falling down holes and such adventures typical in many children's dreams, in Robert's 12 dreams, he visits a bizarre magical land of number tricks with the number devil as his host. For ages 9 to 12.

The Grapes of Math by Gregory Tang (Scholastic Press, 2001; ISBN: 043921033X)—This story offers a series of counting riddles and encourages the reader to find shortcuts to determining the mathematical answers by looking for patterns, symmetries, and familiar number combinations. For ages 9 to 12.

Math Curse by Jon Scieszka and Lane Smith (Viking Books, 1995; ISBN: 0670861944)—An award-winning, very amusing picture book about dealing with numbers in everyday life. For ages 4 to 8.

How Big Is a Foot? by Rolf Myller (Yearling; Reissue edition, 1991; ISBN: 0440404959)—A humorous picture book that begins: "Once upon a time there lived a King and his wife, the Queen...." From there, the book explains to readers ages 4 to 8

> ## What popular movies pitted dinosaurs versus a mathematician?
>
> **B**oth *Jurassic Park* (1993) and *The Lost World: Jurassic Park* (1997) had a mathematician as one of the main (human) protagonists. Jeff Goldblum played chaos theory mathematician Dr. Ian Malcolm, who tried to sound a warning about the inherent instability of the park's experiment with dinosaurs. Unfortunately, no one listened to him, and the dinosaurs soon escaped the control of their human caretakers. In the second movie, Dr. Malcolm had to rescue a colleague/love interest (and his daughter) on a second island.

a concept in measurement—the foot—and why it's necessary to have measurement standards.

What **movies** had mathematics as their primary focus?

The following three movies were either about mathematicians or had a dominant mathematical theme:

A Beautiful Mind (2001)—Based on the book by Sylvia Nasar (also see above), this is Hollywood's version of John Nash's rise, descent into mental illness, and eventual redemption. Directed by Ron Howard, it stars Russell Crowe and Ed Harris.

Pi (1998)—This dark, low-budget, black-and-white film is about a mathematician's obsessive search for patterns in everything, while everyone from Wall Street investors to religious fundamentalists want to find and exploit him. Directed by Darren Aronofsky, it stars Sean Gullette and Mark Margolis.

Infinity (1996)—This movie about the early life of brilliant and eccentric Nobel Prize-winning physicist Richard Feynman (1918–1988), is both a tribute to the scientist and a whimsical romance. Directed by Matthew Broderick, it stars Matthew Broderick and Patricia Arquette.

What **mathematical concept** was featured in the movie *The Andromeda Strain*?

The concept of exponential growth, also known as geometric growth, provided the tension in this movie and the book on which it was based. The basic premise was that microscopic organisms had been inadvertently brought to Earth from outer space by a returning satellite. Once on Earth, they reproduced exponentially, doubling in number every 20 minutes. Very quickly, they spread over the entire planet, killing nearly everyone. (And no, we won't give away the ending!)

457

The mathematical concept of chaos theory is explained in the movie *Jurassic Park* in which geneticists accidentally unleash carnivorous dinosaurs like these upon the modern world. *The Image Bank/Getty Images.*

SURFING THE INTERNET

(Note: These Web sites were active at the time of this writing. Because content on the Internet can change rapidly, some of these sites may no longer be functional, even though they were active at press time. We apologize for any inconvenience this may cause.)

Are there any **online magazines** dealing with mathematics?

The following are a few of the numerous online magazines that are devoted to mathematics:

Convergence
http://convergence.mathdl.org/jsp/index.jsp
Sponsored by the Mathematical Association of America (MAA) in cooperation with the National Council of Teachers of Mathematics. Emphasis is on teaching mathematics using its history.

Journal of Online Mathematics and Its Applications (JOMA)
http://www.joma.org/jsp/index.jsp

A publication of the Mathematical Association of America (MAA). Articles are peer-reviewed, content is published continuously, and there are graphics, hyperlinks, applets, along with audio and video clips.

Pi in the Sky

http://www.pims.math.ca/pi/

A semi-annual periodical designed for high school students in Canada. Published by the Pacific Institute for the Mathematical Sciences.

Plus

http://plus.maths.org/index.html

This online magazine is a part of the Millennium Mathematics Project, a United Kingdom initiative based in Cambridge. Its aim is to introduce readers to the practical applications and beauty of mathematics.

What **other Web-based magazines** also often contain mathematical content?

As with print magazines, there are many Web-based magazines that often contain mathematical news or content. The following lists just a few (notice most of these sites are science-oriented):

Discovery.com

http://www.discovery.com/

A site sponsored by television's Discovery Channel. It often carries stories about mathematicians or applications of mathematics in science.

Eureka Alert

http://www.eurekalert.org

This Web site is a global news service sponsored by the American Association for the Advancement of Science (AAAS). Mathematics is included under its "News by Subject" section.

Science Daily

http://www.sciencedaily.com/

This daily science journal covers the latest in the sciences—and often mathematics. Mathematics is listed under the "Topics" section.

What are some **"essential"** math **Web sites**?

The following are some math Web sites to bookmark, because you'll likely be using them again and again if you are interested in math. These are great, very extensive resources with clear and accurate explanations:

The Math Forum

http://mathforum.org

This site, operating under Drexel University's School of Education, provides math-related materials, resources, activities, and educational products. There is even an opportunity to have your questions personally answered by "Dr. Math®."

MathWorld
http://mathworld.wolfram.com
An extremely comprehensive and interactive mathematics encyclopedia, developed over ten years using input from mathematicians. The site is continuously updated, so everyone from the casual student to the seasoned professional will find something of interest here.

S.O.S Mathematics
http://www.sosmath.com
A math study site containing more than 2,500 pages, ranging from algebra to differential equations. Geared towards high school and college students, but equally useful to adults.

What is a **Web site** that shows how **mathematics** can be **applied to technology**?

The British Columbia Institute of Technology has a great Web site called *Exactly How Is Math Used in Technology?* There are numerous examples of the application of various areas of math to a wide range of technological disciplines, including electronics, nuclear medicine, robotics, and biomedical engineering. The Web site is located at: http://www.math.bcit.ca/examples/index.shtml.

What is a good **reference Web site** if someone is interested in **engineering math**?

460 A site called *eFunda*, which stands for engineering *Funda*mentals, is an online

resource site covering all the basics of engineering mathematics. Located at http://www.efunda.com/math/math_home/math.cfm, mathematical formulas are presented, along with explanations of their usage in the proper context.

What **Web sites** offer a way to **convert units of measurement**?

The conversion of units of measurement are necessary, especially between same types of units (such as feet to inches) or between Standard and metric units (such as miles to kilometers). There are numerous sites on the World Wide Web to make such conversions. In most cases, just type in the number you want converted, hit the return button, and a list of the converted numbers are displayed. Among these sites are:

- http://www.onlineconversion.com
- http://www.convert-me.com/en/
- http://www.sciencemadesimple.com/conversions.html
- http://www.convertit.com/Go/ConvertIt/

What **Web sites** offer links to the **history of mathematics**?

The history of mathematics is an extensive topic—too much to cover in this text. Some of the best sites to explore its history are as follows:

British Society for the History of Mathematics (http://www.dcs.warwick.ac.uk/bshm/)—If you want "one stop shopping" in your search for the history of mathematics, go to this comprehensive link site. This organization promotes research into the history of mathematics at both professional and amateur levels, as well as the use of that history in education.

Clark University's Department of Mathematics and Computer Science (http://aleph0.clarku.edu/~djoyce/mathhist/mathhist.html).

Math Archives—History of Mathematics (http://archives.math.utk.edu/topics/history.html)—A mathematical historical archive with other mathematical topics, too.

MacTutor History of Mathematics (http://www-groups.dcs.st-and.ac.uk/~history/) —This archive is from the School of Mathematics and Statistics at the University of St. Andrews Scotland. It is an extensive resource on the history of mathematics from ancient times to the present.

What are some **Web sites** devoted to **recreational math**?

There are literally hundreds, if not thousands, of recreational math Web sites. The following is just a taste of what's out there:

Interactive Mathematics Miscellany and Puzzles
http://www.cut-the-knot.org/content.shtml
A little bit of everything, from games and puzzles, to fallacies and visual illusions.

461

Mathpuzzle.com
http://www.mathpuzzle.com
Another great site for recreational math and puzzles.

Puzzles.com
http://www.puzzles.com
Illusions, puzzles, tricks and toys can be found here. There is also a gift shop, puzzle links, and online help.

What **Web site** contains some **fun (and interesting) facts** about mathematics?

If you are interested in fun facts about math and don't want to hunt all over the Web for them, go to *Mudd Math Fun Facts*, located at http://www.math.hmc.edu/funfacts/. Created by Francis Edward Su of the Harvey Mudd College Math Department as a "warm-up" activity for calculus courses, these are tidbits from all areas of math. The site is definitely fun, entertaining, and addictive.

Appendix 1:

Measurement Systems and Conversion Factors

The following abbreviations are used for these conversion tables: **atm** = atmosphere; **Btu**$_{IT}$ = British Thermal Unit (international table); **cal** = calorie; **cal**$_{IT}$ = calorie (international table); **cm** = centimeter; **cu ft** = cubic feet; **ft** = feet; **ft-lbf** = foot-pound force; **g** = gram; **gal** = gallon; **hp hr** = horsepower-hour; **in** = inch; **int J** = international Joules; **J** = Joule; **kg** = kilogram; **kgf** = kilogram-force; **kWh** = kilowatt-hour; **L** = liter; **lb** = pound; **lbf** = pound-force; **m** = meter; **mmHG** = millimeters of mercury (also Torr); **m ton** = metric ton; **N** = Newton; **oz** = ounce; **Pa** = Pascal; **qt** = quart; **yd** = yard.

Units of Length

Units	cm	m	in	ft	yd	mile
1 in	2.54	0.0254	1	0.08333...	0.02777...	1.578283×10^{-5}
1 ft	30.48	0.3048	12	1	0.333...	$1.89393939... \times 10^{-4}$
1 yd	91.44	0.9144	36	3	1	$5.68181818... \times 10^{-4}$
1 mile	1.609344×10^5	1.609344×10^3	6.336×10^4	5280	1760	1
1 cm	1	0.01	0.3937008	0.03280840	0.01093631	6.213712×10^{-6}
1 m	100	1	39.37008	3.280840	1.093613	6.213712×10^{-4}

Units of Area

Units	cm²	m²	in²	ft²	yd²	mile²
1 in²	6.4516	6.4516×10^{-4}	1	$6.9444... \times 10^{-3}$	7.716049×10^{-4}	2.490977×10^{-10}
1 ft²	929.0304	0.09290304	144	1	0.111...	3.587007×10^{-8}
1 yd²	8,361.273	0.8361273	1296	9	1	3.228306×10^{-7}
1 mile²	2.589988×10^{10}	2.589988×10^6	4.014490×10^9	2.78784×10^7	3.0976×10^6	1
1 cm²	1	10^{-4}	0.1550003	1.076391×10^{-3}	1.195990×10^{-4}	3.861022×10^{-11}
1 m²	10^4	1	1550.003	10.76391	1.195990	3.861022×10^{-7}

Units of Volume

Units	m³	cm³	in³	L	ft³	qt	gal
1 in³	1.638706×10^{-5}	16.38706	1	0.01638706	5.787037×10^{-4}	0.01731602	4.329004×10^{-3}
1 ft³	2.831685×10^{-2}	28,316.85	1,728	28.31685	1	2.992208	7.480520
1 qt	9.46353×10^{-4}	946.353	57.75	0.946353	0.0342014	1	0.25
1 U.S. gal	3.785412×10^{-3}	3,785.412	231	3.785412	0.1336806	4	1
1 m³	1	10^6	6.102374×10^4	10^3	35.31467	1.056688×10^3	264.1721
1 cm³	10^{-6}	1	0.06102374	10^{-3}	3.531467×10^{-5}	1.056688×10^{-3}	2.641721×10^{-4}
1 L	10^{-3}	1,000	61.02374	1	0.03531467	1.056688	0.2641721

Units of Mass

Units	g	kg	oz	lb	m ton	ton
1 oz	28.34952	0.02834952	1	0.0625	2.834952×10^{-5}	3.125×10^{-5}
1 lb	453.5924	0.4535924	16	1	4.535924×10^{-4}	0.0005
1 m ton	10^6	1,000	35,273.96	2,204.623	1	1.102311
1 ton	907,184.7	907.1847	32,000	2,000	0.907847	1
1 g	1	10^{-3}	0.03527396	2.204623×10^{-3}	10^{-6}	1.102311×10^{-6}
1 kg	1,000	1	35.27396	2.204623	10^{-3}	1.102311×10^{-3}

Units of Density

Units	$g \times cm^{-3}$	$g \times L^{-1}, kg \times m^{-3}$	$oz \times in^{-3}$	$lb \times in^{-3}$	$lb \times ft^{-3}$	$lb \times gal^{-1}$
$1\ oz \times in^{-3}$	1.729994	1,729.994	1	0.0625	108	14.4375
$1\ lb \times in^{-3}$	27.67991	27,679.91	16	1	1,728	231
$1\ lb \times ft^{-3}$	0.01601847	16.01847	9.259259×10^{-3}	5.7870370×10^{-4}	1	0.1336806
$1\ lb \times gal^{-1}$	0.1198264	119.8264	4.749536×10^{-3}	4.3290043×10^{-3}	7.480519	1
$1\ g \times cm^{-3}$	1	1,000	0.5780365	0.03612728	62.42795	8.345403
$1\ g \times L^{-1}, kg \times m^{-3}$	10^{-3}	1	5.780365×10^{-4}	3.612728×10^{-5}	0.06242795	8.345403×10^{-3}

Units of Energy

Units	g mass	kWh	J	hp hr	cal	ft-lbf	cal_{IT}	cu ft-lbf in⁻²	Btu_{IT}	L-atm
1 g mass	1	2.496542×10^{7}	8.987552×10^{13}	3.347918×10^{7}	2.148076×10^{13}	6.628878×10^{13}	2.146640×10^{13}	4.603388×10^{11}	8.518555×10^{10}	8.870024×10^{11}
1 J	1.112650×10^{-14}	$2.777... \times 10^{-7}$	1	3.725062	0.2390057	0.7375622	0.2388459	5.121960×10^{-3}	9.478172×10^{-4}	9.869233×10^{-3}
1 cal	4.655328×10^{-14}	$1.16222... \times 10^{-6}$	4.184	1.558562×10^{-6}	1	3.085960	0.9993312	2.143028×10^{-2}	3.965667×10^{-3}	0.04129287

Units	g mass kWh	J hp hr	cal ft-lbf	cal$_{IT}$ cu ft-lbf in^{-2}	Btu$_{IT}$ L-atm
1 cal$_{IT}$	4.658443×10^{-14} 1.163000×10^{-6}	4.1868 1.559609×10^{-6}	1.000669 3.088025	1 2.144462×10^{-2}	3.968321×10^{-3} 0.04132050
1 Btu$_{IT}$	1.173908×10^{-11} 2.930711×10^{-4}	$1{,}055.056$ 3.930148×10^{-4}	252.1644 778.1693	251.9958 5.403953	1 10.41259
1 kWh	4.005540×10^{-8} 1	$3{,}600{,}000$ 1.341022	$860{,}420.7$ $2{,}655{,}224$	$859{,}845.2$ $18{,}439.06$	$3{,}412.142$ $35{,}529.24$
1 hp hr	2.986931×10^{-8} 0.7456998	$2{,}684{,}519$ 1	$641{,}615.6$ $1{,}980{,}000$	$641{,}186.5$ $13{,}750$	$2{,}544.33$ $26{,}494.15$
1 ft-lbf	1.508551×10^{-14} 3.766161×10^{-7}	1.355818 $5.050505\ldots \times 10^{-7}$	0.3240483 1	0.3238315 $6.9444\ldots \times 10^{-3}$	1.285067×10^{-3} 0.01338088
1 cu ft-lbf in^{-2}	2.172313×10^{-12} 5.423272×10^{-5}	195.2378 $7.272727\ldots \times 10^{-5}$	46.66295 144	46.63174 1	0.1850497 1.926847
1 L-atm	1.127393×10^{-12} 2.814583×10^{-5}	101.3250 3.774419×10^{-5}	24.21726 74.73349	24.20106 0.5189825	0.09603757 1

Units of Pressure

Units	Pa, 1 N × m^{-2} kgf × cm^{-2}	dyne × cm^{-2} mmHg	bar in HG	atm lbf × in^{-2}
1 Pa, 1 N × m^{-2}	1 1.019716×10^{-5}	10 7.500617×10^{-3}	10^{-5} 2.952999×10^{-4}	9.869233×10^{-6} 1.450377×10^{-4}
1 dyne × cm^{-2}	0.1 1.019716×10^{-6}	1 7.500617×10^{-4}	10^{-6} 2.952999×10^{-5}	9.869233×10^{-7} 1.450377×10^{-5}
1 bar	10^{5} 1.019716	10^{6} 750.0617	1 29.52999	0.9869233 14.50377
1 atm	$101{,}325$ 1.033227	$1{,}013{,}250$ 760	1.013250 29.92126	1 14.69595

Units	Pa, N × m⁻² / kgf × cm⁻²		dyne × cm⁻² / mmHg		bar / in HG		atm / lbf × in⁻²	
1 kgf × cm⁻²	98,066.5	1	980,665	735.5592	0.980665	28.95903	0.967841	14.22334
1 mmHg	133.3224	1	1,333.224	0.03937008	1.333224×10^{-3}	0.01933678	1.3157895×10^{-3}	1
1 in Hg	3,386.388	0.0353155	33,863.88	25.4	0.03386388	1	0.03342105	0.4911541
1 lbf × in⁻²	6,894.757	0.07030696	68,947.57	51.71493	0.06894757	2.036021	0.06804596	1

Appendix 2:

Log Table in Base 10 for the Numbers 1 through 10

Number	Log	Number	Log	Number	Log
1.000	0.00000000	1.026	0.01114736	1.052	0.02201574
1.001	0.00043408	1.027	0.01157044	1.053	0.02242837
1.002	0.00086772	1.028	0.01199311	1.054	0.02284061
1.003	0.00130093	1.029	0.01241537	1.055	0.02325246
1.004	0.00173371	1.030	0.01283722	1.056	0.02366392
1.005	0.00216606	1.031	0.01325867	1.057	0.02407499
1.006	0.00259798	1.032	0.01367970	1.058	0.02448567
1.007	0.00302947	1.033	0.01410032	1.059	0.02489596
1.008	0.00346053	1.034	0.01452054	1.060	0.02530587
1.009	0.00389117	1.035	0.01494035	1.061	0.02571538
1.010	0.00432137	1.036	0.01535976	1.062	0.02612452
1.011	0.00475116	1.037	0.01577876	1.063	0.02653326
1.012	0.00518051	1.038	0.01619735	1.064	0.02694163
1.013	0.00560945	1.039	0.01661555	1.065	0.02734961
1.014	0.00603795	1.040	0.01703334	1.066	0.02775720
1.015	0.00646604	1.041	0.01745073	1.067	0.02816442
1.016	0.00689371	1.042	0.01786772	1.068	0.02857125
1.017	0.00732095	1.043	0.01828431	1.069	0.02897771
1.018	0.00774778	1.044	0.01870050	1.070	0.02938378
1.019	0.00817418	1.045	0.01911629	1.071	0.02978947
1.020	0.00860017	1.046	0.01953168	1.072	0.03019479
1.021	0.00902574	1.047	0.01994668	1.073	0.03059972
1.022	0.00945090	1.048	0.02036128	1.074	0.03100428
1.023	0.00987563	1.049	0.02077549	1.075	0.03140846
1.024	0.01029996	1.050	0.02118930	1.076	0.03181227
1.025	0.01072387	1.051	0.02160272	1.077	0.03221570

Number	Log	Number	Log	Number	Log
1.078	0.03261876	1.26	0.1003705	1.64	0.2148438
1.079	0.03302144	1.27	0.1038037	1.65	0.2174839
1.080	0.03342376	1.28	0.1072100	1.66	0.2201081
1.081	0.03382569	1.29	0.1105897	1.67	0.2227165
1.082	0.03422726	1.30	0.1139434	1.68	0.2253093
1.083	0.03462846	1.31	0.1172713	1.69	0.2278867
1.084	0.03502928	1.32	0.1205739	1.70	0.2304489
1.085	0.03542974	1.33	0.1238516	1.71	0.2329961
1.086	0.03582983	1.34	0.1271048	1.72	0.2355284
1.087	0.03622954	1.35	0.1303338	1.73	0.2380461
1.088	0.03662890	1.36	0.1335389	1.74	0.2405492
1.089	0.03702788	1.37	0.1367206	1.75	0.2430380
1.090	0.03742650	1.38	0.1398791	1.76	0.2455127
1.091	0.03782475	1.39	0.1430148	1.77	0.2479733
1.092	0.03822264	1.40	0.1461280	1.78	0.2504200
1.093	0.03862016	1.41	0.1492191	1.79	0.2528530
1.094	0.03901732	1.42	0.1522883	1.80	0.2552725
1.095	0.03941412	1.43	0.1553360	1.81	0.2576786
1.096	0.03981055	1.44	0.1583625	1.82	0.2600714
1.097	0.04020663	1.45	0.1613680	1.83	0.2624511
1.098	0.04060234	1.46	0.1643529	1.84	0.2648178
1.099	0.04099769	1.47	0.1673173	1.85	0.2671717
1.10	0.0413927	1.48	0.1702617	1.86	0.2695129
1.11	0.0453230	1.49	0.1731863	1.87	0.2718416
1.12	0.0492180	1.50	0.1760913	1.88	0.2741578
1.13	0.0530784	1.51	0.1789769	1.89	0.2764618
1.14	0.0569049	1.52	0.1818436	1.90	0.2787536
1.15	0.0606978	1.53	0.1846914	1.91	0.2810334
1.16	0.0644580	1.54	0.1875207	1.92	0.2833012
1.17	0.0681859	1.55	0.1903317	1.93	0.2855573
1.18	0.0718820	1.56	0.1931246	1.94	0.2878017
1.19	0.0755470	1.57	0.1958997	1.95	0.2900346
1.20	0.0791812	1.58	0.1986571	1.96	0.2922561
1.21	0.0827854	1.59	0.2013971	1.97	0.2944662
1.22	0.0863598	1.60	0.2041200	1.98	0.2966652
1.23	0.0899051	1.61	0.2068259	1.99	0.2988531
1.24	0.0934217	1.62	0.2095150	2.00	0.3010300
1.25	0.0969100	1.63	0.2121876	2.01	0.3031961

Number	Log	Number	Log	Number	Log
2.02	0.3053514	2.40	0.3802112	2.78	0.4440448
2.03	0.3074960	2.41	0.3820170	2.79	0.4456042
2.04	0.3096302	2.42	0.3838154	2.80	0.4471580
2.05	0.3117539	2.43	0.3856063	2.81	0.4487063
2.06	0.3138672	2.44	0.3873898	2.82	0.4502491
2.07	0.3159703	2.45	0.3891661	2.83	0.4517864
2.08	0.3180633	2.46	0.3909351	2.84	0.4533183
2.09	0.3201463	2.47	0.3926970	2.85	0.4548449
2.10	0.3222193	2.48	0.3944517	2.86	0.4563660
2.11	0.3242825	2.49	0.3961993	2.87	0.4578819
2.12	0.3263359	2.50	0.3979400	2.88	0.4593925
2.13	0.3283796	2.51	0.3996737	2.89	0.4608978
2.14	0.3304138	2.52	0.4014005	2.90	0.4623980
2.15	0.3324385	2.53	0.4031205	2.91	0.4638930
2.16	0.3344538	2.54	0.4048337	2.92	0.4653829
2.17	0.3364597	2.55	0.4065402	2.93	0.4668676
2.18	0.3384565	2.56	0.4082400	2.94	0.4683473
2.19	0.3404441	2.57	0.4099331	2.95	0.4698220
2.20	0.3424227	2.58	0.4116197	2.96	0.4712917
2.21	0.3443923	2.59	0.4132998	2.97	0.4727564
2.22	0.3463530	2.60	0.4149733	2.98	0.4742163
2.23	0.3483049	2.61	0.4166405	2.99	0.4756712
2.24	0.3502480	2.62	0.4183013	3.00	0.4771213
2.25	0.3521825	2.63	0.4199557	3.01	0.4785665
2.26	0.3541084	2.64	0.4216039	3.02	0.4800069
2.27	0.3560259	2.65	0.4232459	3.03	0.4814426
2.28	0.3579348	2.66	0.4248816	3.04	0.4828736
2.29	0.3598355	2.67	0.4265113	3.05	0.4842998
2.30	0.3617278	2.68	0.4281348	3.06	0.4857214
2.31	0.3636120	2.69	0.4297523	3.07	0.4871384
2.32	0.3654880	2.70	0.4313638	3.08	0.4885507
2.33	0.3673559	2.71	0.4329693	3.09	0.4899585
2.34	0.3692159	2.72	0.4345689	3.10	0.4913617
2.35	0.3710679	2.73	0.4361626	3.11	0.4927604
2.36	0.3729120	2.74	0.4377506	3.12	0.4941546
2.37	0.3747483	2.75	0.4393327	3.13	0.4955443
2.38	0.3765770	2.76	0.4409091	3.14	0.4969296
2.39	0.3783979	2.77	0.4424798	3.15	0.4983106

Number	Log	Number	Log	Number	Log
3.16	0.4996871	3.54	0.5490033	3.92	0.5932861
3.17	0.5010593	3.55	0.5502284	3.93	0.5943926
3.18	0.5024271	3.56	0.5514500	3.94	0.5954962
3.19	0.5037907	3.57	0.5526682	3.95	0.5965971
3.20	0.5051500	3.58	0.5538830	3.96	0.5976952
3.21	0.5065050	3.59	0.5550944	3.97	0.5987905
3.22	0.5078559	3.60	0.5563025	3.98	0.5998831
3.23	0.5092025	3.61	0.5575072	3.99	0.6009729
3.24	0.5105450	3.62	0.5587086	4.00	0.6020600
3.25	0.5118834	3.63	0.5599066	4.01	0.6031444
3.26	0.5132176	3.64	0.5611014	4.02	0.6042261
3.27	0.5145478	3.65	0.5622929	4.03	0.6053050
3.28	0.5158738	3.66	0.5634811	4.04	0.6063814
3.29	0.5171959	3.67	0.5646661	4.05	0.6074550
3.30	0.5185139	3.68	0.5658478	4.06	0.6085260
3.31	0.5198280	3.69	0.5670264	4.07	0.6095944
3.32	0.5211381	3.70	0.5682017	4.08	0.6106602
3.33	0.5224442	3.71	0.5693739	4.09	0.6117233
3.34	0.5237465	3.72	0.5705429	4.10	0.6127839
3.35	0.5250448	3.73	0.5717088	4.11	0.6138418
3.36	0.5263393	3.74	0.5728716	4.12	0.6148972
3.37	0.5276299	3.75	0.5740313	4.13	0.6159501
3.38	0.5289167	3.76	0.5751878	4.14	0.6170003
3.39	0.5301997	3.77	0.5763414	4.15	0.6180481
3.40	0.5314789	3.78	0.5774918	4.16	0.6190933
3.41	0.5327544	3.79	0.5786392	4.17	0.6201361
3.42	0.5340261	3.80	0.5797836	4.18	0.6211763
3.43	0.5352941	3.81	0.5809250	4.19	0.6222140
3.44	0.5365584	3.82	0.5820634	4.20	0.6232493
3.45	0.5378191	3.83	0.5831988	4.21	0.6242821
3.46	0.5390761	3.84	0.5843312	4.22	0.6253125
3.47	0.5403295	3.85	0.5854607	4.23	0.6263404
3.48	0.5415792	3.86	0.5865873	4.24	0.6273659
3.49	0.5428254	3.87	0.5877110	4.25	0.6283889
3.50	0.5440680	3.88	0.5888317	4.26	0.6294096
3.51	0.5453071	3.89	0.5899496	4.27	0.6304279
3.52	0.5465427	3.90	0.5910646	4.28	0.6314438
3.53	0.5477747	3.91	0.5921768	4.29	0.6324573

Number	Log	Number	Log	Number	Log
4.30	0.6334685	4.68	0.6702459	5.06	0.7041505
4.31	0.6344773	4.69	0.6711728	5.07	0.7050080
4.32	0.6354837	4.70	0.6720979	5.08	0.7058637
4.33	0.6364879	4.71	0.6730209	5.09	0.7067178
4.34	0.6374897	4.72	0.6739420	5.10	0.7075702
4.35	0.6384893	4.73	0.6748611	5.11	0.7084209
4.36	0.6394865	4.74	0.6757783	5.12	0.7092700
4.37	0.6404814	4.75	0.6766936	5.13	0.7101174
4.38	0.6414741	4.76	0.6776070	5.14	0.7109631
4.39	0.6424645	4.77	0.6785184	5.15	0.7118072
4.40	0.6434527	4.78	0.6794279	5.16	0.7126497
4.41	0.6444386	4.79	0.6803355	5.17	0.7134905
4.42	0.6454223	4.80	0.6812412	5.18	0.7143298
4.43	0.6464037	4.81	0.6821451	5.19	0.7151674
4.44	0.6473830	4.82	0.6830470	5.20	0.7160033
4.45	0.6483600	4.83	0.6839471	5.21	0.7168377
4.46	0.6493349	4.84	0.6848454	5.22	0.7176705
4.47	0.6503075	4.85	0.6857417	5.23	0.7185017
4.48	0.6512780	4.86	0.6866363	5.24	0.7193313
4.49	0.6522463	4.87	0.6875290	5.25	0.7201593
4.50	0.6532125	4.88	0.6884198	5.26	0.7209857
4.51	0.6541765	4.89	0.6893089	5.27	0.7218106
4.52	0.6551384	4.90	0.6901961	5.28	0.7226339
4.53	0.6560982	4.91	0.6910815	5.29	0.7234557
4.54	0.6570559	4.92	0.6919651	5.30	0.7242759
4.55	0.6580114	4.93	0.6928469	5.31	0.7250945
4.56	0.6589648	4.94	0.6937269	5.32	0.7259116
4.57	0.6599162	4.95	0.6946052	5.33	0.7267272
4.58	0.6608655	4.96	0.6954817	5.34	0.7275413
4.59	0.6618127	4.97	0.6963564	5.35	0.7283538
4.60	0.6627578	4.98	0.6972293	5.36	0.7291648
4.61	0.6637009	4.99	0.6981005	5.37	0.7299743
4.62	0.6646420	5.00	0.6989700	5.38	0.7307823
4.63	0.6655810	5.01	0.6998377	5.39	0.7315888
4.64	0.6665180	5.02	0.7007037	5.40	0.7323938
4.65	0.6674530	5.03	0.7015680	5.41	0.7331973
4.66	0.6683859	5.04	0.7024305	5.42	0.7339993
4.67	0.6693169	5.05	0.7032914	5.43	0.7347998

Number	Log	Number	Log	Number	Log
5.44	0.7355989	5.82	0.7649230	6.20	0.7923917
5.45	0.7363965	5.83	0.7656686	6.21	0.7930916
5.46	0.7371926	5.84	0.7664128	6.22	0.7937904
5.47	0.7379873	5.85	0.7671559	6.23	0.7944880
5.48	0.7387806	5.86	0.7678976	6.24	0.7951846
5.49	0.7395723	5.87	0.7686381	6.25	0.7958800
5.50	0.7403627	5.88	0.7693773	6.26	0.7965743
5.51	0.7411516	5.89	0.7701153	6.27	0.7972675
5.52	0.7419391	5.90	0.7708520	6.28	0.7979596
5.53	0.7427251	5.91	0.7715875	6.29	0.7986506
5.54	0.7435098	5.92	0.7723217	6.30	0.7993405
5.55	0.7442930	5.93	0.7730547	6.31	0.8000294
5.56	0.7450748	5.94	0.7737864	6.32	0.8007171
5.57	0.7458552	5.95	0.7745170	6.33	0.8014037
5.58	0.7466342	5.96	0.7752463	6.34	0.8020893
5.59	0.7474118	5.97	0.7759743	6.35	0.8027737
5.60	0.7481880	5.98	0.7767012	6.36	0.8034571
5.61	0.7489629	5.99	0.7774268	6.37	0.8041394
5.62	0.7497363	6.00	0.7781513	6.38	0.8048207
5.63	0.7505084	6.01	0.7788745	6.39	0.8055009
5.64	0.7512791	6.02	0.7795965	6.40	0.8061800
5.65	0.7520484	6.03	0.7803173	6.41	0.8068580
5.66	0.7528164	6.04	0.7810369	6.42	0.8075350
5.67	0.7535831	6.05	0.7817554	6.43	0.8082110
5.68	0.7543483	6.06	0.7824726	6.44	0.8088859
5.69	0.7551123	6.07	0.7831887	6.45	0.8095597
5.70	0.7558749	6.08	0.7839036	6.46	0.8102325
5.71	0.7566361	6.09	0.7846173	6.47	0.8109043
5.72	0.7573960	6.10	0.7853298	6.48	0.8115750
5.73	0.7581546	6.11	0.7860412	6.49	0.8122447
5.74	0.7589119	6.12	0.7867514	6.50	0.8129134
5.75	0.7596678	6.13	0.7874605	6.51	0.8135810
5.76	0.7604225	6.14	0.7881684	6.52	0.8142476
5.77	0.7611758	6.15	0.7888751	6.53	0.8149132
5.78	0.7619278	6.16	0.7895807	6.54	0.8155777
5.79	0.7626786	6.17	0.7902852	6.55	0.8162413
5.80	0.7634280	6.18	0.7909885	6.56	0.8169038
5.81	0.7641761	6.19	0.7916906	6.57	0.8175654

Number	Log	Number	Log	Number	Log
6.58	0.8182259	6.96	0.8426092	7.34	0.8656961
6.59	0.8188854	6.97	0.8432328	7.35	0.8662873
6.60	0.8195439	6.98	0.8438554	7.36	0.8668778
6.61	0.8202015	6.99	0.8444772	7.37	0.8674675
6.62	0.8208580	7.00	0.8450980	7.38	0.8680564
6.63	0.8215135	7.01	0.8457180	7.39	0.8686444
6.64	0.8221681	7.02	0.8463371	7.40	0.8692317
6.65	0.8228216	7.03	0.8469553	7.41	0.8698182
6.66	0.8234742	7.04	0.8475727	7.42	0.8704039
6.67	0.8241258	7.05	0.8481891	7.43	0.8709888
6.68	0.8247765	7.06	0.8488047	7.44	0.8715729
6.69	0.8254261	7.07	0.8494194	7.45	0.8721563
6.70	0.8260748	7.08	0.8500333	7.46	0.8727388
6.71	0.8267225	7.09	0.8506462	7.47	0.8733206
6.72	0.8273693	7.10	0.8512583	7.48	0.8739016
6.73	0.8280151	7.11	0.8518696	7.49	0.8744818
6.74	0.8286599	7.12	0.8524800	7.50	0.8750613
6.75	0.8293038	7.13	0.8530895	7.51	0.8756399
6.76	0.8299467	7.14	0.8536982	7.52	0.8762178
6.77	0.8305887	7.15	0.8543060	7.53	0.8767950
6.78	0.8312297	7.16	0.8549130	7.54	0.8773713
6.79	0.8318698	7.17	0.8555192	7.55	0.8779470
6.80	0.8325089	7.18	0.8561244	7.56	0.8785218
6.81	0.8331471	7.19	0.8567289	7.57	0.8790959
6.82	0.8337844	7.20	0.8573325	7.58	0.8796692
6.83	0.8344207	7.21	0.8579353	7.59	0.8802418
6.84	0.8350561	7.22	0.8585372	7.60	0.8808136
6.85	0.8356906	7.23	0.8591383	7.61	0.8813847
6.86	0.8363241	7.24	0.8597386	7.62	0.8819550
6.87	0.8369567	7.25	0.8603380	7.63	0.8825245
6.88	0.8375884	7.26	0.8609366	7.64	0.8830934
6.89	0.8382192	7.27	0.8615344	7.65	0.8836614
6.90	0.8388491	7.28	0.8621314	7.66	0.8842288
6.91	0.8394780	7.29	0.8627275	7.67	0.8847954
6.92	0.8401061	7.30	0.8633229	7.68	0.8853612
6.93	0.8407332	7.31	0.8639174	7.69	0.8859263
6.94	0.8413595	7.32	0.8645111	7.70	0.8864907
6.95	0.8419848	7.33	0.8651040	7.71	0.8870544

Number	Log	Number	Log	Number	Log
7.72	0.8876173	8.10	0.9084850	8.48	0.9283959
7.73	0.8881795	8.11	0.9090209	8.49	0.9289077
7.74	0.8887410	8.12	0.9095560	8.50	0.9294189
7.75	0.8893017	8.13	0.9100905	8.51	0.9299296
7.76	0.8898617	8.14	0.9106244	8.52	0.9304396
7.77	0.8904210	8.15	0.9111576	8.53	0.9309490
7.78	0.8909796	8.16	0.9116902	8.54	0.9314579
7.79	0.8915375	8.17	0.9122221	8.55	0.9319661
7.80	0.8920946	8.18	0.9127533	8.56	0.9324738
7.81	0.8926510	8.19	0.9132839	8.57	0.9329808
7.82	0.8932068	8.20	0.9138139	8.58	0.9334873
7.83	0.8937618	8.21	0.9143432	8.59	0.9339932
7.84	0.8943161	8.22	0.9148718	8.60	0.9344985
7.85	0.8948697	8.23	0.9153998	8.61	0.9350032
7.86	0.8954225	8.24	0.9159272	8.62	0.9355073
7.87	0.8959747	8.25	0.9164539	8.63	0.9360108
7.88	0.8965262	8.26	0.9169800	8.64	0.9365137
7.89	0.8970770	8.27	0.9175055	8.65	0.9370161
7.90	0.8976271	8.28	0.9180303	8.66	0.9375179
7.91	0.8981765	8.29	0.9185545	8.67	0.9380191
7.92	0.8987252	8.30	0.9190781	8.68	0.9385197
7.93	0.8992732	8.31	0.9196010	8.69	0.9390198
7.94	0.8998205	8.32	0.9201233	8.70	0.9395193
7.95	0.9003671	8.33	0.9206450	8.71	0.9400182
7.96	0.9009131	8.34	0.9211661	8.72	0.9405165
7.97	0.9014583	8.35	0.9216865	8.73	0.9410142
7.98	0.9020029	8.36	0.9222063	8.74	0.9415114
7.99	0.9025468	8.37	0.9227255	8.75	0.9420081
8.00	0.9030900	8.38	0.9232440	8.76	0.9425041
8.01	0.9036325	8.39	0.9237620	8.77	0.9429996
8.02	0.9041744	8.40	0.9242793	8.78	0.9434945
8.03	0.9047155	8.41	0.9247960	8.79	0.9439889
8.04	0.9052560	8.42	0.9253121	8.80	0.9444827
8.05	0.9057959	8.43	0.9258276	8.81	0.9449759
8.06	0.9063350	8.44	0.9263424	8.82	0.9454686
8.07	0.9068735	8.45	0.9268567	8.83	0.9459607
8.08	0.9074114	8.46	0.9273704	8.84	0.9464523
8.09	0.9079485	8.47	0.9278834	8.85	0.9469433

Number	Log	Number	Log	Number	Log
8.86	0.9474337	9.24	0.9656720	9.62	0.9831751
8.87	0.9479236	9.25	0.9661417	9.63	0.9836263
8.88	0.9484130	9.26	0.9666110	9.64	0.9840770
8.89	0.9489018	9.27	0.9670797	9.65	0.9845273
8.90	0.9493900	9.28	0.9675480	9.66	0.9849771
8.91	0.9498777	9.29	0.9680157	9.67	0.9854265
8.92	0.9503649	9.30	0.9684829	9.68	0.9858754
8.93	0.9508515	9.31	0.9689497	9.69	0.9863238
8.94	0.9513375	9.32	0.9694159	9.70	0.9867717
8.95	0.9518230	9.33	0.9698816	9.71	0.9872192
8.96	0.9523080	9.34	0.9703469	9.72	0.9876663
8.97	0.9527924	9.35	0.9708116	9.73	0.9881128
8.98	0.9532763	9.36	0.9712758	9.74	0.9885590
8.99	0.9537597	9.37	0.9717396	9.75	0.9890046
9.00	0.9542425	9.38	0.9722028	9.76	0.9894498
9.01	0.9547248	9.39	0.9726656	9.77	0.9898946
9.02	0.9552065	9.40	0.9731279	9.78	0.9903890
9.03	0.9556878	9.41	0.9735896	9.79	0.9907827
9.04	0.9561684	9.42	0.9740509	9.80	0.9912261
9.05	0.9566486	9.43	0.9745117	9.81	0.9916690
9.06	0.9571282	9.44	0.9749720	9.82	0.9921115
9.07	0.9576073	9.45	0.9754318	9.83	0.9925535
9.08	0.9580858	9.46	0.9758911	9.84	0.9929951
9.09	0.9585639	9.47	0.9763500	9.85	0.9934362
9.10	0.9590414	9.48	0.9768083	9.86	0.9938769
9.11	0.9595184	9.49	0.9772662	9.87	0.9943172
9.12	0.9599948	9.50	0.9777236	9.88	0.9947569
9.13	0.9604708	9.51	0.9781805	9.89	0.9951963
9.14	0.9609462	9.52	0.9786369	9.90	0.9956352
9.15	0.9614211	9.53	0.9790929	9.91	0.9960737
9.16	0.9618955	9.54	0.9795484	9.92	0.9965117
9.17	0.9623693	9.55	0.9800034	9.93	0.9969492
9.18	0.9628427	9.56	0.9804579	9.94	0.9973864
9.19	0.9633155	9.57	0.9809119	9.95	0.9978231
9.20	0.9637878	9.58	0.9813655	9.96	0.9982593
9.21	0.9642596	9.59	0.9818186	9.97	0.9986952
9.22	0.9647309	9.60	0.9822712	9.98	0.9991305
9.23	0.9652017	9.61	0.9827234	9.99	0.9995655

Appendix 3:

Common Formulas for Calculating Areas and Volumes of Shapes

Area and Perimeter of a Square or Rectangle

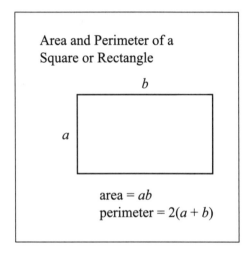

area = ab

perimeter = $2(a + b)$

Area of a Triangle

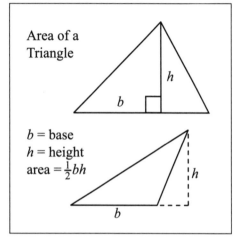

b = base
h = height
area = $\frac{1}{2}bh$

Area and Perimeter of a Parallelogram

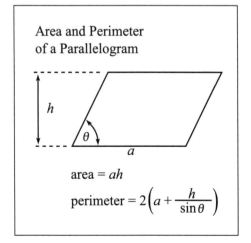

area = ah

perimeter = $2\left(a + \dfrac{h}{\sin\theta}\right)$

Area of a Trapezoid

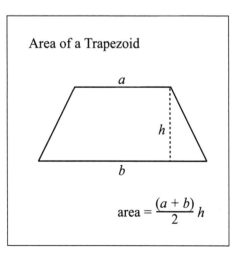

area = $\dfrac{(a + b)}{2}h$

Area and Circumference of a Circle

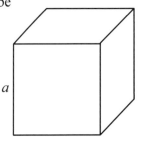

r = radius
d = diameter = $2r$

$$\text{area} = \pi r^2 = \frac{\pi d^2}{4}$$

circumference = $2\pi r = \pi d$
π = circumference/diameter

Area of an Ellipse

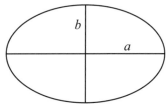

$$\text{area} = \pi\,ab$$

Surface Area and Volume of a Cube

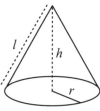

volume = a^3
lateral surface area = $6a^2$

Surface Area and Volume of a Cone

surface area = πrl

$$\text{volume} = \frac{1}{3}\pi r^2 h$$

Volume of a Pyramid and Irregular Cone

$$\text{volume} = \frac{1}{3}\,\text{base area} \times h$$

Volume of a Parallelepiped

A = area of base

h = height

volume = Ah

480

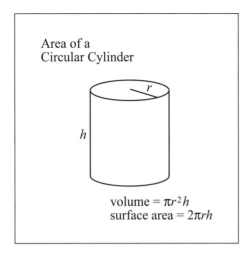

Area of a
Circular Cylinder

volume = $\pi r^2 h$
surface area = $2\pi rh$

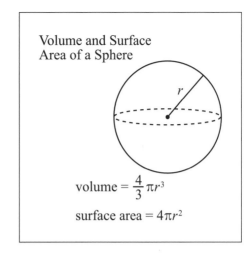

Volume and Surface
Area of a Sphere

volume = $\dfrac{4}{3}\pi r^3$

surface area = $4\pi r^2$

Index

(ill.) indicates photos and illustrations.

483

Hurricanes, 310–11 (table)
Hutchings, Michael, 367
Huygens, Christiaan, 252–53, 253 (ill.)
Hydrogen, 285–86
Hydrostatic principle, 19
Hypatia of Alexandria, 18
Hyperbolic functions, 204–5
Hyperbolic geometry, 205, 206
Hypotenuse, 182

I

IBM, 354, 359
Identities (algebraic), 135
Identities (trigonometric), 203, 204
Identity equations, 141–42
Identity matrix, 159–60
Identity transformation, 191
"If and only if" (iff), 221
Image processing, 245, 245 (ill.)
Imaginary calculus, 213
Imaginary numbers, 75, 76–78, 78 (ill.), 340
Imperial gallon, 49
Implicit differential equations, 232
Improper fractions, 98–99
Improper integral, 231
Inch, 44
Inches of mercury, 307
Incompleteness: The Proof and Paradox of Kurt Gödel (Goldstein), 453
Indefinite integral, 230
Independent axiomatic system, 116
Independent events, 250
Independent sampling, 257
Independent variables, 135, 136, 259
Indeterminate number, 91
Indirect proofs, 118, 178
Induction, 118–19
Inductor, 340
Industrial engineering, 342–44
Inequality, 134

Infant mortality rate, 381
Inferential statistics, 258
Infinite series, 215
Infinitesimal calculus, 220–21
Infinity, 80, 129, 129 (ill.), 218 (ill.), 218–19
Infinity (movie), 457
Information theory, 270–71
Initial conditions, 233–34
Initial term, 216
Initial value problems, 234
Institute and Museum of the History of Science, 449
Insurance tables, 247
Integers, 71, 72, 72 (table)
Integral calculus, 209, 213. *See also* Calculus; Differential calculus; Differential equations; Mathematical analysis; Vectors
 antiderivatives, 231
 antidifferentiation, 231
 common integrals, 228
 definite integral, 229–30
 definition of, 228
 double/triple integrals, 231–32
 Fundamental Theorem of the Calculus, 231
 improper integral, 231
 indefinite integral, 230
 integration, 229, 229 (ill.)
Integrated-circuit machines, 363
Integration, 229, 229 (ill.)
Interactive Mathematics Miscellany and Puzzles (Web site), 461
Intercept, 195
Interest, 388–89
Interest rate, 412, 413
Interior angles, 174
International Bureau of Weights and Measures, 50, 54, 58–59
International Business Machines (IBM), 354, 359
International System of Units (SI), 50
International units, 53
Interpolation, 324

Intersection, 125
Intuitionism, 107, 107 (ill.)
Inverses, 94, 141, 222–23
Ion, 284
Ionians, 16
"Iron Ulna," 46
Irrational numbers, 74–75
Irregular cone, 480
Irregular polygons, 179
Islamic patterns, 375–76, 376 (ill.)
Isometry, 191
Isosceles trapezoid, 183
Isosceles triangle, 182 (ill.), 182–83
Iteration, 143

J

Jacquard, Joseph-Marie, 353–54
Jainism, 384–85
James, Bill, 434
Japanese numerals, 418 (ill.)
Jiuzhang suanshu, 21
John I (pope/saint), 61
Jones, William, 40
Journal of Online Mathematics and Its Applications (online magazine), 458–59
Julian calendar, 62, 63
Julius Caesar, 62
Jurassic Park (movie), 457

K

Kappa Mu Epsilon, 446
Kasner, Edward, 73
Kassites, 10
Kelvin scale, 55
Kepler: A Novel (Banville), 455
Kepler, Johannes, 210, 288–89, 383
Kepler's Laws of Planetary Motion, 288–89
Khuwarizmi, Muhammad ibn Musa al-, 21, 22, 112–13, 131
Kindi, al-, 21

499